Jennifer Cramer and Chris Montgomery (Eds.)

Cityscapes and Perceptual Dialectology

Language and Social Life

Editors
David Britain
Crispin Thurlow

Volume 5

Cityscapes and Perceptual Dialectology

Global Perspectives on Non-Linguists' Knowledge of the Dialect Landscape

Edited by
Jennifer Cramer
Chris Montgomery

DE GRUYTER
MOUTON

ISBN 978-1-61451-477-0
e-ISBN (PDF) 978-1-61451-008-6
e-ISBN (EPUB) 978-1-5015-0026-8
ISSN 2364-4303

Library of Congress Cataloging-in-Publication Data
A CIP catalog record for this book has been applied for at the Library of Congress.

Bibliographic information published by the Deutsche Nationalbibliothek
The Deutsche Nationalbibliothek lists this publication in the Deutsche Nationalbibliografie; detailed bibliographic data are available on the Internet at http://dnb.dnb.de.

Typesetting: Meta Systems Publishing & Printservices GmbH, Wustermark
Printing and binding: CPI books GmbH, Leck
♾ Printed on acid-free paper
Printed in Germany

www.degruyter.com

Contents

List of illustrations

List of tables

Preface

The notion of *place* has been one of the main concerns of traditional dialectology throughout its history as a field. This type of research has typically aimed to present a picture of some dialect landscape, using phonological and lexical items as the basis for dialect groupings, with place being a physical, objective, and bounded entity. In doing so, dialectology has ignored a "socially rich" (Britain 2009: 142) interpretation of space and place, with the primary "role of space largely reduced to that of a canvas onto which dialectological findings could be painted" (Britain 2009: 144). Traditional dialectology has also tended to ignore the perceptions, attitudes, and beliefs non-linguists have about connections between language, space, and place.

Perceptual Dialectology (PD) is a branch of folk linguistics that attempts to redress the balance somewhat by focusing on what non-linguists think, say, and understand about language and linguistic variation. This includes where they think variation comes from, where they think it exists, and why they think it happens. This field has a rather long tradition, with its earliest roots in Dutch and Japanese traditions, though more recent work in the American tradition, led by the work of Dennis Preston, has had more widespread success. From Preston's "Visions of America" (Preston 1989) type of studies, we have gained an enormous amount of information about speaker perceptions – information that had been previously missing in many accounts of linguistic variation. The development of PD tasks, like the draw-a-map task (used extensively in this volume), wherein non-linguists are asked to draw lines around areas of a blank map where they believe people have a certain way of speaking, have provided a framework for understanding those folk beliefs, and the use of Geographic Information Systems (GIS) by several of the authors in this collection in the analysis of these maps helps solidify this connection between language, space, and place.

Thus, in this tradition, place becomes less objective; it is not only geographic or political facts that guide non-linguists' ideas about language but also their social realities. PD allows researchers to question the assumptions often made about the close connections between place, language, and identity. It can also provide support or contradicting evidence for the ways in which traditional dialectology divides the linguistic landscape and ask "the question of just how much dialectologists' supposedly scientific determination of dialect areas may be artifacts of the dialectologists' own cultural bias" (Butters 1991: 296).

As a field of inquiry, then, PD provides the right tools for understanding how place conditions non-linguists' thoughts about language. PD can aid in our understanding of how people in cities view the linguistic landscape in ways that are often different from but sometimes similar to people outside of city limits. For example, previous research in New York City (Preston 1989; Blake et al. 2008), Memphis (Fridland, Bartlett, and Kreuz 2004, 2005; Fridland and Bartlett 2006), Reno (Fridland and Bartlett 2006), Boston (Hartley 2005), and Louisville (Cramer 2010) in the United States, as well as research conducted in cities and other locations across the globe (e.g., Romanello 2002; Boughton 2006), has exemplified the ways in which the tools of PD can be used to determine how city dwellers view their own way of speaking as well as that of nearby communities. Perceptual tasks of the type used in some PD investigations (e.g., Montgomery 2007) are increasingly being used to investigate salient social and linguistic features in city-based populations (Kerswill et al. 2008), highlighting the utility of such methods.

Research in the UK outside of cities has highlighted the importance of urban areas in perception (Montgomery 2007; Montgomery and Beal 2011), something that had not previously been observed (Inoue 1996). The important role that these urban areas play in the perceptual picture of the country appears to tally with the increasing role of such areas in the development of supra-regional linguistic features (see Watt 2002; Kerswill 2003). As countries become increasingly urbanized, the role of cities and their inhabitants will be of further interest to scholars in sociolinguistics, and perceptual studies should not neglect this important development.

This book brings together a collection of research that addresses the perception of dialects in and from the vantage point of urban areas. Drawing together examples of studies conducted in the United States, Korea, Austria, Germany, France, Ireland, and Great Britain, the volume offers the reader an opportunity to make cross- and intra-linguistic comparisons between the perceptions of dialect in quite different locations. The book is organised in order to review the state of the art in perceptual dialectology and survey the latest methodologies and novel processing techniques in the field at the outset of the collection. There then follows three sections, in which papers are collected in order to deal with particular issues relating to the nature of the perception of dialect. The first section of the volume deals with the differences in the perception of rural and urban areas, a matter of longstanding interest to those interested in the ways in which non-linguists perceive the linguistic landscape. The book's second section brings together work that considers speech perception and the ways in which evaluations are arrived at by listeners. The final section of the collection draws together chapters that investigate the extent to which non-linguists' perceptions match the 'reality' of the dialect situation and explores the important factors that contribute to the ways in which non-linguists perceive their linguistic environment. It is hoped that this volume, whilst demonstrating some of the latest research in perceptual dialectology, acts as a motivation for others to make connections between the chapters and spot possible future avenues for their own research.

Jennifer Cramer and Chris Montgomery

References

Blake, Renee, Elizabeth Coggshall, Daniel Erker & Michael Taylor. 2008. New York City English: Perceptual Dialectology and Research Design. (Paper Presented at NWAV 37, Houston, TX).

Boughton, Zoe. 2006. When perception isn't reality: Accent identification and perceptual dialectology in French. *French Language Studies* 16. 277–304.

Britain, David. 2009. Language and Space: the variationist approach. In Peter Auer & Jürgen Erich Schmidt (eds.), *Language and Space: an international handbook of linguistic variation*, 142–162. Berlin: Mouton de Gruyter.

Butters, Ronald R. 1991. Review of Dennis Preston, Perceptual Dialectology. *Language in Society* 20. 294–299.

Cramer, Jennifer. 2010. "The Effect of Borders on the Linguistic Production and Perception of Regional Identity in Louisville, Kentucky". Unpublished PhD dissertation, University of Illinois at Urbana-Champaign. *https://www.ideals.illinois.edu/bitstream/handle/2142/18426/Cramer_Jennifer.pdf?sequence=1.* (accessed 29 July 2014).

Fridland, Valerie & Kathryn Bartlett. 2006. Correctness, Pleasantness, and Degree of Difference Ratings across Regions. *American Speech* 81(4). 358–386.

Fridland, Valerie, Kathryn Bartlett & Roger Kruez. 2004. Do you hear what I hear? Experimental measurement of the perceptual salience of acoustically manipulated vowel variants by Southern speakers in Memphis, TN. *Language Variation and Change* 16(1). 1–16.

Fridland, Valerie, Kathryn Bartlett & Roger Kruez. 2005. Making Sense Of Variation: Pleasantness and education ratings of southern vowel variants. *American Speech* 80(4). 366–386.

Hartley, Laura C. 2005. The Consequences of Conflicting Stereotypes: Bostonian Perceptions of U.S. Dialects. *American Speech* 80(4). 388–405.

Inoue, Fumio. 1996. Subjective Dialect Division in Great Britain. *American Speech* 71(2). 142–161.

Kerswill, Paul. 2003. Dialect Levelling and Geographical Diffusion in British English. In David Britain & Jenny Cheshire (eds.), *Social Dialectology: In honour of Peter Trudgill*, 223–243. Amsterdam: John Benjamins.

Kerswill, Paul, Eivind Torgerson, Arfaan Khan & Sue Fox. 2008. Perceiving ethnicity and place in Multicultural London English. (Paper presented at the NWAV 37, Houston, TX).

Montgomery, Chris. 2007. Northern English dialects: A Perceptual Approach. *Unpublished PhD dissertation, University of Sheffield. http://etheses.whiterose.ac.uk/1203/.* (accessed 30 July 2014).

Montgomery, Chris & Joan C. Beal. 2011. Perceptual Dialectology. In Warren Maguire & April McMahon (eds.), *Analysing Variation in English*, 121–148. Cambridge: Cambridge University Press.

Preston, Dennis R. 1989. *Perceptual Dialectology: Non-linguists' view of aerial linguistics.* Dordrecht: Foris.

Romanello, Maria Teresa. 2002. The perception of urban varieties: Preliminary studies from the south of Italy. In Daniel Long & Dennis R. Preston (eds.), *Handbook of perceptual dialectology*, 329–349. Amsterdam: John Benjamins.

Watt, Dominic. 2002. "I Don't Speak with a Geordie Accent, I Speak, Like, the Northern Accent": Contact-induced levelling in the Tyneside vowel system. *Journal of Sociolinguistics* 6(1). 44–63.

Acknowledgements

This book has its roots in a panel presentation titled "Dialect Perceptions in the City" presented at the *Sociolinguistic Symposium 19* in Berlin, Germany in 2012. Many members of that panel, as well as some who presented their work in similar panels at the same conference, came together to produce a volume that we believe is timely and much needed within the field of perceptual dialectology. Thank you to audience members for their insightful comments and suggestions.

We would also like to thank the many people at Mouton de Gruyter who helped make this volume a reality. Thanks to Emily Farrell, who met with us during the *Sociolinguistic Symposium* to discuss the possibility of turning this panel into an edited collection. Thanks to Lara Wysong, our project editor, for working so closely with us to put this collection together, and Wolfgang Konwitschny, for his role as production editor. Thanks to series editors David Britain and Crispin Thurlow, and the founding editor of the Language and Social Life series, Richard J. Watts, for the positive feedback they have all provided on the concept for this volume and for support throughout the process.

We must also send a very big thank you to the contributing authors for their hard work, dedication, and patience as we put this volume together. We sincerely believe a book can only be as great as its contributors, and this group is outstanding. We especially want to thank Dennis Preston, as he has worked so diligently to support the work and careers of both editors, as well as many of our other contributing authors. His work has inspired us to push the limits of what had heretofore been possible in perceptual dialectology, and that inspiration is prevalent throughout this book.

Finally, we gratefully and humbly acknowledge the tremendous support and encouragement we have received from our own families in our careers. Jennifer would like to thank Aaron, her husband, as well as Isabelle, Nathan, and Benjamin, her children, for their unending love and care. Chris would like to thank Emma and his daughter Lara for their love, support, and patience over the course of this project.

Notes on contributors

Natalie Braber is Senior Lecturer in Linguistics at Nottingham Trent University. She received her PhD in 2001 from St. Andrews University. Her research focuses on language variation in the East Midlands, including the perception of language in the region and senses of local and regional identity. Natalie's research has also included the study of identity and language change in Glasgow – both for Glaswegians living in Glasgow and those now living in England. She is also interested in the role of emotion in language use and how this can be examined. She has presented her work at numerous national and international conferences and has published articles in journals such as *English Today, Identity, Journal of Pragmatics* and *Journal of Germanic Linguistics*. She has also contributed chapters to edited book collections and co-authored *Exploring the German Language* with Sally Johnson. Email: natalie.braber@ntu.ac.uk

Jennifer Cramer is Assistant Professor of Linguistics in the Department of English at the University of Kentucky. She received a PhD in linguistics in 2010 from the University of Illinois. Her research focuses on the perception and production of linguistic variation at dialect and regional borders, with a specific interest in the dialects spoken in Kentucky, located at the border between the Southern and Midwestern regions of the United States. Her research utilizes the tools of Perceptual Dialectology and traditional dialectology to investigate connections between language and identity. She has presented her work at numerous national and international conferences, has published papers in *American Speech, Discourse & Society, English World-Wide, Southern Journal of Linguistics*, and *Studies in the Linguistic Sciences*, and has co-authored a book chapter in Marina Terkourafi's (ed.), *The Languages of Global Hip Hop*. Email: jennifer.cramer@uky.edu

Patricia Cukor-Avila is Associate Professor and Director of the Linguistics Program in the College of Information at the University of North Texas. Her primary research is on the study of linguistic variation and change, specifically grammatical change over time in African American Vernacular English (AAVE). Her longitudinal panel study (1988-present) of AAVE in a rural Texas community has provided much of the data for presentations and publications that center on approaches to sociolinguistic fieldwork, transmission and diffusion, as well as documenting innovations in African American English. She is coeditor (with Guy Bailey and Natalie Maynor) of *The Emergence of Black English: Text and Commentary* (1991) and author of several journal articles and book chapters. She also researches regional variation and perceptual dialectology in American English and has collaborated on a similar project in Korea. She is currently working with Lisa Jeon (Rice University) conducting cross-disciplinary research that incorporates the analytical methods of variation studies into traditional perceptual dialectology through the use of a GIS and R. Email: pcavila@unt.edu

Betsy E. Evans holds a PhD from Michigan State University and is currently Associate Professor of Linguistics at the University of Washington. Her research concentrates on linguistic variation and how that relates to the functions of language in marking identity, status, group solidarity, and cultural values and draws heavily on perceptions and attitudes of language variation. Her recent work focuses on spatial perceptions of language using digital technology for the analysis and interpretation of results. Email: evansbe@uw.edu

Médéric Gasquet-Cyrus is Assistant Professor in Linguistics, co-director of the Department of Linguistics at the University of Aix-Marseille, and an affiliate of the Laboratoire Parole et Langage (CNRS). He received his PhD in Linguistics in 2004. His research focus is in the field of Sociolinguistics, with a specific interest in the accents and the varieties of French, including the French spoken in Marseille and the Occitan area. His works are based on ethnographic approaches with a large part devoted to language attitudes. His research also concerns fields such as language policy, endangered languages, lexicography, verbal humor, and the epistemology of sociolinguistics, with a specific interest in methodology and ethics. He has presented his work at numerous national and international conferences, has published papers in the *International Journal of the Sociology of Language*, *Journal of French Language Studies, Cahiers de Linguistique*, and *Langage et Société*. He has co-edited four books and one issue of a journal and has co-authored papers in Mari Jones and David Hornsby (eds.), *Language and Social Structure in Urban France*; Mari Jones and Christopher Connolly (eds.), *Endangered Languages and New Technologies*; Mari Jones and Sara Ogilvie (eds.), *Keeping Languages Alive. Documentation, Pedagogy and Revitalisation*. Email: mederic.gasquet-cyrus@univ-amu.fr

Lisa Jeon is currently a PhD candidate in the Linguistics Department at Rice University. Her research concerns the perception of linguistic variation, the relationship between language, place, and society, and methodological approaches to the study of Perceptual Dialectology. Her research has focused on locations in South Korea and Texas and has examined the intersection of perceived linguistic variation and social factors, migration patterns, demography, and language ideology. Lisa has presented her work at numerous national and international conferences and has published articles in the *Texas Linguistic Forum, Rice Working Papers in Linguistics,* and *Dialectologia*. Email: lisa.r.jeon@rice.edu

John Lonergan is a Data Specialist Manager at Evi (an Amazon company). He received an MA in Applied Linguistics from Macquarie University, Australia in 2008 and a PhD in Linguistics from University College Dublin, Ireland in 2013. His research to date has focused on variation and change in the phonology of Dublin English. This work has combined acoustic phonetics with perceptual dialectology. He has co-authored chapters in New Perspectives on Irish English (John Benjamins, 2012) and the *Proceedings of the Australian Linguistic Society Conference 2008*. He has also presented at a number of Irish and international conferences, including an invited presentation at the 2013 Interdisciplinary Linguistics Conference in Queen's University Belfast. Email: johndavidlonergan@gmail.com

Stephen L. Mann is Assistant Professor of Linguistics in the Department of English at the University of Wisconsin-La Crosse. He earned his PhD in linguistics from the University of South Carolina in 2011. His research addresses attitudes toward and perceptions of varieties of American English used or believed to be used by gay men in the U.S. He is especially interested in how attitudes and perceptions are shaped by regional affiliations of speakers and listeners. He has presented his work at key national and international conferences and has published papers in the *Journal of Homosexuality* and the *Journal of Language and Sexuality*. His vignette on ethics in fieldwork is included in Mallinson, Childs, and Van Herk (eds.), *Data Collection in Sociolinguistics: Methods and Applications*. Email: smann@uwlax.edu

Chris Montgomery is a Lecturer in Dialectology at the University of Sheffield. He gained his PhD in 2007 and subsequently held an ESRC postdoctoral fellowship at the University of Edinburgh. His research focus is in the field of Perceptual Dialectology, with a specific interest in methodological approaches to the study of non-linguists' perceptions. His research has focused on locations in the north of England and southern Scotland, and has discussed the role of (real and imagined) borders in perception. Chris has presented his work at numerous national and international conferences, and has published articles in the *Journal of Sociolinguistics* and *Studies in Variation, Contacts and Change in English*. His other contributions include book chapters in Terttu Nevalainen and Elizabeth Closs Traugott (Eds.), *The Oxford Handbook of the History of English*, Alfred Lameli et al (Eds.), *Language and Space – An International Handbook of Linguistic Variation, Volume 2: Language Mapping*, and Warren Maguire and April McMahon (Eds.), *Analysing Variation in English*. Email: c.montgomery@sheffield.ac.uk

Sylvia Moosmüller is Associate Professor in the Acoustics Research Institute of the Austrian Academy of Sciences, Vienna. She received her PhD in General and Applied Linguistics in 1984 from the University of Vienna. In 2007, she finished her post-doctoral thesis (Habilitation) on the *Vowels in Standard Austrian German* and has been qualified Associate Professor in Applied Linguistics, Phonetics, and Phonology since 2008. Her main research focuses on the phonetics, phonology, and sociolinguistics of the varieties spoken in Austria. Another focus of research is aimed at the description of selected insufficiently described languages. She is the author of two monographs – *Soziophonologische Variation im gegenwärtigen Wiener Deutsch. Eine empirische Untersuchung* and *Hochsprache und Dialekt in Österreich. Soziophonologische Untersuchungen zu ihrer Abgrenzung in Wien, Graz, Salzburg und Innsbruck* (1991) – and has published articles in various journals and books. Email:sylvia.moosmueller@oeaw.ac.at

Nicole Palliwoda is a project assistant in the research project "Perceptual Dialectology – The German language area from the perspective of linguistic laymen" [Wahrnehmungs-dialektologie – Der deutsche Sprachraum aus der Sicht lingustischer Laien] at the University of Kiel, funded by the *Deutsche Forschungsgemeinschaft* (DFG). She is currently working toward her PhD with a focus on whether the construct of *Mauer in den Köpfen* ['wall in mind'] still prevails within East and West German citizens and how it is composed. Within her research project, which includes qualitative and quantitative methods, she is analysing autobiographical-narrative interviews and working with the Priming method. One central aim of her research is to find out whether primes influence language perception of informants. She has presented her work at numerous conferences, has published a book chapter in Yvonne Hettler et al. (Eds.), *Variation, Wandel, Wissen. Studien zum Hochdeutschen und Niederdeutschen*, and has written a paper for *Linguistik Online* with Christina A. Anders and Saskia Schröder. Email: palliwoda@germsem.uni-kiel.de

Dennis R. Preston is Regents Professor and Co-Director of the Center for Oklahoma Studies, Oklahoma State University and University Distinguished Professor Emeritus, Michigan State University. He has been a visiting professor at numerous US and overseas institutions and was Director of the 2003 LSA Institute at Michigan State. He was President of the American Dialect Society and has served on the Executive Boards of that society and many others. His work focuses on sociolinguistics and dialectology, and he has directed four recent NSF grants, two in folk linguistics and two in language variation and change. His most recent book-length publications are, with James Stanford, *Variation in Indigenous Languages* (2009); and, with Nancy Niedzielski, *A Reader in Sociophonetics* (2010). He is a fellow of the Linguistic Society of America and the Japan Society for the Promotion of Science and holds the Officer's Cross of the Order of Merit of the Polish Republic. Email: dennis.preston@okstate.edu

Saskia Schröder is a project assistant in the research project "Perceptual Dialectology – The German language area from the perspective of linguistic laymen" [Wahrnehmungs-dialektologie – Der deutsche Sprachraum aus der Sicht lingustischer Laien] at the University of Kiel, funded by the *Deutsche Forschungsgemeinschaft* (DFG). She is currently working toward her PhD which deals with the data ascertained in the research project. It focuses on the results of the pilesort-task which is a central part of the guided interviews that are used to reveal the laylinguists' mental maps of the German language area. She has presented her work at numerous conferences, has published an article in Yvonne Hettler et al. (Eds.), *Variation, Wandel, Wissen. Studien zum Hochdeutschen und Niederdeutschen*, and has co-authored a paper in Helen Christen and Evelyn Ziegler (Eds.), *Die Vermessung der Salienz (forschung)/Measuring (the Research on) Salience* (*Linguistik Online*). Email: schroeder@germsem.uni-kiel.de

Christoph Hare Svenstrup is a PhD student at the universities of Freiburg and Copenhagen. His research area is language attitudes to varieties within the Dialect-Standard continuum. His PhD project focuses on adolescents' attitudes to the (German) varieties spoken in the Stuttgart area in Southwest Germany and combines quantitative methods with qualitative methods in the investigation. Furthermore, Christoph is a part of the European language attitudes research network the SLICE project (Standard Language Ideology in Contemporary Europe), and he has published in the first two of the projects' publications, *Standard Languages and Language Standards in a Changinge Europe* and *Language (de)standardisation in Late Modern Europe: Experimental Studies*. Email: christophhare@gmail.com.

Dennis R. Preston

Chapter 1
Introduction

Just when we thought it would be uncontroversial, perhaps even innovative, to focus on cities rather than larger geographic expanses in the enterprise of perceptual (or folk) dialectology,[1] Britain (2012) tells us that the countryside has all the requisite detail that one needs to engage in modern, even sociolinguistically-oriented, dialectology.[2] Here is the concluding part of his introductory rationale:

> ... [W]e should put speakers at the centre of our theorisation about language, and if speakers move and come into contact with other speakers, the likely outcomes of that contact are typologically similar, but are insensitive to urban or rural location. (Britain 2012: 13)

I will not quibble with "typologically similar," for, if I understand it correctly, it would be absurd to claim that the underlying, cognitive conditions of human language differ in cities and the countryside. It is the "and if" condition that might trouble those who have focused on the city as an important and even distinctive site for the study of variation and change.[3] In this volume, the city is treated as a place where movement and contact are more intense and variable and provide, therefore, a no more or less valid but interestingly different environment for the study of perception and attitude.

This book focuses on such perceptions of and attitudes towards varieties of language, as well as the triggers of them and their influence on performance in urban settings. I will not pursue here either the history of folk linguistic and perceptual studies[4] or further justification of such work. The field is now well-

---

1 I do not mean to suggest that the chapters presented here are the first to report on perception in and of cities. Evans (2002a) focuses on Montréal, Hartley (2005) on Boston, Romanello (2002) on Lecce and Brindisi in Puglia, and substantial parts of Labov (1966) deal with perceptual data gleaned from his New York City respondents.

2 Britain as well as this introduction take seriously the dictum that "Dialectology without sociolinguistics at its core is a relic" (Chambers and Trudgill 1998: 188); Britain certainly does not suggest that a focus on rural areas would do away with sociolinguistic considerations.

3 And, therefore, I would assert, the city is an important place to do not only perceptual dialectology but also folk linguistics in general (broadly defined) because the study of variation and change risks faulty interpretation and may even miss explanatory opportunities without it (e.g., Weinreich et al. 1968; Preston, to appear).

4 The history of and early work in perceptual dialectology may be consulted in Preston 1999a. A more recent survey is Montgomery and Beal 2011. There is, unfortunately, no current compre-

established in the literature and further evidenced by its presence in all major sociolinguistic conferences and even in conferences devoted exclusively to it. I provide here only a brief review of the chapters, with the understanding that they illustrate an area of linguistic concern that is both well-established and of proven value[5].

In the chapter following this, Montgomery and Cramer outline methodological practices and recent advances. Particularly noteworthy are recent advances in GIS-based software that allow graphic and statistical generalizations of large numbers of hand-drawn maps and the possibility of comparison of these generalizations with other geographic-demographic facts made available in the large databases associated with these systems. The chapters by Cramer, Evans, Jeon and Cukor-Avila, and Montgomery all illustrate various uses of this technique. It is also the case that the more recent uses of discourse-analytic techniques in the study of language attitudes and ideologies have influenced perceptual studies. The chapters by Gasquet-Cyrus, Lonergan, Mann, Schröder and Palliwoda, and Svenstrup focus on interactions between respondents and between respondents and interviewers making use of these modes of investigation. Finally, Braber anchors her perceptual study of Nottingham in media representations, another developing technique in the study of public regard for language varieties, and Montgomery makes use of similar resources in arguing for the growing salience of the Manchester dialect. In Chapter 2, Montgomery and Cramer provide both a historical review of the field and a more detailed look at methods as they are employed in this volume.

The first section, "Differences in the perception of rural and urban areas," focuses on the first issue raised in this introduction. More directly than any other, it points to the production bias of Britain's position that little separates the city from the country. This production bias would seem to exclude perceptual studies, and, while one of the aims, perhaps the earliest, of perceptual dialectology was to compare professional versus folk representations of variety, the value of folk impressions that do not correspond to linguistic reality are just as important as those that do in the study of variation and change, and, as noted above, often form the backbone of explanatory interpretations.

Cramer's chapter (3) contrasts the perception and performance of northern versus southern features in Louisville, Kentucky, a city on the Ohio River, taken

---

hensive bibliography, but this volume contains extensive documentation in its various chapters.

5 Some may be convinced by its presence in Wikipedia: http://en.wikipedia.org/wiki/Perceptual_dialectology, but, alas, not the one for folk linguistics (http://en.wikipedia.org/wiki/Folk_linguistics), which is sorely lacking in breadth, depth, and documentation.

by many to be the major dividing line between these two US speech areas (e.g., Labov et al. 2006). Cramer first determines by means of a map-drawing task where her Louisville respondents believe dialect differences lie on a map of 13 nearby states, 7 of which actually border Kentucky, resulting in 11 distinctions, 3 within the state itself: Louisville-Lexington (the major urban areas), Eastern Kentucky (an Appalachian area), and the rest of the state, this last clearly regarded as rural. Cramer goes on to study the attitudes of her Louisville respondents to these areas along both status and solidarity dimensions. Finally, Cramer studies the perception-production interface by establishing a baseline of /ai/-monophthongization (from a media source), a well-known and stereotypical feature of U.S. Southern speech in general (e.g., Plichta and Preston 2005), and then compares her evaluation data with production data, making the perception-production link suggested above and illustrating the importance of the perceived urban-rural distinction in the way these border-Southerner respondents perform their own identities.

In Chapter 4 Evans studies a similar urban-rural divide in the Pacific northwestern US state of Washington. Evans had her respondents carry out a map-drawing task, only for the state, and classified the labels written by the respondents on the map that were associated with specific areas, making use of a technique in the study of area language perceptions introduced by Hartley and Preston (1999). Since these labels can be entered in the GIS described above, along with any other geographic and demographic material already available in the system, Evans is able not only to map where Washingtonians believe different speech areas are in the State but also to show precisely the areas that reflect labels of such urban-rural dichotomies as "hick" (rural) versus "gangsta" (urban). But this chapter is not just an exercise in beautiful map-making. Evans goes on to show that, no matter what the linguistic differences are, the urban-rural dichotomy is an extremely strong divide in the folk linguistic mentality of her respondents, a fact surely worth showing and, as suggested above, one important to any study of language variety in the area.

Mann's study is the first I know of in which gay speech is considered in geographical terms, but it is not surprising to find relevance here. Stereotypes of Southern culture in the US strongly embed notions of "manliness" and such "manly" pursuits as hunting, fishing, auto racing, fighting, and sports, which, along with such other stereotypes as the predominance of conservative Christian beliefs, allow Mann to explore the apparent cognitive folk linguistic dissonance involved in "talking gay" and "being Southern." His interviews with gay respondents, however, reveal odd correlations (vowel length) that appear to exist in both gay and Southern speech, causing one respondent to suggest that such a fact "... throws your gaydar all completely out of whack." Although

Mann conducts experiments (voice sample evaluations and identifications), the main focus of this chapter is on the focus group interview settings and the elaboration of the details obtained. Its fit in this section is confirmed by the interesting confirmation of the rural stereotype of Southern US varieties.

The last chapter in this section focuses on the perception of the urban-rural divide in Korea, using 436 respondent hand-drawn maps obtained from both inside and outside the major city of Seoul. Although one goal of the chapter was to compare folk and perceptual dialect boundaries, the goal that justifies the chapter's placement in this section was that of comparing insider and outsider perceptions of the urban and rural areas of Korea, utilizing, as in Evans in Chapter 4, recent GIS map-making technology, and, also as in Evans' work, identifying qualitative comments on varieties written on the maps by the respondents. The chapter focuses on the urban-rural split through the prism of *standardness*, a concept used in labeling one or more areas by more than 96 % of the respondents. These sophisticated maps allow the authors to show that indeed Seoul is most associated with the "standard" labels but that all rural areas are not equal for labels associated with "nonstandardness": the South and Southeast are most intensely identified as such, and, more importantly, these nonstandard rural areas are much more intensely identified as nonstandard than the urban areas are as standard, perhaps a regional commonplace (e.g., see the intense recognition of the stereotypically nonstandard U.S. South in Preston 1996: 305). These findings are further elaborated on by separating and mapping urban and rural respondent performances, showing a much wider area of standardness for the urban than for the rural respondents.

The second section of the volume, "Processes of perception and language change," explores the cognitive and speech perception characteristics of perception, particularly in the area of identification, as well as potential influences from perception on production and vice versa, focusing on areas where rapid, ongoing change appears to be taking place.

Moosmüller's contribution (Chapter 7) deals with "dark" /l/ in Vienna and is a model of sophisticated acoustic work on the production/perception interface. Local knowledge clearly identifies dark /l/ as a stereotype of lower status use in the city. Authentic working class Viennese recordings and actor-imitations of the Viennese were used and the actor performances confirmed public or folk knowledge of the dark /l/ stereotype, although a strong sex difference showed up in the authentic performances – men were much more frequent users of the stigmatized form than women, but the female actors used the same high rate as the male actors and native speakers, suggesting an all-or-nothing character in at least some folk perceptions. At least partly as a result of this failure, in a perception test that asked about "Viennese dialect authenticity," respondents rated the authentic female speakers lower than the female actors, further confirming the importance of the feature, although native speakers of the dialect are by no means required to use it categorically, and the "best" raters were not fooled by this variability, and, perhaps not surprisingly, they were all from Vienna. They were also the ones who most consistently distinguished between the actor and authentic performances in their assignments of authenticity, a finding earlier pointed out for studies of imitations of Hessian in which all the imitators were identified as such by dialect-speaking local respondents (Purschke 2011). Moosmüller concludes that indeed both insiders and outsiders know the status-linked stereotypicality of dark /l/ but that only a few "best" raters know the sociolinguistic variability associated with it.

Chapter 8 directly explores Kristiansen's (2009) contention that the direction of variety change is represented best in nonconscious attitude sampling rather than in conscious labeling or interview tasks. Svenstrup's work complicates this in Southwest Germany; indeed, his respondents find Hochdeutsch (the national standard) highly evaluated in a vocal guise task, but it is also highly rated and, more importantly, not always seen as distinct from the local variety in conscious tasks. This evaluation of the local as a "local standard" complicates the findings from Denmark in a way that reflects the local "destandardization" process seen in much of Europe (e.g., Kristiansen and Grondelaers 2013). But the meat of the offering here is in the sensitive and careful analysis of conversations with respondents about the boundaries and meanings of local varieties. The group interview settings clearly reveal conscious understandings of the mixed varieties that are emerging, at least in part of Germany, as Svenstrup's respondents indicate by the use of such labels as "neuschwäbisch" and "youth-schwäbisch." The sequential interaction methodology adopted here (Auer 1995) proves very useful for the investigation of metalinguistic commentary on language variety.

In Chapter 9, Gasquet-Cyrus offers a look at Marseilles, France's second largest city, a site often perceived as having a single, local accent, but we find that locals know better, and this chapter explores an intensive ethnolinguistic project devoted to uncovering local perception, one largely carried out through the investigation of metalinguistic ("epilinguistic" in the French tradition) commentary by nonlinguist respondents, although the data form only a part of the larger investigation of variety in the city. Gasquet-Cyrus first establishes the enregisterment of the Provençal dialect in all of France; it is, as Southern US English has been shown to be (e.g., Preston 1996), the most salient variety in the country, and, like Southern US English, some of its attributes are positive, but local speakers still apparently suffer from linguistic insecurity; at the regional level, in fact, the dialect of the city is regarded negatively compared to that of the surrounding area, a clear contrast between "tough" city folk and "pleasant" countryside people, just as Evans found in Washington in Chapter 4. The heart of this chapter lies in its sensitive enquiry into the varieties of French in the city (tripartite it turns out) and the way in which the respondents associate territoriality and social groups, not surprisingly assigning masculine value to the oldest, most centrally located, vernacular variety, but at the same time paying careful attention to the designation of the emerging standard-influenced and immigrant-based varieties. The deployment of rich, ethnographic interpretive methods in the collected interview data stand out in this chapter as well as do the integrative perspectives (uses of media, literature, etc.) taken on the various means of data interpretation.

The final section of the volume, "The relationship between perception and 'reality'," returns to the oldest question in perceptual dialectology: How good are the folk? Do their boundaries correspond to those drawn by professional dialectologists? In Chapter 10, Montgomery offers first a sophisticated GIS-based restudy of production studies based on line density analysis, resulting in a comprehensive map that I suspect will be much-used in future discussions of areal differences in Great Britain. With this map in hand, Montgomery goes on to compare it to three perceptual studies: Inoue ([1996] 1999a) and two of his own (2007 and 2012), allowing him to also suggest whether there has been any change in perception over time. Montgomery suspects that in spite of his denial, Inoue's perceptual map is not such a bad fit, at least for major areas, but the most important reflections here deal with perceptual change and the proximity effect, i.e., the idea that geographically closer areas are perceived with greater detail and perhaps even accuracy, although Montgomery notes that "cultural salience" may also be an important factor in the recognition of an area, even a distant one. He makes use of this notion of salience in his suggestion that the leveling of regional speech areas in Great Britain has been supplemented by the rise of the salience of cities, the major difference between his maps and Inoue's older one, and another argument for the importance of perceptual work in urban areas.

In Chapter 11 Braber studies a city, Nottingham, that is not particularly salient in the overall scheme of English urban areas but one that reveals interesting internal perceptions: her respondents are not good at recognizing local speech but they are nearly uniform in attesting to its nonstandardness. Braber first asks them to carry out a typical hand-drawn map task but adds to it a recognition task, in which the respondents were to indicate on the map where they believed 14 regional speech samples were from. In the map-drawing task, there is confirmation of Montgomery's claim that cities have grown in importance since the most frequently identified areas are Liverpool, Newcastle, and

Birmingham (though not Manchester); Nottingham itself was in 10th place, identified by only 29 %. Interestingly, the accuracy in the voice placement task mirrored this perceptual salience: Liverpool, Newcastle, and Birmingham voices were accurately identified, but the local voices were not, suggesting a much greater importance for salience than proximity for these respondents. Although this chapter also includes very interesting data from a "mind map" of the local and surrounding areas, Braber's interpretation of the lack of local salience is gleaned from media resources and makes the interesting proposal that the lack of linguistic salience (or the denial of it) may stem from local insecurity about Nottingham (e.g., culture, crime, etc.) itself in spite of locally distinct linguistic practices, an interesting case of cultural insecurity overwhelming any that might derive from local linguistic practices.

Lonergan's study of Dublin in Chapter 12 is another example of a careful comparison between perception and actual linguistic production. Like Gasquet-Cyrus in Chapter 9, Lonergan wants to know how many varieties in the city are perceived by locals and, again, based on a hand-drawn map study, comes up with three: "inner-city," "northside," and "D4" (the last a postcode designation for a prosperous southern area of the city), but professional study does not support the distinctions. In-depth interviews revealed much more nuanced characterizations of the varieties, which include the covert prestige for their own variety among the inner-city respondents, and an odd association with "American accents" for D4; a ranking of the perceived areas shows a strong correlation between the "correct" and "pleasant" designations (as in Hartley 1999 for the state of Oregon in the US). The chapter concludes with copious and interesting conversational data that support and supplement the drawing and assessment tasks.

Schröder and Palliwoda address the question of salience directly in the final chapter of this collection (13), drawing their data from a larger study of German perceptual dialectology (or *Wahrnehmungsdialektologie*). Their work, however, focuses in particular on the salience of specific features and goes about investigating the relative salience by means of a quiz and extensive interviews. The quiz involved associating a voice sample (a read folktale) with a city, and the cities were derived from a previous perceptual task (the pile-sort method, e.g., Tamasi 2003). Schröder and Palliwoda then deployed a method that identified the time of city assignment with the point in the reading when the assignment was made, the assumption being that the linguistic feature that was most responsible for the assignment could be identified in this way. Although that finely-grained study is not a part of this chapter, it is an extremely compelling approach. Included here, however, are identification correctness and evaluative scores, in which, interestingly, standard German was least well identified but rated "most pleasant" and "most correct." Most interestingly in this chapter, however, the authors derive the most common comments about varieties from their extensive interview data and compile them in easy-to-read pie charts which reveal salient contrasts: the Eastphalian sample is characterized by "standardness" and "clarity," while the High Alemannic variety is most often singled out for its pronunciation characteristics. Once these folk characterizations are combined with the identification scores and the timed response data, this long-range study will prove to be a valuable methods handbook for future research.

This volume shows us that the city and the city-countryside divide loom large in the folk understanding of language variety. Perhaps we could learn everything that is to be learned about the linguistic outcomes of contact and mobility by studying only the countryside, but the modern landscape and peoplescape make me doubt that. If what people believe about language and its distribution is important, then this focus on the city is a most welcome addition.

This volume also highlights the maturing of perceptual studies with experimental, technical, and discoursal detail in particular. New techniques in implicit judgment tests, the ability to present resynthesized samples, and GIS-mapping all contribute to the breadth and depth of the enterprise, and there is no doubt that the careful investigation of respondent interactive and discoursal representations form a rich part of the modes of enquiry. Although the volume focuses on the city and the urban-rural split, there is much here that should engage any scholar interested in what the folk say and feel about languages and varieties.

Chris Montgomery and Jennifer Cramer

Chapter 2 Developing methods in Perceptual Dialectology

1 Introduction

In many traditional dialectology studies, speakers' perceptions, attitudes, and beliefs about linguistic varieties and variation were not included. In general, many dialectologists, particularly those from the American structuralist tradition (more particularly, those heavily influenced by Bloomfield) have not been interested in the overt opinions of non-linguists, as non-linguists' perceptions and attitudes have been generally assumed to be secondary to the analysis of "real" data, such as the phonetic and lexical variables used in traditional dialectology (Preston 1989; Niedzielski and Preston 2000; Benson 2003).

Despite the prevalence of the Bloomfieldian perspective in the 1960s, Hoenigswald (1966) incited interest in the beliefs of speakers, or *the folk*, in linguistic research. He suggested that linguists should be concerned not only with language as production but also with how people react to language and how people represent language in talk about language. Thus, a field referred to as *folk linguistics* was established, and work done by Dennis Preston, among others, in the 1980s and later, emphasized the importance of language attitudes and perceptions in the study of linguistic variation.

But, as Preston has noted[1], this was not the actual beginning of the field. Perceptual Dialectology[2], a branch of folk linguistics that has its focus in what non-linguists say about language variation, including where they think it comes from, where they think it exists, and why they think it happens, has its earliest roots in the Dutch and Japanese traditions. In a 1939 Dutch dialect survey, respondents were asked to identify areas in which people spoke either similarly or differently. Rensink's work in this survey ([1955] 1999) utilized the

1 For a more detailed account of the history of the field, as well as early research in previous traditions, see the Introduction (Preston 1999b) and Parts I and II of the Handbook of Perceptual Dialectology (Preston 1999a).

2 A terminological note: Preston uses the term *Perceptual Dialectology*, as opposed to *folk dialectology*, because the word *folk* is often understood to mean "false" (Preston 1999b). We follow this same practice throughout the collection.

little-arrow method developed by Weijnen (1946, cited in Preston 1999a). This method linked respondents' home areas to the locations they described as linguistically similar.

Similarly, in Japan, a tradition for accounting for the beliefs people held about language was developed, amid some controversy. Sibata ([1959] 1999) undertook a study in which respondents were asked to list villages in which people spoke differently than people in their own villages. Not aware of the little-arrow method, Sibata processed his data using increasingly thick lines to delineate 'difference boundaries'. Having done this, Sibata found that the perceived boundaries did not match the production boundaries and therefore determined them to be uninteresting. Noting this, Grootaers (1959, [1964] 1999), much like Bloomfield, complained that speaker perceptions were too subjective and, therefore, not very valuable. However, Weijnen ([1968] 1999), whose little-arrow method had been successful in the Netherlands, responded to these claims, heightening the debate about how (or if) non-linguists' perceptions can inform those interested in language variation.

This controversy, however, did not spell the end for Perceptual Dialectology. Preston's own work, which he refers to as the "modern" trend in Perceptual Dialectology (Preston 1999b: xxxiii), has produced a wealth of knowledge in the subject matter. Typically, the methods in this tradition have included the following: a draw-a-map task, which, drawing heavily on methods in cultural geography (Gould and White 1986), is designed to elicit where non-linguists believe dialect boundaries exist (see below); a degree-of-difference rating task, wherein a participant rates how different a variety is with respect to his or her own variety; pleasantness and correctness evaluations, in which participants rate how pleasant or how correct a certain way of speaking is; voice placing tasks, which require participants to estimate, given an audio sample, from where a particular voice comes; and qualitative analysis of other data, including labels used in the draw-a-map task, focus group reactions to voices, and interviews in which participants overtly share language beliefs. These methods have been employed not only in the United States, which has served as the center for modern Perceptual Dialectology studies (e.g., Preston 1989; Hartley 1999; Fought 2002; Benson 2003; Fridland, Bartlett, and Kreuz 2004; Fridland, Bartlett, and Kreuz 2005; Hartley 2005; Fridland and Bartlett 2006; Bucholtz et al. 2007; Bucholtz et al. 2008; Cramer 2010), but also in numerous locations around the globe, including Brazil (Preston 1989), France (Kuiper 1999), Germany (Dailey-O'Cain 1999; Diercks 2002), Great Britain (Inoue [1996] 1999b; Montgomery 2007), Italy (Romanello 2002), Japan (Inoue [1995] 1999a; Long 1999a, 1999b), Korea (Long and Yim 2002), Spain (Moreno and Moreno 2002), Turkey (Demirci and Kleiner 1999; Demirci 2002), Wales (Coupland, Williams, and Garrett 1999; Williams, Garrett, and Coupland 1999), among many others.

With his work leading the way, Preston has shown linguists why the perceptions of language users matter for linguistics. Preston indicated that

> [w]ithout knowledge of the value-ridden classifications of language and language status and function by the folk, without knowledge of where the folk believe differences exist, without knowledge of where they are capable of hearing major and minor differences, and, most importantly, without knowledge of how the folk bring their beliefs about language to bear on their solutions to linguistic problems, the study of language attitudes risks being: 1) a venture into the investigation of academic distinctions which distort the folk reality or tell only a partial truth or, worse, 2) a misadventure into the study of theatrically exaggerated speech caricatures. (Preston 1993: 252)

Thus, the work of folk linguists/perceptual dialectologists can serve to bridge the gap left by linguists and social psychologists in bypassing the interrelatedness of ideologies, society, and linguistic practices (Milroy and Preston 1999). In what follows, we further explore advances in the methods used in Perceptual Dialectology, which seek to discover more about the different ways in which non-linguists perceive the linguistic landscapes in which they live. Recent years have seen development in both the theory and methods of Perceptual Dialectology, and this chapter provides a survey of such research (i.e., Anders 2010; Montgomery 2011a; Purschke 2011a) in order to provide an overview of the current state of the field.

2 The draw-a-map task

Forming part of the suite of methods in Perceptual Dialectology noted above, the draw-a-map task (Preston 1982) is perhaps the best known of all the methods of Perceptual Dialectology. It involves directly accessing respondents' mental maps (Gould and White 1986) of language variation. At the core of the draw-a-map task is a blank map, into which respondents are invited to add lines which indicate the boundaries of dialect areas in a task that lasts between 10 and 20 minutes (Montgomery 2007). These respondent-added dialect areas are given names by the map drawers, and other information about the areas may also be added to the map. This can include a range of linguistic and nonlinguistic information (see Long 1999b for a method of classifying such information). Thus, via the draw-a-map task, data relating to the names, placement, and extent of dialect areas can be collected. In addition to this, further information about linguistic features of the dialects and attitudes towards the speakers who live in the areas can also be gathered.

The draw-a-map task's evolution owes a good amount to research methods developed by perceptual geographers in the 1960s and 1970s. Such research was initially developed to understand the way in which residents of cities thought about major landmarks and the travel routes they used (Lynch 1960; Goodey et al. 1971). The research by both Lynch and Goodey's team involved respondents producing "quick, unaided impression[s]" (Gould and White 1986: 12) of the cities in which they lived. The hand-drawn impressions of the city were then combined in order to understand respondents' collective perception of the environment. Such collective impressions are useful in applied contexts such as town planning, but can also shed light on the differential perceptions that groups might have of their environments.

Such perceptual differences were noted by Orleans (1973) in his research in Los Angeles. The combined results from his mapping task demonstrated a link between socioeconomic class and perceptual information, with white, upper class informants from a well-off suburb of Los Angeles (Westwood) providing "a well-formed, and generalised image of the entire Los Angeles Basin," (Orleans 1973: 118), in contrast to Spanish-speaking informants from the centre of the city whose knowledge was "confined to a few city blocks" (Orleans 1973: 118). Goodchild (1974) found a similar socioeconomic class effect in the English town of Market Drayton, further demonstrating the way in which different groups might experience their environment differently. Socioeconomic factors were found to impact the way in which people experience the world around them, with residential location also having an effect, as demonstrated by Gould and White's (1986: 42) "local dome of preference" effect. Such an effect showed that respondents tended to view their own familiar area in a positive light despite rating other areas as desirable.

Perceptual geographers' findings, such as those outlined above, were made possible by the ability of the varied researchers working in the field to generalise the maps their respondents had drawn, although researchers such as Ladd (1970) used individual maps to explore qualitative aspects of the way in which respondents viewed their environments. As we discuss below, the draw-a-map task in Perceptual Dialectology has primarily been used quantitatively, with map generalisation a central aim of the discipline (Preston and Howe 1987). Although the benefits of qualitative aspects of data from draw-a-map tasks should not be overlooked, the remainder of this section and the next will deal with the way in which the draw-a-map task has developed over time as a quantitative method.

As we mention above, at its most basic level, the draw-a-map task consists of a blank map onto which lines are drawn by respondents. However the development of the "blank" draw-a-map task has seen various levels of detail applied to the base maps in an effort to address a lack of geographical knowledge amongst respondents (see the discussion in Wales 2006: 58–59). Although

Preston's (1982) first foray into the field using the draw-a-map task used a completely blank map of the United States, "the resulting confusion was so great that it became necessary to use a map state lines" (Niedzielski and Preston 2003: 46). In further research, Preston (1986a) used such a state line map, which itself is not without its problems (Carver 1991: 434). However, as "folk dialectological research is confounded with folk geography" (Niedzielski and Preston 2003: 47) to some extent, the use of some geographical information on maps is generally regarded as necessary (although see Hofer 2004) in order to provide some means of comparison with other datasets (both linguistic and non-linguistic).

Lameli et al. (2008) have investigated the effect of the inclusion of various levels and types of detail on base maps used in draw-a-map tasks in Germany. Montgomery (2007) and Jeon (2011) have also examined this in England and Korea respectively, and Cukor-Avila et al.'s (2012) research in Texas used different map types with the hope of reporting on the impact of the different levels of detail. Lameli and his colleagues used seven different map types which were completed by between 23 and 26 respondents each. The seven maps contained an increasing level of detail, from a basic state outline to a map with a large amount of geographical data including cities, federal state boundaries, and rivers. In analysis of the results from the completed tasks, the map with the largest amount of data resulted in a greater mean number of areas being added to the maps and the map with simply the state outline the least (Lameli, Purschke, and Kehrein 2008: 62), although the greatest number of areas added by one individual were added to the state outline map. Despite these results, there were no statistically significant differences between the results for each map type save for a map including only relief data, which produced significantly fewer areas than maps with other detail levels. These findings indicate that the level of detail included on maps does not have a significant impact on the number of areas drawn.

These findings are slightly different from those of Montgomery (2007), who trialed different levels of detail whilst developing his draw-a-map task in England and Wales. Montgomery's research trials involved pilot studies using different panels of respondents responding. The first panel of respondents was given a map containing the outlines of England and Wales and asked to draw dialect areas with no additional assistance. The second and third pilot study's group of respondents were supplied with the same outline map, but were shown a separate map with the location of major towns and cities whilst they completed the task. The final pilot study allowed respondents to consult the location map for the first half of the task and replace the wholly blank map with one that contained dots indicating the location of major cities in England.

Tab. 2.1: Development of the draw-a-map method in Montgomery's (2007) research (darker shading indicates greater number of times areas recognised across pilot studies). Respondent recognition (RR) percentage indicated for each dialect area.

Pilot study 1 (*n* = 130) Mixed locations		**Pilot study 2 (*n* = 16) Newcastle**		**Pilot study 3 (*n* = 14) Sheffield**		**Pilot study 4 (*n* = 14) Hull**	
Area	**% RR**	**Area**	**% RR**	**Area**	**% RR**	**Area**	**% RR**
Geordie	27.7	Scouse	81.3	Geordie	100	Yorkshire	66.7
Yorkshire	20.8	Cockney	81.3	Yorkshire	92.9	Scouse	63.3
Scouse	20	Yorkshire	75	Midlands	78.6	London	53.3
Cornish	19.2	Geordie	68.8	Scouse	71.4	Brummie	46.7
Manc	19.2	Manc	43.8	Cockney	71.4	Geordie	46.7
London	16.2	Cornish	43.8	Manc	57.1	Midlands	26.7
Brummie	16.2	Brummie	37.5	Brummie	57.1	West Country	26.7
Cockney	11.5	Northumbrian	31.3	Cornish	35.7	Manc	23.3
Midlands	8.5	North East	25	Southern	28.6	East Anglia	20
Norfolk	8.5	Cumbria	25	East Anglia	28.6	Bristol	13.3

Respondents were requested to add the names of the cities to the dots before adding dialect area lines and labels. Table 2.1 shows the result of the pilot studies, in terms of areas added to the maps.

Table 2.1 shows the top ten areas recognised in each pilot study, and demonstrates that similar areas were recognised across each pilot study. The table demonstrates that the recognised areas are very similar, with only 12.5 % of areas appearing in only one study, and 50 % of the areas appearing in all four of the pilots. The table also reveals the impact of the changing method on the way in which respondents completed the task, and in particular that the additional support given to respondents had the effect of increasing respondent confidence. The first pilot study's recognition rates are particularly low, with no area receiving a higher rate than 27.7 % and a mean rate of 16.8 % (over the top ten areas). This rises to a mean of 51.3 % in the second study and 62.1 % in the third. The fourth study saw a drop in the recognition rates to a mean of 38.7 %. This drop coincided with the introduction of the city location dots, which were included in order to provide some 'anchor' points in the country which would help to orient respondents when completing their maps. Such anchor points were needed as the geographical accuracy of the respondents' placement of areas in the previous two pilot studies was in some cases very poor, resulting in the need to discard some of the maps. The use of the location dots and the location map therefore improved recognition rates from 16.8 % to 38.7 % whilst ensuring that the areas added to maps had some geographical underpinning.

Montgomery's (2007) and Lameli et al.'s (2008) tests of the draw-a-map method both demonstrate that the draw-a-map task method itself does not influence the results, as Carver (1991: 434) has contended. Instead, the task allows respondents to discuss their perceptions of dialect variation in ways that they had not previously been able to do. Adding various levels of detail to the map will either have no effect on the number of areas added to maps, as in the research in Germany, or will increase the confidence of respondents when adding areas, as in the research in England and Wales. In neither case does the addition of data influence the areas that are added, suggesting that the draw-a-map task is a robust research method that can be used to shed light on the way in which the dialect landscape is perceived.

Of course, the scale of the research area is something that is an important consideration when detail is being added to the map onto which areas will be drawn. Generally, the larger the area, the fewer details can be included. Thus, country-wide surveys have tended to rely on state outlines or maps with federal boundaries indicated (e.g., Preston 1986a; Williams, Garrett, and Coupland 1996; Hartley 2005). Studies that have looked at the perception in parts of countries or single federal states have similarly used basic outline maps (e.g., Evans 2002a; Benson 2003; Cramer 2010), although at this scale more detailed maps have also been used. Major roads and the locations of cities were used in Evans' (2011a) research in Washington State, whilst Purschke (2011b: 97) used the locations of cities in his research in Germany, as did Demirci (2002). City dots were used in Bucholtz et al.'s (2007; 2008) research in California, (although in this research, location dots were not to be labelled as in Montgomery's (2007) research), whilst Inoue's (1999b) research in England used county boundaries. In smaller areas still, more detail has typically been added to maps, as in Romanello's (2002) research in two Italian cities and Stoeckle's (2012) work in the South West of Germany, where cities, roads, and administrative boundaries were included.

Despite the proliferation of studies using different map types at different scales, the research questions that underpin the use of the draw-a-map task remain the same. Specifically, these concern what non-linguists think about:

a) The similarities of their own speech to, and differences from, the speech of other areas.
b) What [where] they believe dialect areas to be.
c) The characteristics of regional speech. (Preston 1988: 475–6)

The draw-a-map task has been, and continues to be, a valuable tool that goes some way to addressing these questions. The following section deals with the ways in which draw-a-map task data can be processed, and what this means for the future of the task.

3 Processing and aggregating draw-a-map data

As noted above and in Montgomery and Stoeckle (2013: 52), aggregating draw-a-map data is, and always has been, an important part of the approach. It is surprising, therefore, that a reliable, robust, and widely-useable method of aggregating lines drawn on maps by respondents proved such an elusive goal for perceptual dialectologists for so long. Perceptual geographers were able to generalize their data as a result of what they were asking of their respondents. Research dealing with the perception of cities tended to focus on the routes respondents used to get around, along with prominent landmarks (e.g., Goodey et al. 1971; Orleans 1973). Other perceptual geographers, such as Gould and White (1986), were concerned with perceptions of the desirability of locations as residential areas. They instituted ranking tasks based on pre-existing geographical boundaries (something adopted by Preston 1999c in his correctness, pleasantness, and degree-of-difference tasks). In both of these instances, aggregation of data was relatively straightforward as respondents could only complete tasks in specified ways, and superfluous data could be ignored.

As draw-a-map tasks typically ask respondents for completely unaided maps of dialect areas[3], there is less control over the maps that respondents will complete. Such a lack of prescription about the ways in which a respondent can complete the task means that maps will often provide very rich, detailed, and varied data that reflects each respondent's individual experience of the dialect landscape. This is an undoubted benefit of the method, but it does also mean that the data types gathered are quite varied. The areas added to the map can be counted to assess a rank-order of perceptual prominence (see Table 2.1 and Bucholtz et al. 2007; Bucholtz et al. 2008). Attitudinal data can also be extracted from comments on maps, along with other commentary data which can be classified according to methods outlined in Long (1999b), as noted above. Although these methods shed light on aspects of perceptions of the dialect landscape, they neglect the key geographical component of the data gathered draw-a-map tasks, namely the information relating to perceived placement and extent of dialect areas.

---

3 The use of the word 'dialect' is often avoided in instructions for respondents, as the use of this term might lead respondents to activate concepts other than those aimed for. Instead, many studies ask for perceptions of different or similar speech (e.g., Benson 2003: 312).

It is data relating to the placement and extent of dialect areas that proved difficult to work with for some time. Preston's first use of the draw-a-map task (1982) trailed a generalization technique that examined clusters of lines (Preston and Howe 1987: 363), although this technique was less suitable for instances in which lines did not coincide, as observed by Carver (1991: 434). Lack of line coincidence does not make generalization "meaningless" (Carver 1991: 434), as despite a lack of agreement, "cognitive 'landscapes' of regions and dialects do exist [...] influenced by a mish-mash of cultural factors, [which] therefore have their own kind of 'reality'" (Wales 2006: 59) for each respondent engaged in a draw-a-map task. A method of reflecting the "reality" of perceptual dialect areas whilst also acknowledging that there is more or less agreement over the location and extent of areas was therefore needed. Preston and Howe (1987: 371) introduced an early attempt to do this using shaded maps. Lines were traced with a digitizing pad in order to enter them into a computer. Once all lines had been added for a given area, the user was then able to output the extent of a given area at various levels of "agreement". This method was the first to provide a generalization technique that acknowledged the particular type of geographical data captured by the draw-a-map task.

There were limitations to the method however, not least a relatively poor graphical resolution and the need to specify agreement levels to be outputted in advance. It was clear however that using shaded maps was the best solution to the problem posed by draw-a-map task data. Onishi and Long (1997) developed *Perceptual Dialectology Quantifier for Windows* (PDQ) which was able to build on Preston and Howe's method. It again relied on tracing lines using a digitizing pad but also allowed various types of data to be attached to each line. Such data included basic biographical data (sex, and age), which permitted results to be queried after processing. The output from PDQ displayed perceptual dialect areas in relation to agreement, just as Preston and Howe's method allowed. However, instead of having to instruct the programme to output an agreement level PDQ displayed all of the data with different colours or patterns showing different levels of agreement. For the first time it was possible to see on the same map the extent of agreement about the placement and extent of perceptual dialect areas.

Both of these early attempts to process data from the draw-a-map task permitted greater understanding of the perception of the dialect landscape and were excellent uses of the technology available at the time. Neither method was designed to be portable, however, and required expensive and increasingly obsolete equipment to input the line data[4]. Furthermore, both of the

4 Indeed, PDQ only ran on a small number of increasingly elderly computers housed at Tokyo Metropolitan University, and data processing using this method necessitates a trip to Japan.

methods allowed the display of only one dialect area at a time and used rather a crude graphical resolution which made publication of the data difficult. In addition to these issues, both methods' outputs were graphical only. This means that they were not geographically "anchored" in any way, and bore no relationship to real-world coordinates. This meant that comparisons with other types of data (both linguistic and non-linguistic) were made difficult. A new method that addressed these problems was needed.

Although there have been recent innovations using overlay techniques in vector graphics programmes, such as CorelDraw or Adobe Illustrator, (e.g., Purschke 2011b) the most widely used method for processing draw-a-map task data now uses Geographical Information Systems (GIS). A GIS is defined as a system which integrates the three basic elements of hardware, software, and data "for capturing, managing, analysing, and displaying all forms of geographically referenced information" (ESRI 2011). A GIS is particularly well-suited for working with draw-a-map data as it works with layers of data that can be combined in order to produce composite maps, which is precisely what perceptual dialectologists have strived to do with their map-based data. There are currently a number of GIS approaches to working with draw-a-map data, although most use the proprietary software package ArcGIS. Montgomery and Stoeckle (2013) outline one such method, although slightly different approaches have been taken by Jeon (2011), Cramer (2010), and Evans (2011a; Evans 2011b). In essence, all of the methods convert the line data from maps to area data that can be overlaid and then calculate the levels of agreement in order to produce gradient shaded areas that display the extent of agreement over the placement and extent of perceptual dialect areas. On the face of it, this is not a much of an improvement on the PDQ method, although it does address some of the problems with graphical resolution and the ability to display more than one dialect area at a time. In addition, results are able to be queried more effectively, resulting in greater understanding of the data. A major strength of using GIS is that the processed data is geographically meaningful, which in turn means that it can be compared with other geographically referenced data. For example, in the UK, this means that comparisons between perceptual area data and data from the BBC Voices survey processed using GIS (Holliday et al. 2013) can be examined together, as can any other geographically referenced datasets. Across the globe, the adoption of GIS by perceptual dialectologists means that their data can be shared, compared, and used alongside other geographical data, permitting ever increasing amounts of "crosstalk" (Nerbonne 2007) within and between disciplines and further helping to explain the patterns in draw-a-map data.

4 Research in language regard

In addition to the draw-a-map task, studies conducted within a Perceptual Dialectology framework also employ methods of analysis that deal more generally with language perception and metalinguistic beliefs about language, or *language regard* (Preston 2013). This area has long been fertile ground for perceptual dialectologists and those interested in non-linguists' reactions to spoken and conceptual data (Giles 1970; Williams, Garrett, and Coupland 1999; Evans 2002b; Goeman 2002; Coupland and Bishop 2007). The study of how people perceive language has been variably referred to as the study of *language beliefs*, *language attitudes*, or *language ideologies*, and this area of focus has a long history in anthropology and social psychology.

Silverstein (1979) defined language ideologies as a "sets of beliefs about language articulated by users as a rationalization or justification of perceived language structure and use" (1979: 193). Similarly, Irvine (1989) describes language ideologies as "the cultural system of ideas about social and linguistic relationships, together with their loading of moral and political interests" (1989: 255). Such ideologies are seen as imbued with the political, social, and moral issues prevalent within a community (Irvine and Gal 2000). A group has multiple ideologies, which are context-specific and constructed over time through the experiences of individuals (Kroskrity 2004). But our social constructions are based on more than direct sensory input; in fact, ideologies can come about from our perceptions of certain sensory information (Edwards 1999). These ideologies are seen as dynamic entities, not static ones (Woolard 1992; Kroskrity 2004).

A great deal of research in this domain has served to showcase how language ideologies, linguistic form, and social use of said forms are interconnected. Each one is thought to shape and inform the others, and within this, "language ideology is a mediating link between social structures and forms of talk" (Woolard 1992: 235). Preston (2010) presents a schematic of how language and reactions to language, both conscious and unconscious, are interconnected, as in Figure 2.1.

Some have found it quite difficult, however, to focus on all three variables at once (Woolard 2008). For instance, while variationist studies and the study of language attitudes have often seemed to be separate ventures within linguistics (Milroy 2004), social psychologists who do explore language attitudes tend to also ignore linguistic variation (Edwards 1999; Milroy and Preston 1999). Milroy claims that there is a need for "a framework for incorporating into mainstream variationist work an account of language attitudes, treated as manifestations of locally constructed language ideologies" (2004: 161). Her approach,

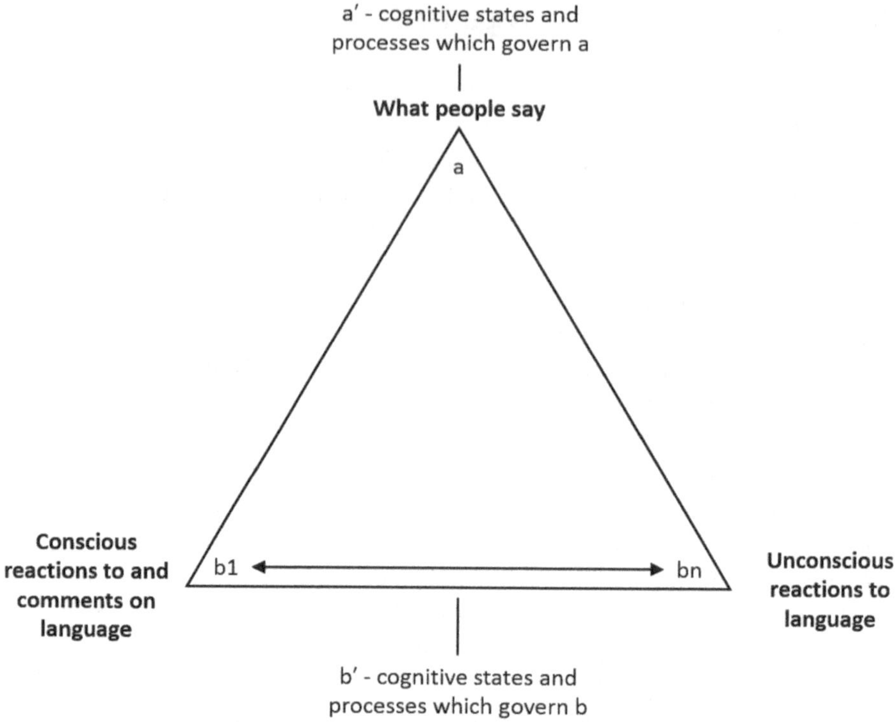

Fig. 2.1: Connections between language use and reactions to language use (based on Preston 2010, adapted from Niedzielski and Preston 2000: 26).

which draws on Labov's (1963) classic study of variation in Martha's Vineyard and more recent work by Eckert (2000), emphasizes ethnographic detail and indexicality in identity and ideology research. Such a framework encourages an understanding of locally relevant social categories before conducting research in and on a place, allowing the researcher to better understand the ideological motivations of group members to affiliate with a particular group.

Perceptual Dialectology, as a whole, provides a framework similar to Milroy's in goal, though perhaps typically slightly more concerned with attitudes and beliefs than with variation. Traditionally, in earlier Perceptual Dialectology projects aimed at understanding language attitudes, the focus had been on how certain *places* and the linguistic varieties associated with them are perceived in terms of degree-of-difference, pleasantness, and correctness.

In the American tradition, the degree-of-difference task often entails having respondents rank each of the 50 states in terms of difference from how the respondents themselves speak. The task usually involves a scale of one to four, where one means the variety is the same as the respondent's and four means the variety is unintelligibly different from the respondent's. When a state receives an overall low mean score, it is suggested that people from the respondents' home area believe that way of speaking to be similar to their own.

For example, in Preston's earlier work (reviewed in Niedzielski and Preston 2000: 77–82), Michiganders view their neighboring states (Ohio, Indiana, Illinois, and Wisconsin), as well as Iowa and Minnesota, as rather similar to their way of speaking. States that Gould and White (1986) termed the *Southern Trough*, namely Louisiana, Mississippi, and Alabama, are viewed as the most different from the way people speak in Michigan. On the other hand, Southerners in Preston's studies (mostly from Alabama, Georgia, and South Carolina) view Georgia and South Carolina as most similar, with a secondary similarity zone in the surrounding Southern states, while they see Wisconsin, Delaware, and all states northeast of Pennsylvania as unintelligibly different.

In addition to understanding how similar a respondent believes a variety to be, Perceptual Dialectology studies have often attempted to capture how respondents view linguistic variation in terms of certain social characteristics. For Preston, over the course of many studies, the most prominent social characteristics to surface for respondents dealt with notions of correctness and pleasantness. This is similar to findings in other work, including Ryan, Giles, and Sebastian (1982), where speakers judged audio samples on status and solidarity, and Inoue (1999a), where respondents identified characteristics associated with a particular "dialect image" as intellectual and emotional.

Like in the degree-of-difference task, American respondents are asked to rate all 50 states with respect to these characteristics. These studies typically use a scale from one to nine, where one indicates a variety that is least correct or least pleasant, while nine indicates a variety that is most correct or most pleasant. This is similar to the practices of cultural geographers, like Gould and White (1986), where respondents were asked to indicate areas of the country that were most desirable with respect to residence.

More examples from Preston's work (reviewed in Niedzielski and Preston 2000: 63–77) show that the Michigan respondents discussed above have a rather high opinion of themselves (as also revealed in other studies, like Niedzielski 2002). They rank their own variety alone as the most correct variety and only rate four other states as high as Michigan on pleasantness. They rate Southern states, particularly Alabama, rather low on correctness and relatively low on pleasantness. On the other hand, Indiana respondents, who also rate Southern states as low on correctness, tended to give high scores for pleasantness in those same states[5]. Southern respondents, being rather linguistically

5 These respondents, however, were pretty generous with their pleasantness ratings; a large majority of the country was rated at a six or higher.

insecure, rated their own varieties (at least for the Georgia and Alabama respondents) as lower in terms of correctness than several other northern, eastern, and western states. They made the further distinction that Texas, Louisiana, and Mississippi speak the least correct varieties, confirming the generalization made about the mental maps that regional locals make more distinctions within their home area. The Southern respondents also rated Alabama as the most pleasant, with coastal southern states ranking as a secondary locus for pleasantness. They rank northeastern states, like New York and New Jersey, as relatively low on the pleasantness scale.

Interesting work has also been conducted outside of the United States using these methods. For example, Dailey-O'Cain's (1999) work in post-unification Germany showed that, given a list of different varieties of German, residents of the former West Germany view their and other western varieties as significantly more correct and pleasant than eastern varieties. For residents of the former East Germany, there was no significant difference between eastern and western varieties in terms of correctness, but they perceived western varieties to be more pleasant. While this does not exactly mirror the north-south division in the United States, it is clear that former East Germans experience some level of linguistic insecurity similar to that felt in the American South.

Based on the results of the many studies that have utilized these methodologies, Preston (1999c) has suggested some generalizations about how people rank other ways of speaking. He claims that respondents from areas with high levels of *linguistic security*, a term used in reference to Labov's (1966) definition of *linguistic insecurity*, like Michigan, where speakers believe their variety is the same as Standard American English (Niedzielski 2002), tend to rate varieties that they classify as least correct and pleasant as most different, even unintelligibly different from their own way of speaking. Respondents from areas that are more linguistically insecure, like southern Indiana, where "respondents put part of their state in the Midwest, part in the North, and, curiously, the part where they live, nowhere at all" (Preston 1997: 321), rate varieties that they found to be high or low on the correctness and pleasantness scales as rather different.

More recent work in language regard has taken these traditional methods and modified them in varying ways. For example, some studies using Perceptual Dialectology tasks have also made connections between the perceptions of respondents and linguistic production. Fridland, Bartlett, and Kreuz (2004, 2005) examined the folk perceptions of Southern and non-Southern vowel variants among participants in Memphis, Tennessee. Using synthesized tokens from native Memphians as the stimuli, respondents were asked to identify vowels representative of the Southern Vowel Shift (e.g., Feagin 1986; Fridland

1998). The authors found that the front vowels, especially /ei/, were more likely to be identified as Southern. In previous work (e.g., Fridland 1998), Fridland showed that /ei/ is the most actively participating vowel in the Southern Shift in Memphis, and the fact that it was the vowel most commonly identified as Southern in this study suggests a strong connection between perception and production, indicating that "the ability of participants to accurately rate differences between vowel variants and assign scores appears to vary, depending on whether the local community speech norms involve those particular variants and whether those variants are shared with other regions" (Fridland, Bartlett, and Kreuz 2004: 13).

Memphians were also asked to rate these particular vowel variants in terms of education and pleasantness. The authors found that respondents rated the non-Southern variants front vowels in the Southern Shift vowels to be significantly more educated and pleasant and favored the traditionally positioned back vowels to the shifted ones. Ultimately, the authors found that the less regional salience a vowel had for respondents, the more positive ratings it received.

Cramer (2010) has explored making explicit connections between the traditional draw-a-map task and the ratings of difference, pleasantness, and correctness. While traditional Perceptual Dialectology projects have left these two tasks separate, in that participants draw on maps but then rate individual states, Cramer asked respondents to complete a language attitudes survey based on the labels used on their maps. Respondents rated whichever varieties they delimited in terms of degree of difference, pleasantness, and correctness, as well as a few other social characteristics like standardness, formality, beauty, and education. This method more accurately shows which varieties respondents hold in high esteem and which areas are not seen as desirable. It also aligns well with the bottom-up approach of dialect identification that is typical of Perceptual Dialectology studies.

5 In this volume

As a field of inquiry, Perceptual Dialectology in its current state, with advances in both the kinds of questions explored and the technologies used to do so, provides the right tools for understanding how the notion of place conditions non-linguists' views about language and linguistic variation. This edited collection includes many chapters relating to the state of the art in Perceptual Dialectology research. The collection takes a global view of the field of Perceptual Dialectology, broadly defined, to assess the similarities and contrasts in non-linguists' perceptions of the dialect landscape.

With its global focus, the volume includes chapters which discuss data gathered in the United States, the United Kingdom, the Republic of Ireland, France, Germany, Austria, and South Korea. The common methods used by many of the contributors means that readers should be able to draw comparisons from the breadth of the volume. For example, as Preston highlighted in the introduction to this collection, several of the contributions showcase the advances in data processing that the use of GIS in Perceptual Dialectology research has brought. These chapters demonstrate how this technology can be used to ask further questions of the draw-a-map data and how it can be used to produce high quality graphical outputs. Other contributors focused on attitudes toward language and linguistic variation. Still other studies use more traditional dialectological methods, but they do so with the intent to explore the beliefs of non-linguists. All in all, the combination of these many diverse methods has helped to create a rather complete and complex view of the perspectives non-linguists use to characterize the linguistic landscapes in and of cities.

The goal of this volume is to address the specific theoretical and methodological issues associated with the examination of dialect perceptions in and of cities, drawing on current advances in Perceptual Dialectology to examine how the urban setting influences those perceptions. In the course of examining the connections between place and perceptions, the authors in this collection explore interrelated themes of linguistic variation, including the differences in the perception of rural and urban areas, the processes of perception and language change, and the relationship between perception and "reality". Therefore, we have divided the book into these three sections. It is hoped that, in the course of reading this collection, readers will come to understand the importance of including the non-linguist in the research process as a valuable source of pertinent linguistic information. Such respondents play an essential role in our understanding of the social processes that impact language and linguistic variation.

Part I: Differences in the perception of rural and urban areas

Jennifer Cramer

Chapter 3
Rural vs. urban: Perception and production of identity in a border city

1 Introduction

In even the earliest of the large-scale American Dialectology projects (e.g., Kurath et al. 1939–1943; Kurath 1949; Atwood 1953; Kurath and McDavid 1961), both urban and rural areas were investigated, in hopes of better understanding not only the linguistic variation evident across the country but also the relationship that exists between these two types of communities. As Kurath (1931) said, in his original planning of the *Linguistic Atlas of the United States and Canada*,

> If the inhabitants of a village or a countryside have intimate contacts with a neighboring community, they may abandon some of their own dialectal peculiarities and adopt speech habits of their neighbors, especially if the community with which they entertain relations is of greater importance in the economic, political, religious, or cultural life of the region. (483)

Despite a current emphasis on linguistic variation in and of urban areas, many recent local studies have been conducted to examine the impact of the urban/rural distinction in communities like Texas (Thomas 1997; Koops 2010; Cukor-Avila 2012), Michigan (Ito 1999; Gordon 2001), Arizona (Hall-Lew 2005), Ohio (Durian 2007), Kentucky (Irons 2007), Minnesota (Nguyen 2011), and others.

And while this research has been informative for our understanding of how urban and rural varieties influence each other in terms of linguistic production, little has been said within this framework of analysis about how the speakers of these varieties perceive the relationships between the rural and urban areas within their regions and about how those perceptions condition their linguistic acts of identity (Le Page and Tabouret-Keller 1985). Indeed, in much of the work of traditional dialectology, non-linguists' perceptions of dialect variation have often been considered secondary to the analysis of phonetic and lexical variables (see Preston, this volume).

Work in perceptual dialectology (e.g., Preston 1989, 1999a; Long and Preston 2002), however, has shown that perceptual data can provide important insights into dialect variation that are complementary to production data and that provide an empirical basis for our delimitations of rural and urban linguistic areas that are based in the speaker's reality, not in geographical, historical, or statistical definitions. For example, despite media representations of California as urban and coastal, Bucholtz et al. (2007) found that Californians labeled a large portion of the state, mostly in the north and inland, using words associated with rural dwellers like *hick*, *hillbilly*, and *redneck*, labels which had not been previously used to describe California in other perceptual studies (Fought 2002). These labels were most commonly used to describe Northern California speech, and interestingly, subsequent work in the area (Bucholtz et al. 2008) showed that, despite this categorization of the north as rural, respondents marked Northern California as the location of the "best" speech in the state. A combination of this research with linguistic production data that highlights the distinction perceived by non-linguist respondents would paint a more complete picture of the dialect landscape of California.

Within the southern United States, the variety of English most widely spoken "has long been regarded as a conservative variety preserved in large part by the rural, insular character of the region" (Tillery and Bailey 2003: 159). Thus, much of the focus in studying the varieties of speech in this region has been on the historical development of those varieties, with the greatest attention being paid to rural speech. But urban centers in the South have been growing. Between 1950 and 2000, the population of the South as a whole increased by 130.6 %, with major metropolitan areas growing by 361 %, accounting for a total of 63 % of the entire Southern growth (McDonald 2013). Thus, it is important to recognize how the changing face of the Southern dialect region (with respect to urbanity) influences the linguistic production and perception of a Southern regional identity.

Linguistic research related to this change in demography must account for how, in the American South, urban speakers distinguish themselves from rural speakers in terms of not only production but also perception. In this chapter, Louisville, Kentucky (see Figure 3.1), which is located within the American South and is the only major urban area within the largely rural state, is presented as an example of how perceptual and production data can be used together to understand the connections between urbanity and linguistic acts of identity (Le Page and Tabouret-Keller 1985). This chapter addresses the ways in which Louisvillians[1] constantly negotiate and contest both their level of

1 I use the term "Louisvillian" to refer only to people from Jefferson County, Kentucky, all of which is included in the consolidated city-county government of Louisville Metro. The larger metropolitan statistical area (MSA) includes portions of Southern Indiana, but this population has been purposely excluded in this project, as the original focus was on the Ohio River as a boundary for people who live on the Kentucky/southern side of river.

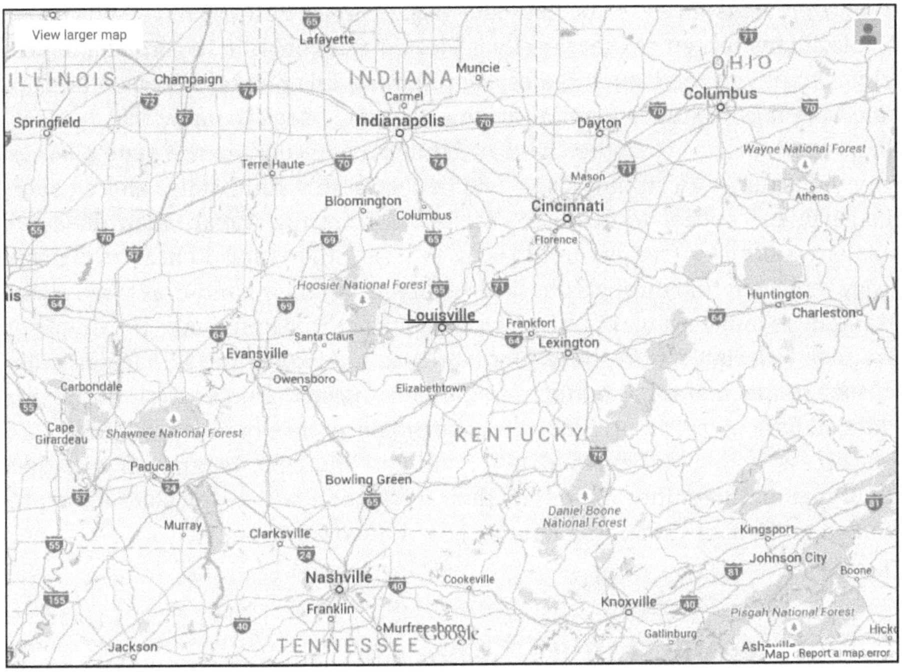

Fig. 3.1: Louisville's location in the larger region (Map credit: Google Maps).

Southernness and their urbanity, in the active and agentive expression of their amplified awareness of belonging brought about by their geopolitical position in the American landscape.

Following the models of mental mapping utilized in much of the perceptual dialectology research (cf. Preston 1989; Montgomery and Cramer, this volume), this chapter presents the perceptions non-linguists in Louisville have about variation in the region, with a specific focus on how Louisvillians distinguish themselves from the rest of the state. Additionally, production data from a reality television show, *Southern Belles: Louisville*, are presented to show how the perceptual data complement the realities of production. Labov, Ash, and Boberg's (2006) Southern dialect region is defined by the presence of monophthongization of /ai/ before voiced consonants and in final position, a hallmark of Southern speech. This definition led them to classify Louisville as a Southern city. Recalling this classification, I examine the production data for examples of /ai/ monophthongization using an optimization-based curve fitting procedure developed in Cramer (2009, 2010). An analysis of the speakers' variable use of this feature is presented.

In previous analyses of this data (e.g., Cramer 2010), I have shown that Louisvillians recognize that the city is located at a border in the broader regional geography of the United States, representing this fact in their mental maps and language attitudes surveys. In this chapter, I seek to show that Louisvillians are located at another kind of border. Louisville, as the state's largest city, appears as an urban island within the largely rural state, and as such, Louisvillians appear to distance themselves from the Appalachian dialect of Eastern Kentucky (see Figure 3.1), as well as other rural Kentucky dialects. They represent this through physical separation in their hand-drawn maps and through negative evaluations in their language attitudes surveys. The linguistic features associated with those rural varieties are likely the same features discussed in much of the literature on Southern speech[2]. The analysis of the production data reveals that, likely because of Louisville's border position, speakers access both Southern and non-Southern linguistic features. Thus, their fluid regional identities as well as their distinction within the state are represented in their seemingly random selection of vowel variants.

2 Data collection and analysis

The data for this project were collected in 2009. The data for the perceptual part of the project come from 23 participants living in Louisville, most of who claim to have spent all or nearly all of their lives in Louisville. These participants completed a mental mapping task, using the map in Figure 3.2[3], and were asked to provide labels for the varieties they distinguished. Participants were then asked to complete a language attitudes survey, where they listed the labels used on their maps and, using a four-point scale, rated these varieties in terms of difference (with respect to their own variety), correctness, pleasantness, standardness, formality, beauty, and education.

In section 3, I present several individual maps that highlight the rural/urban distinction being addressed in this paper. Individual maps alone, however, cannot create an image of how the larger community views linguistic variation. Therefore, composite maps were also created through a process by

2 Such features include, but are not limited to /ai/ monophthongization, the Southern Shift, pin/pen merger, vowel breaking, etc. See Labov, Ash, and Boberg (2006) for a full description of Southern dialectal features.

3 Maps with state lines only are typically used in Perceptual Dialectology studies, though some studies have given other information as well. See Montgomery and Cramer, this volume, for a discussion of the tools of Perceptual Dialectology.

Fig. 3.2: Map of region for Louisville draw-a-map task.

which individual maps were scanned and the most commonly delimited regions were digitized using ArcGIS 9 (see Montgomery and Cramer, this volume). To determine which regions were most commonly delimited, I analyzed the specific labels used by participants to determine which areas were most frequently identified. The composite map created from these individual maps shows 11 major regions defined by Louisvillians and presents an overall picture of how Louisvillians perceive these regions.

With the 11 major regions established, I examine how participants perceive the varieties of English they have delimited through an analysis of language attitudes surveys. Using a post hoc Tukey HSD (Honestly Significant Difference) test[4], I compare the scores for different varieties, to determine which varieties are perceived as better than or worse than others, given the social descriptors in the language attitudes survey listed above. Specifically, this chapter is interested in how participants perceive a Louisville dialect in relation to a larger Kentucky dialect or to an Appalachian dialect.

4 A post hoc analysis is necessary, as the categories used by participants were not determined before the experiment. This test is a variation of the *t* distribution, using instead the studentized range distribution *q*. This method is used to compare all pairs of means of every treatment and is used instead of multiple standard t-tests to reduce the likelihood of type I errors. This method was also selected because it can handle unequal sample sizes, as found in my data.

Finally, randomly selected individual tokens of the /ai/ vowel (ten per speaker) were extracted from the production data, which consists of more than seven hours of broadly transcribed video, and were subjected to spectral analysis using Praat version 5.0.35 (Boersma and Weenink 2008). For each word, I hand-selected the vowel in Praat through visual inspection of the spectrogram. The boundaries of the vowel were determined by listening to the speech sample, zooming in to the spectrogram, looking for the higher energy bands typical of vowel formants, and identifying the end of the preceding phoneme and the closure of the following phoneme.

In American English, diphthongs typically exhibit two steady-states: one in the onset, which is followed by a transition, and another in the offset (Lehiste and Peterson 1961). To analyze the speakers' use of the monophthongal variant, I exported the formant values from Praat for each token of /ai/, and, using MATLAB, I applied an optimization-based curve fitting procedure to the F_1 and F_2 values[5]. This procedure allowed me to fit the data to the ideal diphthong pattern using four parameters: transition beginning time, transition end time, beginning frequency, and end frequency. A vowel was considered to be monophthongal if the change in frequency (in F_1 or F_2) from the initial point of the vowel to the end point was less than 25 %[6].

To get an understanding of the level of participation in /ai/ monophthongization, certain statistical models are necessary. The null hypothesis states that because of Louisville's location in the South (Labov, Ash, and Boberg 2006), a speaker will produce a monophthong with at least as great a probability as that of a "real" Southern speaker, using the 33 % mark established in Cramer (2009). I test the null hypothesis that Louisville speakers produce monophthongs at this same rate. In this case, a one-tailed binomial test, rather than the standard normal approximation, is used because of the smaller sample sizes involved. Also, 95 % confidence intervals constructed using Wilson score intervals are provided for the probability that a speaker will produce a monophthong in a particular word.

3 Mental maps

I begin by examining some of the individual maps created by Louisville participants. The map in Figure 3.3 most precisely represents the notion that Louis-

5 F_1 and F_2 were analyzed separately.

6 Hewlett and Beck (2006) claim that monophthongs and diphthongs are not discrete categories but points on a continuum. Thus, vowels that are perceived to be monophthongs may exhibit a dynamic phase. My definition accounts for such monophthongs.

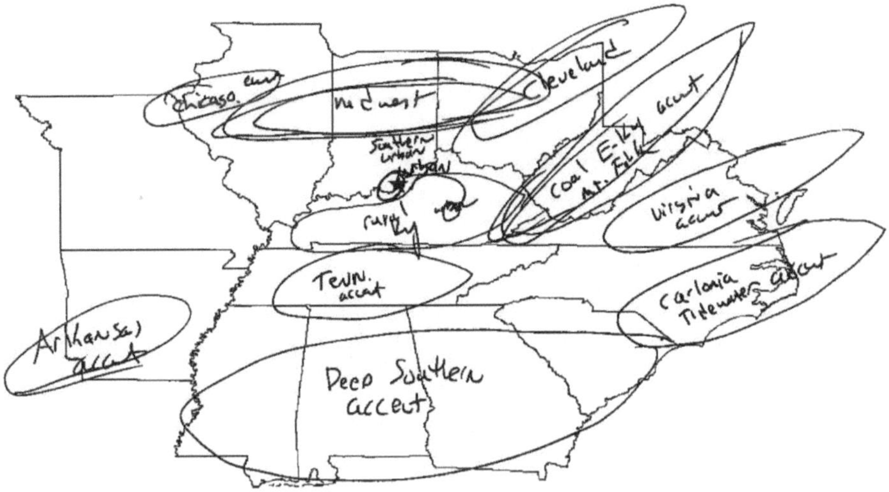

Fig. 3.3: Map drawn by a 40-year-old white female Louisvillian.

ville, represented in these maps with a star[7], along with the second largest city in the state, Lexington, exist as little *Southern urban*[8] islands inside *rural Ky.* A third variety is distinguished within the state: in the far eastern part of the state, where the Appalachian Mountains can be found, this participant has circled an area called *Coal E. Ky accent Mt. Folk*, which includes not only eastern Kentucky but also parts of western Virginia as well as almost the entire state of West Virginia. This tripartite division of Kentucky has been replicated in ongoing research focusing specifically on variation within the state (see Cramer and Hardymon 2014 for preliminary results), where the most common divisions line up with the urban/rural/mountain rural distinction depicted here.

Similarly, the participant who drew the map in Figure 3.4 drew a distinction between the urban areas of Kentucky (Louisville and Lexington), which are further linked to some urban area in Ohio, likely Columbus, and the Appalachian area in the eastern part of the state. Several participants seemed to recognize only Louisville/Lexington and eastern Kentucky as distinct, while the rest of the state remained unlabeled. While not commenting on the rest of the state may have to do with travel history (i.e., they have never visited those regions) or some other form of unfamiliarity, it is interesting to note that the

7 This star was not provided on the map when participants were creating their mental maps.
8 All labels (presented in italics) are written exactly as they appeared on the participant's individual maps.

Fig. 3.4: Map drawn by a 23-year-old white male Louisvillian.

areas they did commonly delimit were their own location and the area that is most derogated. Such a finding in a state that is known widely in the United States for its "backward" mountain population (Billings, Norman, and Ledford 1999) is not surprising; Louisvillians use their maps to mark a physical and linguistic distinction between themselves and the mountain population they perceive to bring negative stereotypes to their state[9].

Many of the labels used in Figure 3.3 and Figure 3.4 tend to be more neutral, featuring geographic labels sometimes attached to the non-descript language label of "accent". Other maps included some broader and possibly more derogatory labels. For example, one participant used labels like *Poor South Hick* and *Cleveland/New Brunswick Nasal*, both of which (based on the participant's discussion of the labels) appear to be negative. This participant also chose the label *Hillbilly Twang* to describe the variety in the eastern portion of the state. As we will see in the language attitudes section, derogatory labels often go hand in hand with negative evaluations of the variety spoken in those locations.

Sometimes, however, participants made broader categorizations, choosing to group much of the state under one label. The map in Figure 3.5 indicates

9 For more on the specific relationship between Louisville and Appalachia, see Cramer 2012.

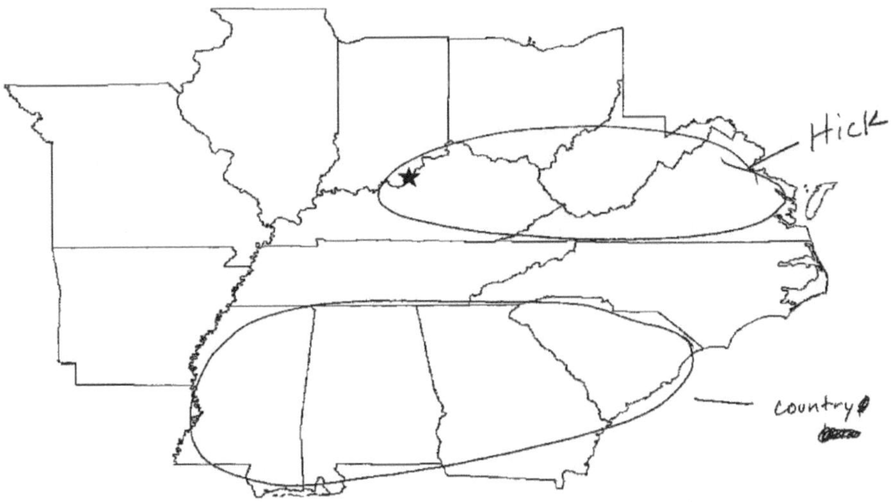

Fig. 3.5: Map drawn by 30-year-old white male Louisvillian.

that this participant only makes two distinctions within the whole region, *Country*, which is spoken across much of the Deep South, and *Hick*, which appears to have some Appalachian connection, though it extends from the Atlantic coast of Virginia to an area in Kentucky just west of Louisville. As such, this participant has marginally included Louisville in an Appalachian variety. Again, however, the participant has chosen a label that is mildly derogatory. This particular participant also identified this way of speaking as being his way of speaking[10], which might suggest the participant has reappropriated (Chen 1998) the term for more acceptable/positive usage.

Unlike the preceding maps, the map in Figure 3.6 does not appear to categorize Louisville as belonging to any particular region, though the outline of *RK* is quite close to the bottom edge of the star marking the location of Louisville. In the key provided by the participant, the area labeled *RK* is rural Kentucky, and the area labeled *EK* is eastern Kentucky, which are distinct varieties for this participant. Thus, while the urban distinction is not produced here, the participant has suggested that all rural varieties are not the same; for him, the mountainous portion of the state speaks differently than the non-mountainous rural regions.

10 The language attitudes survey also included three open-ended questions, which is where this participant indicated his affiliation with this variety. These questions, however, will not figure prominently in the analysis presented in this chapter.

Fig. 3.6: Map drawn by 66-year-old white male Louisvillian.

Finally, the map in Figure 3.7, like the map in Figure 3.6, makes some distinction between eastern Kentucky and a majority of the rest of the state. Still, Louisville is left uncategorized[11], and we can start to see some kind of border running through Louisville. While it is unclear what might be on the other side of the line, it is clear that this participant connects the rest of Kentucky to some kind of Southern dialect. This dialect differs, however, from a *Deep southern accent*; the major distinction this participant makes in Kentucky is yet again between mountain rural and non-mountain rural. She does indicate, in her answer to open-response questions, that the main difference between these two varieties has to do with education, calling the *Hill-billy* variety "comical, uneducated," while describing the *Southern* variety as "not highly educated but not uneducated, pretty widely accepted". This added commentary echoes the results of the quantitative analysis of the language attitudes survey below.

Ultimately, across all completed maps, Kentucky, Louisville/Lexington, and Appalachian varieties emerged as three of the most commonly delimited areas, thus making them categories for the overall composite map featured in Figure 3.8. One of the issues in creating this kind of map is determining what level of agreement to show. As Preston (1989: 28) indicates, "neither the mini-

11 Similar results have been found in the identification of production data, as in Braber's work in Nottingham (this volume).

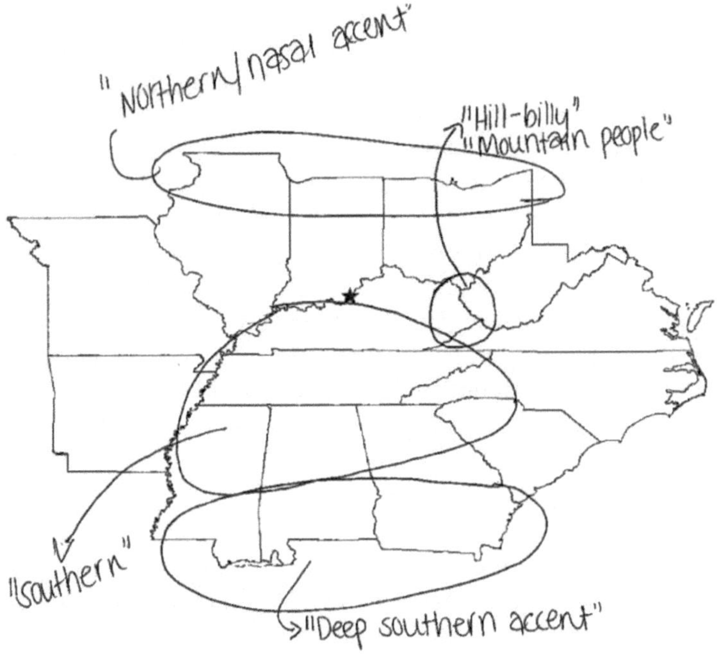

Fig. 3.7: Map drawn by a 23-year-old white female Louisvillian.

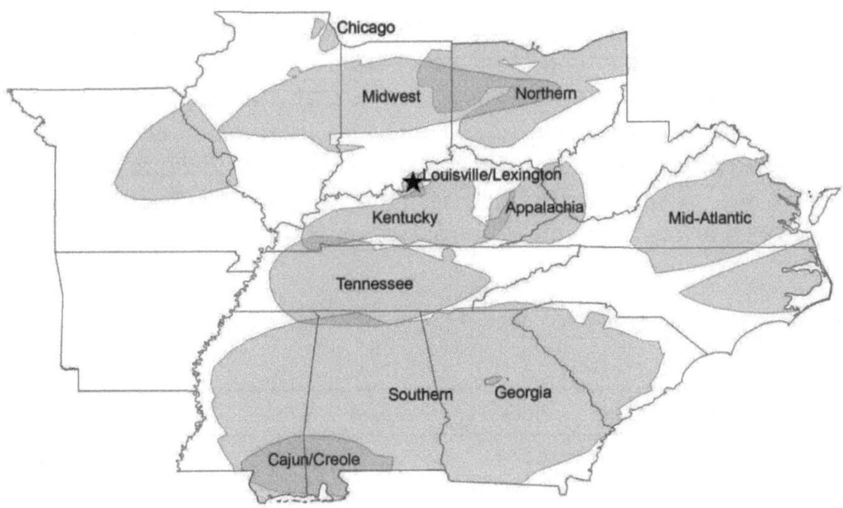

Fig. 3.8: Overall composite map of 50 % agreement among Louisvillians.

mal boundaries [...] nor the maximal ones [...] will do as a generalization" of the perceptions speakers have about varieties of English. Thus, in Preston's work, and in much of the perceptual dialectology research following from it, it has been necessary to select an intermediate level of agreement. The map in Figure 3.8 represents where 50 % of participants agreed on particular regions. Here, the tripartite division in Kentucky is clear; there is slight overlap between the Kentucky region and the Appalachia region, as well as between the Louisville/Lexington region (which in this level of agreement encompasses only Louisville) and the Kentucky region. The prominence of this division in the mental maps of Louisvillians suggests that the urban/rural/mountain rural distinction is real for these speakers. In what follows, I explore how this division is further enacted through the respondents' attitudes toward these varieties.

4 Language attitudes

The composite map in Figure 3.8 clearly shows that Louisvillians recognize three distinct varieties in their state. Therefore, in addition to the physical separation represented in their maps, how do Louisvillians differentiate among these varieties? In this section, I examine how these three varieties are distinguished in terms of attitudes regarding the following social characteristics: degree of difference, pleasantness, correctness, standardness, formality, beauty, and education.

The values provided in Table 3.1 are the mean scores for each of the three Kentucky regions delimited by these Louisville participants. Each score is based on the individual ratings of the varieties on a scale of 1–4, and the lower scores in this table indicate the variety that best fits the category label listed at the top. As is clear from this table, Louisville/Lexington is rated lowest on each of the seven categories. It is perceived to be the most similar to the participants' own way of speaking (a likely result, as all participants are from this region) as well as the most correct, pleasant, standard, formal, beautiful, and

Tab. 3.1: Mean scores for Appalachia, Kentucky, and Louisville/Lexington for seven social characteristics.

Label	Same	Correct	Pleasant	Standard	Formal	Beautiful	Educated
Appalachia	2.857	3	2.143	2.929	3.357	2.357	3.143
Kentucky	2.667	2.5	2.167	2.333	3.5	2.667	3
Louisville/ Lexington	1.375	1.875	1.75	1.75	2	2.125	2.25

Tab. 3.2: Analysis of pairings for level of difference.

	n	Mean	Rank	*q*	*q*-critical	Significant
Appalachia	14	2.857143	9	5.986716475	4.65	TRUE
Louisville/Lexington	8	1.375	1			
Kentucky	6	2.666667	6	4.281618887	4.65	FALSE
Louisville/Lexington	8	1.375	1			

educated. Perceptions of Kentucky and Appalachia were split: Kentucky ranked last in pleasantness, formality, and beauty, while Appalachia ranked last in degree of difference, correctness, standardness, and education.

But when these areas are compared to the entire group of 11 overarching categories (see Figure 3.8), some interesting trends arise. In terms of degree of difference, Louisvillians perceive a Cajun/Creole dialect to be the most different from their own way of speaking and a Louisville/Lexington dialect as most similar to their own way of speaking. The Tukey HSD statistical test reveals that Louisvillians believe the Cajun/Creole, Tennessee, Appalachian, Northern, and Southern dialects to be statistically significantly more different from their own way of speaking than a Louisville/Lexington dialect. In focusing on how Louisvillians position themselves at a larger regional level, I have attributed (Cramer 2010) these results to a distancing from Southern speech and an alliance with Midwestern speech, which was rated second most similar to Louisvillian's own way of speaking. The difference between the Louisville/Lexington variety and the Kentucky dialect was not statistically significant. The results of the statistical tests for these three overarching categories can be seen in Table 3.2.

In terms of correctness, we find that Louisvillians rate an Appalachian variety as the least correct overall, while Chicago and Midwest dialects are tied as the most correct way of speaking. The statistical results indicate that Louisvillians believe the Appalachian dialect to be statistically significantly less correct than the Chicago, Midwest, Northern, Louisville/Lexington, and Southern dialects. The results of the statistical test between Appalachian and Louisville/Lexington can be found in Table 3.3.

The three Kentucky varieties were not ranked in a way that marked them as statistically different from one another with respect to pleasantness. As in many perceptual dialectology studies conducted in the past (e.g., Preston 1989; Preston 1999a; Long and Preston 2002), the Southern dialect region was rated as the most pleasant variety. For these Louisville participants, however, subvarieties in the South, like Tennessee and Georgia, are ranked rather low in pleasantness. The Louisville/Lexington variety is rated just below the Southern

Tab. 3.3: Analysis of pairings for level of correctness.

	n	Mean	Rank	*q*	*q*-critical	Significant
Appalachia	14	3	11	5.302166	4.65	TRUE
Louisville/Lexington	8	1.875	4			

Tab. 3.4: Analysis of pairings for level of standardness.

	n	Mean	Rank	*q*	*q*-critical	Significant
Appalachia	14	2.928571	10	6.219051	4.65	TRUE
Louisville/Lexington	8	1.75	1			

variety, in a way aligning participants in Louisville with the Southern variety. Appalachia is ranked toward the middle, while Kentucky is rated slightly lower. The rankings perhaps indicate a slight three-way distinction among the urban, rural, and mountain rural varieties in Kentucky, but the difference is not significant.

In rating how standard a variety is, a Louisville/Lexington variety is seen as the most standard way of speaking. Kentucky is rated toward the middle, while Appalachia is ranked next to last, with only the Cajun/Creole variety rated lower. Perhaps in this category more than in correctness, a three-way distinction is established: Louisville/Lexington speech is very standard, Appalachian speech is not standard, and Kentucky speech is somewhere in between. The rating of Appalachian speech as very low in standardness is not surprising, as this variety suffers a great deal of derogation with respect to its approximation of a standard, both in the results of the open-ended questions in this study as well as in the larger popular discourse about Appalachia (see Billings, Norman, and Ledford 1999).

In terms of significance, the Appalachian variety is rated statistically significantly less standard than the Louisville/Lexington variety by these Louisville participants. The results of the statistical test can be seen in Table 3.4.

In rating the formality of the varieties in Kentucky, Louisvillians marked a sharp contrast between their own variety and that spoken in Appalachia and the rest of Kentucky. Louisville/Lexington was rated as the most formal way of speaking, while Appalachia and Kentucky were rated at the bottom of the scale. Louisville/Lexington is statistically significantly more formal than those other two varieties, as seen in Table 3.5.

This difference in formality between the speech of Louisville/Lexington and Kentucky is somewhat surprising, given that both dialects were considered

Tab. 3.5: Analysis of pairings for level of formality.

	n	Mean	Rank	*q*	*q*-critical	Significant
Kentucky	6	3.5	11	6.115546	4.65	TRUE
Louisville/Lexington	8	2	1			
Appalachia	14	3.357143	10	6.742339	4.65	TRUE
Louisville/Lexington	8	2	1			

to be statistically significantly different in the level of difference question. It suggests a distancing among Louisvillians between their variety and the variety in the rest of the state, which is achieved by claiming a difference in level of formality. The differences between these two dialects are not this great in the other social categories, suggesting that level of formality is the main factor distinguishing urban and rural varieties for Louisvillians, which would explain the drastic distinction.

Neither beauty nor education provided any statistically significant results, not only for the three Kentucky varieties but also for the entire set of overarching categories. The beauty category seemed to confuse participants, as there was a four-way tie for most beautiful: Cajun/Creole, Mid-Atlantic, Midwest, and Southern. Louisville/Lexington was rated next most beautiful, with Chicago and Appalachia not far behind. And while Kentucky was ranked next to last in this category, the range of mean scores for this category was only between 2 and 2.75, a rather small range of difference located near the middle of the scale which suggests that beauty is not a category Louisvillians use to distinguish varieties of American English.

The lack of statistically significant results for the category of education is somewhat surprising, given the amount of attention paid, for example, to the notion that Appalachians sound "uneducated" (again, both in terms of the answers to open-ended questions as well as in the larger discourse about Appalachia). Indeed, Appalachia, Tennessee, and Kentucky were rated as the least educated varieties for Louisvillians, though the Louisville/Lexington variety was rated below the areas stereotypically associated with education (Northern and Midwest)[12].

12 It is possible that this lack of significance has more to do with methodological issues; Montgomery (personal communication) has indicated that his participants in Britain seem less happy to rate education on a quantifiable scale, possibly because educational achievement is already quantifiable, thus making the task redundant.

Louisvillians rated the Louisville/Lexington dialect as third most educated, echoing the results of the level of correctness analysis above. Participants also clearly make some distinction between the Louisville/Lexington dialect and the Kentucky and Appalachian ones, as they yet again separated these varieties from Louisville/Lexington in their rankings by a fairly large margin. This also points to the urban/(mountain) rural divide within the state, with Louisvillians choosing to support the stereotype of educated city-dwellers versus uneducated rural residents. As Evans (this volume) has found with her Washington respondents, despite the fact that sociologists like Lichter and Brown (2011) proclaim the rural/urban distinction to be obsolete, participants from Louisville seem to still utilize this dichotomy (or, in this case, trichotomy) as a useful categorization method.

5 Production of /ai/

Having explored the perceptions Louisvillians have about their own variety as it compares to others in the state, it is additionally important to see if those perceptions are enacted linguistically through the production of certain linguistic variables. The expectation, based on the results of the perceptual part of this study, is that Louisvillians will also produce linguistic variables in a way that differentiates their variety from the other varieties they perceive. Using production data from a reality television show situated in and about Louisville, I show how the perceptual data complements the realities of production[13].

The specific linguistic variable under consideration is the production of the American English diphthong /ai/, which in the Southern dialect region is often monophthongal before voiced consonants and in final position (Labov, Ash, and Boberg 2006). The data for this part of the project come from *Southern Belles: Louisville*. This show, described as a "docusoap" or "docudrama," follows the lives of five Louisville women in their 20s and 30s, detailing their experiences as friends, as professionals, and as bachelorettes. This section features an acoustic analysis of the production of /ai/ for each of the five women.

I examine each speaker individually, as an overall analysis of five women would not likely yield very interesting generalizations about Louisville as a whole. But the analysis of these individual women does provide insight into

13 It is reasonable to compare the production data of these speakers (who did not participate in the perceptual experiment) because the characters in this show are depicted as reliable representations of Louisvillians and, therefore, of Louisvillians' speech patterns.

the identity processes available for Louisvillians. An analysis of each speaker's speech patterns can show individual alignments in terms of regional identity (i.e., Southern, Midwestern, Northern). Each subsection below discusses the findings for each of the speakers in further detail. The participants are presented here in alphabetical order by first name[14].

Each table in this section features the individual words, the ratios of the end frequencies to the beginning frequencies (Z_2/Z_1), and whether the ratio meets the requirements of a monophthong as defined by the study, showing both F_1 and F_2 separately. The bolded entries are complete monophthongs.

5.1 Emily

In F_1, Emily used the monophthongal variant in four of ten cases. In F_2, she produced a monophthong in all but two words. All four cases where F_1 featured the monophthong also featured a monophthong in F_2, making these vowels complete monophthongs[15]. Table 3.6 shows Emily's level of monophthong use.

An example of a complete monophthong can be seen in Figure 3.9 in Emily's production of the word *kind*. In these figures, the data points exported from Praat are the *x* marks while the best fitting curve for those points is represented as a line through those *x* marks. Here, the line clearly features only a

Tab. 3.6: Level of monophthong use for Emily.

Word	F_1		F_2	
	Z_2/Z_1	**Monophthong?**	Z_2/Z_1	**Monophthong?**
I	0.720361	FALSE	1.14424	TRUE
Invited	0.737399	FALSE	1.452226	FALSE
I'm	**1.063218**	**TRUE**	**1.110029**	**TRUE**
Time	**1.13499**	**TRUE**	**0.857714**	**TRUE**
Kind	**0.952238**	**TRUE**	**0.955882**	**TRUE**
Exciting	0.507763	FALSE	1.032022	TRUE
Assignment	0.563025	FALSE	1.142991	TRUE
My	**1.118502**	**TRUE**	**1.200418**	**TRUE**
I've	0.604966	FALSE	0.74093	FALSE
Decide	0.69919	FALSE	1.140123	TRUE

14 For more information about these participants, see Cramer (2010).

15 Cramer (2009) found some instances where either F_1 or F_2 but not both were monophthongal.

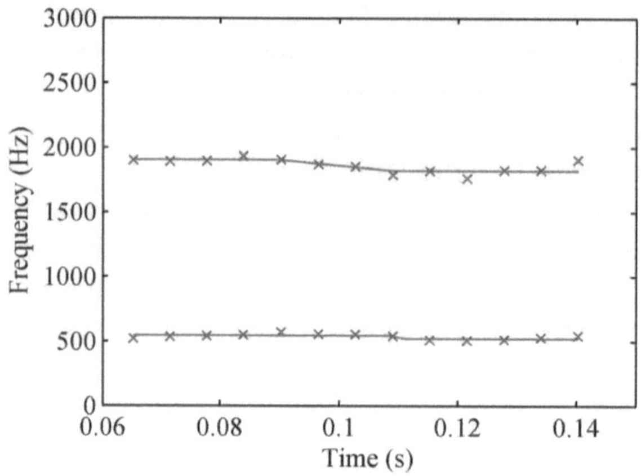

Fig. 3.9: F_1 and F_2 of /ai/ in *kind* as produced by Emily.

Tab. 3.7: Statistical analysis for Emily.

	F_1	F_2
p-value (> 0.05)	0.431632	0.000326
Observed Proportion	40 %	80 %
95 % CI – Lower Bound	17 %	49 %
95 % CI – Upper Bound	69 %	94 %

small amount of transition in F_2 and seemingly no transition in F_1. My definition of a monophthong as consisting of at most a 25 % change in frequency across the vowel is met here, and the graphical representation of this vowel shows rather plainly that this vowel is, in fact, a monophthong.

A statistical analysis of the data, presented in Table 3.7, yields the result that Emily's use of monophthongs in F_2 is statistically significant, assuming a level of significance at 0.05, though her use in F_1 is not. These results, along with the 95 % confidence intervals, which show that even though the interval for F_1 contains a 50 % probability of exhibiting a monophthong pattern at least as often as the Southern pattern, indicate that only the values in the interval for F_2 are greater, with a lower bound quite close to 50 %.

5.2 Hadley

Information about Hadley's level of monophthong use can be found in Table 3.8. In F_1, Hadley used the monophthongal variant in three of ten words. In F_2,

Tab. 3.8: Level of monophthong use for Hadley.

Word	F_1		F_2	
	Z_2/Z_1	Monophthong?	Z_2/Z_1	Monophthong?
I	**0.889346**	**TRUE**	**1.160482**	**TRUE**
Idea	0.659524	FALSE	1.220066	TRUE
I've	**0.880338**	**TRUE**	**1.158072**	**TRUE**
My	0.760456	TRUE	1.33141	FALSE
I'm	0.676849	FALSE	1.146317	TRUE
Kind	0.474486	FALSE	0.90086	TRUE
I'll	0.664927	FALSE	0.724204	FALSE
Finding	0.491478	FALSE	1.792759	FALSE
Guys	1.704143	FALSE	0.951574	TRUE
Guy	1.616764	FALSE	1.129567	TRUE

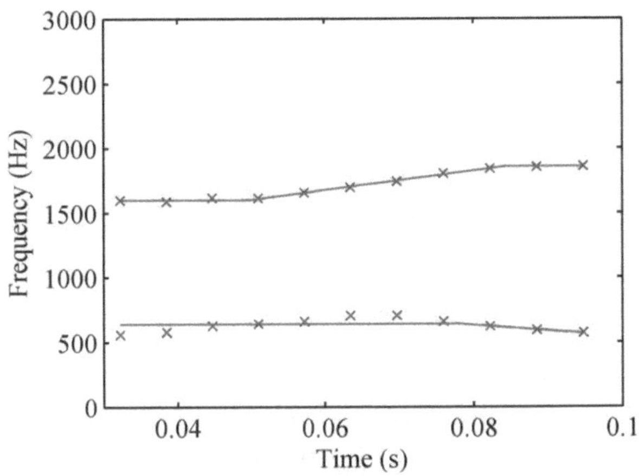

Fig. 3.10: F_1 and F_2 of /ai/ in *I* as produced by Hadley.

she produced a monophthong in seven of ten words. There were only two instances where both F_1 and F_2 were determined to be monophthongs.

An example of a complete monophthong produced by Hadley can be seen in Figure 3.10. This figure represents Hadley's production of the word *I*. It is not unexpected that a common word like *I* might appear in the monophthongal form in a possibly marginally Southern accent because, as Feagin (2000) notes, "the monophthongal unglided vowel in *I* and *my* symbolizes all Southerners' identification with the South" (342–343). But as can be seen in Table 3.8, Hadley uses the monophthongal variant in *I* and *I've*, uses a monophthong in F_2 of *I'm*, and no monophthong in *I'll*, so it cannot be generalized that Hadley

Tab. 3.9: Statistical analysis for Hadley.

	F_1	F_2
p-value (> 0.05)	0.692997	0.018549
Observed Proportion	30 %	70 %
95 % CI – Lower Bound	11 %	40 %
95 % CI – Upper Bound	60 %	89 %

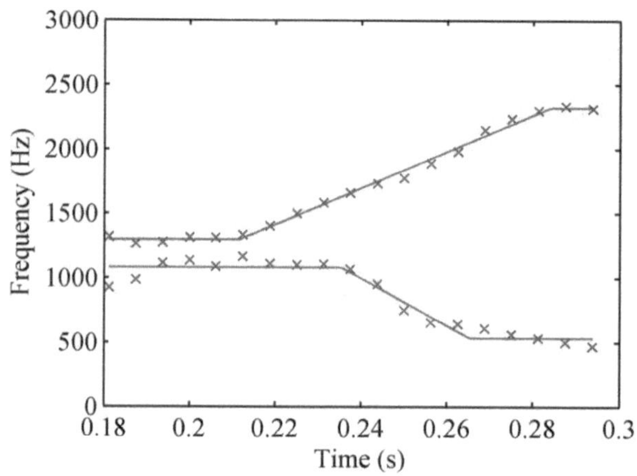

Fig. 3.11: F_1 and F_2 of /ai/ in *finding* as produced by Hadley.

always uses the monophthongal variant in variations of *I*. What the graphical representation shows again is that the vowel, perceived as a monophthong by the author, still has a slight amount of shift across the vowel, though not larger than the 25 % established here.

The images of complete monophthongs alone might not show the entire picture. A comparison of Figure 3.10 and Figure 3.11, which is a representation of Hadley's production of the word *finding*, shows how different the monophthongs appear relative to vowels that feature the typical diphthong pattern. Here, there is a large transition phase between two steady-states, which is lacking in Figure 3.10. Also, F_1 and F_2 both start between 1,000 and 1,500 Hz, and while F_1 drops, F_2 increases, indicating the shift from the low, back /a/ to the high, front /i/.

A statistical analysis of the data, presented in Table 3.9, yields the result that Hadley's use of monophthongs in F_2 is statistically significant, though her use in F_1 is not. Again, while the intervals for both F_1 and F_2 contain a 50 %

probability of exhibiting a monophthong pattern at least as often as the Southern pattern, more of the values for the interval for F_2 are above 50 % than below it, while the majority of the range is located below 50 % for F_1.

5.3 Julie

Table 3.10 shows Julie's level of monophthong use. This table shows that there are four instances of monophthongs in F_1 and seven in F_2. There are three instances where both F_1 and F_2 were determined to be monophthongs.

An example of a complete monophthong can be seen in Figure 3.12. Again, a slight transitional phase can be detected in both F_1 and F_2, but the amount of change in frequency across the vowel is less than 25 %. Similar to Hadley's case, this figure represents Julie's production of *I'm*, which, like *I*, is likely to become monophthongal for Southern speakers. Additionally, Julie's production of *I'll* and *my* were the other two instances of complete monophthongs. Figure 3.13 features the graphical image of Julie's production of *I*, which has the shape of a typical diphthong. Again, a generalization based on this information is not appropriate.

A statistical analysis of the data, presented in Table 3.11, yields the result that, like Emily and Hadley, Julie's use of monophthongs is statistically significant only in F_2. The intervals for both F_1 and F_2 contain a probability of exhibiting a monophthong pattern at least as often as the Southern pattern of greater than 50 %, but again, the values are higher in F_2.

Tab. 3.10: Level of monophthong use for Julie.

Word	F_1		F_2	
	Z_2/Z_1	Monophthong?	Z_2/Z_1	Monophthong?
I	0.626654	FALSE	1.42316	FALSE
Memorized	1.915488	FALSE	1.485803	FALSE
I'm	**1.131347**	**TRUE**	**0.881645**	**TRUE**
I'll	**0.890328**	**TRUE**	**0.935806**	**TRUE**
Why	0.877389	TRUE	1.66882	FALSE
Kinda	1.431471	FALSE	1.167569	TRUE
Guys	1.326222	FALSE	0.940086	TRUE
Sometimes	0.633463	FALSE	1.238234	TRUE
Time	0.743768	FALSE	1.201671	TRUE
My	**0.79565**	**TRUE**	**1.074404**	**TRUE**

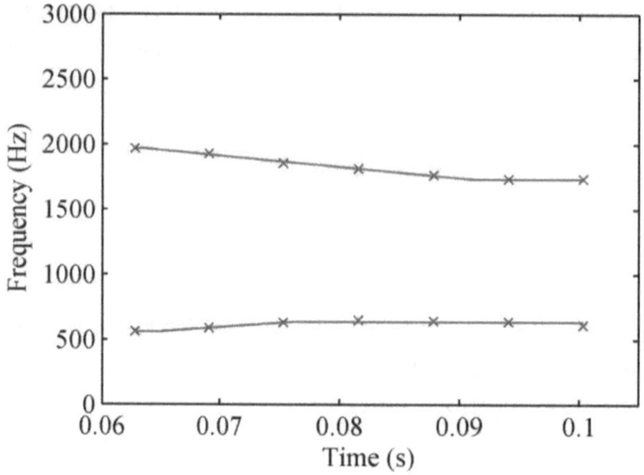

Fig. 3.12: F_1 and F_2 of /ai/ in *I'm* as produced by Julie.

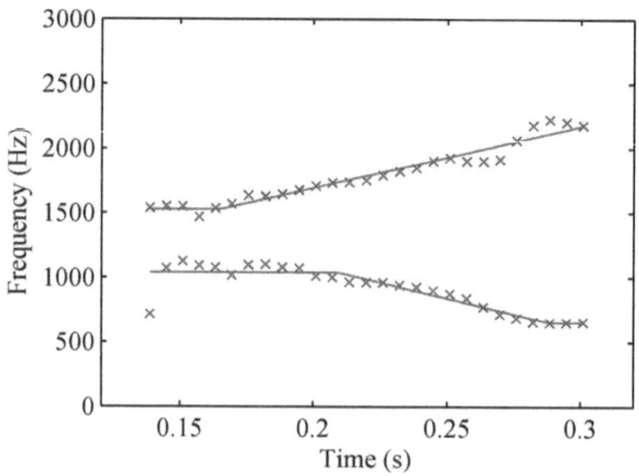

Fig. 3.13: F_1 and F_2 of /ai/ in *I* as produced by Julie.

Tab. 3.11: Statistical analysis for Julie.

	F_1	F_2
p-value (> 0.05)	0.431632	0.018549
Observed Proportion	40 %	70 %
95 % CI – Lower Bound	17 %	40 %
95 % CI – Upper Bound	69 %	89 %

5.4 Kellie

Kellie's overall level of monophthong use can be seen in Table 3.12. In F_1, the monophthong pattern is exhibited in four of the ten words. In F_2, all but two vowels were determined to be monophthongal. Like Emily, all four vowels where F_1 was determined to be a monophthong also featured a monophthong in F_2, making these vowels complete monophthongs. Kellie and Emily are tied for the most complete monophthongs, both with four instances where both F_1 and F_2 exhibit the monophthong pattern.

Tab. 3.12: Level of monophthong use for Kellie.

Word	F_1		F_2	
	Z_2/Z_1	Monophthong?	Z_2/Z_1	Monophthong?
I	**0.806248**	**TRUE**	**0.935804**	**TRUE**
Kind	**0.873164**	**TRUE**	**0.918893**	**TRUE**
My	1.333607	FALSE	0.939343	TRUE
Child	0.383851	FALSE	0.658798	FALSE
Meantime	**0.88388**	**TRUE**	**1.207033**	**TRUE**
I'm	1.554271	FALSE	0.989759	TRUE
Find	**1.135358**	**TRUE**	**0.940683**	**TRUE**
By	1.885134	FALSE	0.872578	TRUE
Lifestyle	1.336664	FALSE	1.364695	FALSE
Kinda	0.371074	FALSE	0.901266	TRUE

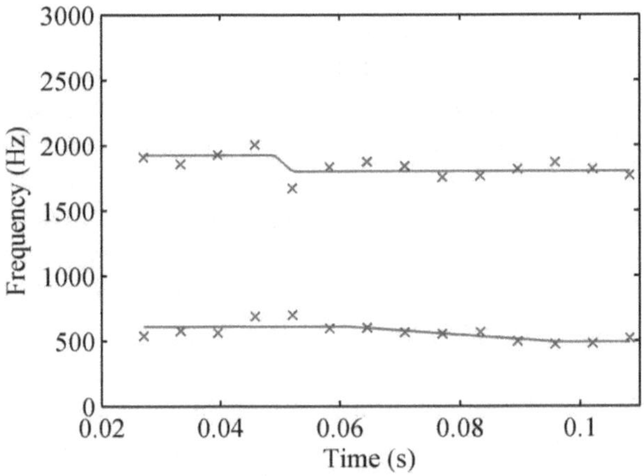

Fig. 3.14: F_1 and F_2 of /ai/ in *I* as produced by Kellie.

Tab. 3.13: Statistical analysis for Kellie.

	F_1	F_2
p-value (> 0.05)	0.431632	0.000326
Observed Proportion	40 %	80 %
95 % CI – Lower Bound	17 %	49 %
95 % CI – Upper Bound	69 %	94 %

An example of a complete monophthong can be seen in Figure 3.14. As with Hadley and Julie, an example of *I* from Kellie is revealed as a complete monophthong. One data point seems to be slightly out of place (at about 0.05 seconds), which causes the somewhat drastic transitional phase between two fairly straight steady states. Despite this issue, the graphical representation exhibits a rather clear monophthongal pattern. Yet, as with the previous women, while this instance of *I* is a monophthong, Table 3.12 indicates that Kellie's production of *my* and *I'm* only featured a monophthong in F_2.

A statistical analysis of the data, presented in Table 3.13, yields the result that Kellie's use of monophthongs is statistically significant only in F_2. The intervals for both F_1 and F_2 contain a probability of exhibiting a monophthong pattern at least as often as the Southern pattern of greater than 50 %, but again, the values are higher in F_2.

5.5 Shea

Shea's level of monophthong use can be seen in Table 3.14. In F_1, three of the curves produced were determined to be monophthongal. In F_2, eight out of ten were determined to be monophthongs. Two of the three monophthongal patterns in F_1 also aligned with monophthongs in F_2 for complete monophthongs.

An example of a complete monophthong can be seen in Figure 3.15, which is a graphical representation of Shea's production of the word *decide*. Here, there is only a slight transition in both F_1 and F_2, both of which cause a change in frequency of less than 25 % across the vowel.

A statistical analysis of the data, presented in Table 3.15, yields the result that, like the others, Shea's use of monophthongs is statistically significant only in F_2. While the intervals for both F_1 and F_2 contain a 50 % probability of exhibiting a monophthong pattern at least as often as the Southern pattern, the values for the interval for F_2 are higher than those for F_1.

Tab. 3.14: Level of monophthong use for Shea.

Word	F_1		F_2	
	Z_2/Z_1	Monophthong?	Z_2/Z_1	Monophthong?
I've	**1.032057**	**TRUE**	**0.880429**	**TRUE**
Decide	**1.192914**	**TRUE**	**0.966586**	**TRUE**
Liner	0.953645	TRUE	1.417447	FALSE
I	2.059912	FALSE	0.900058	TRUE
Kind	0.534898	FALSE	0.882047	TRUE
Lives	1.691689	FALSE	1.158704	TRUE
Time	0.678032	FALSE	1.05476	TRUE
I'm	0.706303	FALSE	1.018932	TRUE
My	0.672218	FALSE	1.315383	FALSE
Find	0.613377	FALSE	1.052925	TRUE

Tab. 3.15: Statistical analysis for Shea.

	F_1	F_2
p-value (> 0.05)	0.692997	0.003167
Observed Proportion	30 %	80 %
95 % CI – Lower Bound	11 %	49 %
95 % CI – Upper Bound	60 %	94 %

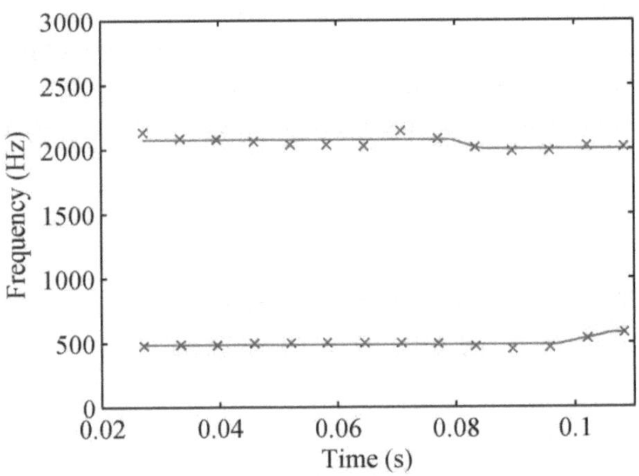

Fig. 3.15: F_1 and F_2 of /ai/ in *decide* as produced by Shea.

5.6 Summary

This section reveals the varying degrees to which these five women utilize the Southern feature of the monophthongal variant of /ai/. Emily and Kellie lead the way with the most instances of the monophthong in their speech (12 of the 20 total curves defined as monophthongs, counting F_1 and F_2 separately, and four complete monophthongs each). Julie and Shea were next, with 11 of 20 curves defined as monophthongs, and Hadley had the least amount, with 10 of 20 curves defined as monophthongs. That said, the total number of monophthongs for all of the women was at least 50 %, a rather medial number for speakers representing a city in the Southern dialect area which is defined by /ai/ monophthongization.

In fact, the number echoes the result for Louisville in Labov, Ash, and Boberg's (2006) atlas. If we turn to the other locations in Kentucky represented in the ANAE, we see that, in fact, variable use of the monophthongal variant is one way in which Louisvillians can distinguish themselves from others in the state. Only Lexington (the second largest city in the state) and Ashland (a small Appalachian town located near the border with West Virginia and Ohio) are featured in the ANAE, but both cities are represented as fully monophthongal in the pre-voiced and open environments. However, as I have argued elsewhere (Cramer 2009, 2010), Louisvillians' variable use of the monophthongal variant may have more to do with the fact that Louisville is on a border between a dialect that participates in /ai/ monophthongization and one that does not, making it just as likely that the speakers will produce either variant. For the purposes of this analysis, the results suggest that Louisvillians' lack of full participation in the Southern feature serves as a way for urban Louisvillians to distinguish themselves from the rural and mountain varieties they perceive in the rest of the state.

6 At the intersection of perception and production

In perceptual dialectology studies, the focus is typically on how non-linguists perceive dialect variation, not on how they produce it. What has been noted, through comparing results of perceptual and production studies, is that the two lines (perceived dialects and dialect boundaries based on production data) do not always align (Benson 2003). Indeed, in this data, I have shown that Louisvillians perceive more variation in their state than standard production studies like the ANAE have by including both a separate Kentucky region and a Louisville one. Combining the analysis of perceptual data and production data shows also that Louisvillians have at least three ways in which they can differentiate themselves from these other varieties in the state: Louisvillians can physically separate the varieties in the drawing of dialect maps, thus showing that they believe such varieties exist and are geographically distinct; they can produce attitudes about those varieties in a way that marks them as different; and they can produce linguistic variables at different rates than speakers of the other varieties[16].

As I have shown in previous analyses (e.g., Cramer 2010), Louisvillians appear to exhibit both Southern and non-Southern identities linguistically in their production and perception of regional identity through varying alignments (similar to those found in Mann, this volume), revealing the contested and dynamic nature of identity in the borderlands. But, on a more local level, the distinction seems to be more about the urban nature of Louisville as it compares to the rural and mountain rural regions perceived in the rest of the state, possibly suggesting an alignment of urban with northernness and rural with southernness.

This research project gives further credence to supplementing large-scale dialectology projects with local perceptual and production studies as a way of better understanding internal variation. It also indicates that more research on how people in urban areas mark themselves linguistically as distinct from rural residents, both in terms of production and perception, would greatly benefit our understanding of the changing state of linguistic variation, especially in areas like the American South where the focus has been firmly situated on the rural. Studies that combine the methods of perceptual dialectology with traditional dialectology can provide important insights into the realities of linguistic variation.

16 Of course, further research on the linguistic production of other Kentuckians would greatly enhance this claim.

Betsy E. Evans

Chapter 4
City talk and *Country talk*: Perceptions of urban and rural English in Washington State

1 Introduction

Woods (2011: 13) asserts, "The distinction between urban and rural, between the city and the country, is one of the oldest and most pervasive of geographical binaries". While there may be a common sense perception that the distinction between rural and urban is obvious and salient, this view has come under pressure and social geographers argue that notions of urban/rural are more nuanced than what a simple binary opposition suggests (e.g., Vanderbeck and Dunkley 2003; Woods 2005; Cloke 2006). Lichter and Brown (2011: 566) have argued that modern society is undergoing a "new urban/rural interdependence" which renders the urban/rural dichotomy "obsolete". They assert that the boundaries between rural and urban areas are being blurred at an accelerated pace by features of modern society such as migration, government devolution and technology. In addition to the blurring of boundaries, they also point to a symmetrical, rather than asymmetrical, relationship between rural and urban. That is, it is not the case, as is often assumed, that the urban influences the rural in a one-way direction such that the rural has no impact on the urban[1]. In other words, rural and urban communities are interdependent with mutual flows of people, information, technology and commodities. This perspective suggests that the urban/rural dichotomy as usually conceived may become irrelevant or perhaps nonexistent. The perceptual dialect data presented here suggests otherwise, at least for the lay population. The socio-cultural and linguistic distinctions of urban and rural are very salient for the respondents in this study. For example, when asked for a label for areas where they perceive English to be "different" in Washington (hereafter WA), respondents' ($N = 178$) most frequent category label (86/336 labels or 25 %) related to notions of rural life (Evans 2011, 2013a). This finding suggests that these respondents

1 Linguistic research such as Bailey et al. (1983) has also found contradictions to the assumption of a flow of influence that starts from urban areas and later reaches rural ones.

have a fairly strong perception of rural life as "different". This raises questions about language perceptions and space. That is, will this dichotomy become less productive as a categorization mechanism as the "reality" of the disintegration of the urban/rural interface, as Lichter and Brown describe it, becomes more apparent to lay people? Or, will the dichotomy continue to be a productive categorization mechanism regardless of "reality"? The present data alone cannot answer this question; however, it suggests that urban vs. rural is currently a salient dichotomous categorization and that while the academic descriptive characterization may be more objective, it must coexist and interact with the lay characterization.

1.1 Lack of attention to rural

The exploration of linguistic variation in rural contexts has received less attention since the shift to the study of urban areas accelerated by Labov (1966). So in turning attention to urban areas, sociolinguistics has largely overlooked rural communities (with some exceptions, e.g., Frazer 1983; Lippi-Green 1989; Hazen 2000; Ito 2000; Marshall 2004; Hall-Lew and Stephens 2012; Cukor-Avila et al. 2012; Cramer 2013). Britain (2009) argues for more attention to rural populations in sociolinguistic research noting that the patterns of variation sociolinguists seek are not only found in urban areas: "the very same cultural, economic, social and political processes and conflicts can affect rural areas as affect urban [areas]" (238). With regard to Perceptual Dialectology (hereafter PD) research, attention to perceptions of rural speech can also provide important details about production and perception of speech in those areas.

Although not directly solicited, a few PD research projects have found perceptions of language related to urban/rural inhabitants. For example, Bucholtz et al. (2007: 345) found in their PD study of California (CA) that rural labels like *hick* and *farmer talk* were the most common (17.9 percent) social labels attributed to speech varieties by Californian raters of language in CA. Interestingly, rural residents with non-standard speech are not salient to non-Californian raters who have been found to rate CA highly on correctness (Preston and Niedzielski 2003: 64). In addition, Fridland and Bartlett (2006), reporting on ratings of US English by speakers in Memphis, Tennessee, found that while Memphians rated their region (the US south) as less correct than other US regions, they rated their own state as more correct and pleasant than their more rural neighboring states (e.g., Mississippi, Arkansas). Fridland and Bartlett suggest that Memphians downgraded their rural neighbors and upgraded themselves because of Memphis' position in the region as an urban center.

More recently, some scholars, such as Hall-Lew and Stephens (2012) have explicitly solicited respondents' reflections on the notion of rural. They interviewed 17 residents living in the border area of Texas and Oklahoma (Texoma) in order to explore the social meaning of *country talk* in that community. Analyses of interviews and PD maps revealed that the notion of *country talk* for Texomans is an important tool for managing Texoma identity. For their respondents, *country talk* indexes a very nuanced meaning that includes particular aspects of rural life, regional distinction and social persona that are uniquely Texoman and facilitates identity creation in a border region where multiple varieties of English are present (261).

1.2 Defining rural

An important question is how to define urban and rural in WA. In this paper, perceptions of rurality in WA are the primary focus, but also more objective data from the US census that may confirm those perceptions is considered. With regard to perceptions, there is much popular consensus on the notion of which areas of WA are rural and urban. This consensus is reflected in the notion of the "Cascade Curtain". The Cascade Curtain is a phrase that refers to the Cascade mountain range which runs north and south through the state and literally divides it in two (see Figure 4.1). It is generally perceived that the two "halves" of the state are culturally different. Seattle is located in the western "half" while the eastern region is populated by several smaller cities and large areas of land devoted to farming or forests. This perception is widely circulated and discussed in a variety of media outlets. One such example is a Seattle Times editorial titled *Parting the Cascade curtain: Rethinking the state's cultural fault line* (Gastil 2006). Although this cultural divide might be thought of as simply a stereotype, and one that the PD data described below reflects strongly, it is possible to compare these perceptions with rather more tangible and objective evidence. For example, Gastil (2006) examined public opinion data and found that there is a sort of cultural divide in WA but that, spatially and culturally, the opposition consists of the city of Seattle and the rest of WA. That is to say when Seattle was excluded from western WA, it was not statistically different from eastern WA on the cultural dimensions measured.

In terms of actual population distribution, we can turn to US Census data, which confirms the urbanity of a large part of WA west of the Cascades and the rurality of a large part east of the Cascades. Figure 4.2 shows a map of the US census rural-urban commuting areas (RUCA) of WA. RUCA uses measures of population density, urbanization, and daily commuting to determine levels of rurality/urbanity. In Figure 4.2, we can see that urban areas in WA are the

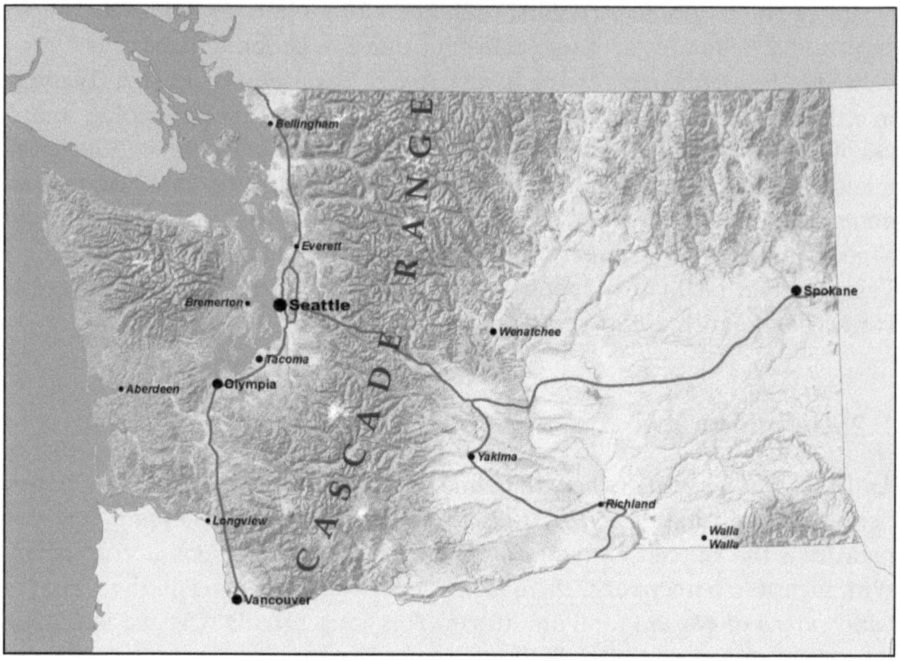

Fig. 4.1: Map of Washington State.

Seattle metro area along the Interstate 5 corridor and Vancouver on the southern border of WA (west of the Cascades) and a few areas in the east, namely, Spokane, the Tri-Cities region, Yakima, and Asotin county (the southeast corner of WA). The RUCA map also shows the rurality of part of western WA that seems to be rather overlooked in discussions of the "Cascade Curtain" and highlights the spatial nuances of the cultural "divide" described by Gastil (2006). While modern cultural geographers critique the use of demographics for the purpose of identifying rural space or culture (e.g., Halfacree 1993; Woods 2006, 2011), the apparent correlation between RUCA maps and the perceptual dialect maps in this data is worth noting.

It is important to note the possible yet unquantifiable (in this data) contribution of the actual spatial divide created by the Cascade mountain range to the dichotomous nature of Washingtonians' perceptions. Further research in WA is necessary to assess what role the specific geography of WA plays in respondents' perceptions.

The data discussed below are from a larger PD project exploring Washingtonians' perceptions of English in the state (Evans 2011, 2013a, 2013b). Data was collected using a draw-a-map technique (Preston and Howe 1987; Chapter 2

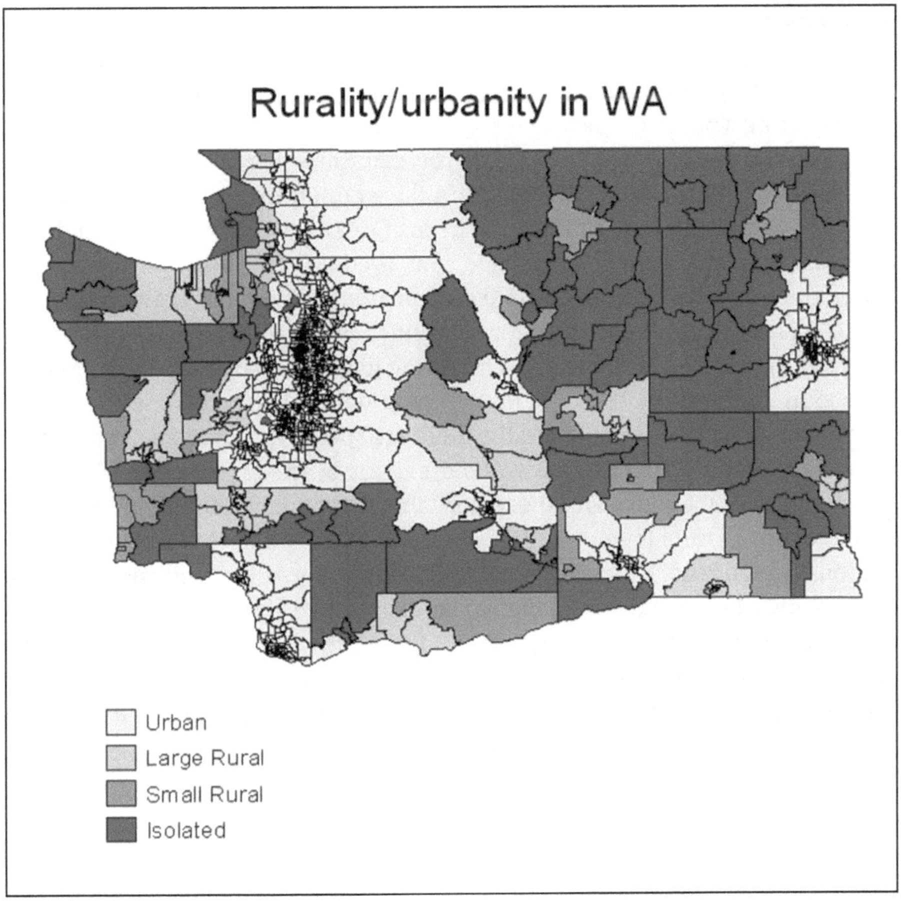

Fig. 4.2: Map of US Census Bureau Rural/Urban Commuting Areas (RUCA) in Washington State[2].

in this volume) in which respondents were asked to indicate on a map of WA where they believe people speak "differently". The sample discussed here is drawn from responses from 178 long-term residents of WA. Below, the methods for data collection and results bearing on these categorizations are described and discussed.

2 Map generated using ArcGIS and a data set consisting of 2000 US Census commuting data and 2004 ZIP codes compiled by the WWAMI Rural Health Research Center. Data available at http://depts.washington.edu/uwruca/ruca-data.php.

2 Methods

Respondents were instructed to draw on a map of WA: "draw a line around places where you think people's English sounds different. Next, write down what you'd call that way of talking, if you can think of a label for it". (Preston and Howe, 1987 Preston 1999b). A full description of the data collection methods and results can be found in Evans (2011, 2013a, 2013b). Respondents' hand-drawn maps were analyzed and compared using ArcGIS 10.0 Geographic Information System (GIS) software (see also Montgomery and Cramer, this volume). This software allowed the areas identified on respondent maps to be quantitatively aggregated and queried, permitting detailed analysis of the spatial information captured by the hand-drawn map data. There are various types of maps created by this method. Firstly, there are spatial associations with "difference" indicated by the lines drawn on the maps by the respondents. Secondly, the qualitative labels given for the regions that respondents identified, including maps generated from them, comprise a different set of data relating to what associations respondents have with those spatial differences they indicated on their map. While GIS is usually used to map more objective "facts", here maps of perceptions are created. Pavlovskaya (2009: 18) argues that "it is vital to articulate GIS as a strategy for mixed methods research that transgresses the established epistemological boundaries". The current study and contributions in this volume demonstrate how this can be achieved in the field of linguistics.

2.1 Procedure

Each map was scanned and saved as a digital image file so that it could be read by ArcGIS. Hand-drawn shapes on each map were traced using a procedure known as *heads-up* or *on-screen digitizing*, resulting in a GIS polygon *feature*. Thus, each shape that respondents drew on their maps is represented by a single polygon in the GIS. Demographic information about the respondents and the labels they provided for each shape were added to each polygon as attributes. Each of these polygon features was combined to create a composite map consisting of all the respondents' maps (see Figure 4.3).

2.2 Qualitative labels

The labels extracted from the maps were analyzed using a content analysis (Bauer 2002; Johnstone 2000) drawing on other PD work (Garrett, Williams and Evans 2005; see also Braber, this volume). This procedure examines the con-

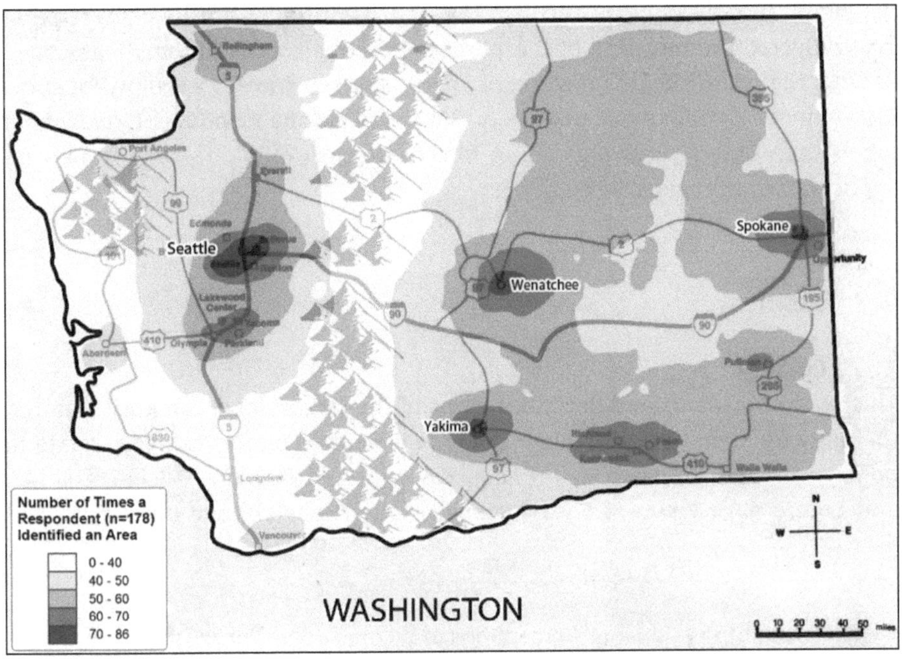

Fig. 4.3: Composite of all responses indicating overall agreement on circled regions.

Tab. 4.1: Label categories and examples.

Category	Number of comments	Examples
Country	63	*rednecks, farmers*
Spanish	40	*Spanglish, spanish influence*
Slang	24	*a lot of slang*
Pronunciation	24	*warshington, word pronunciation*
Gangster	14	*Ghettogangsta slang*
City	14	*city talk, urban*

tent of each label in order to group similar labels and to find emerging themes. Similar words/phrases were collated with others that were in the same semantic field (see Table 4.1). For example, *hick* and *southern* were combined with *cowboy* to form one category called Country. If more than one different word or phrase was associated with a single polygon, each one was considered as a separate response. A noun phrase such as *high society proper* was coded as belonging to only one category, however, and attributed to the category that best reflected the meaning of the phrase. Thus *high society proper* was coded as High Status. While this method of analysis of keywords does allow for some respondents' opinions to be overrepresented (e.g., one respondent gives more labels than others), it avoids being forced to select only one label from a respondent who provides several.

2.3 Results

2.3.1 Country

This section presents and describes maps of the largest label category Country, the most frequent category label. Comments in the Country category relate to notions of rural life such as *farming*, *cowboys* and *boondocks* (see Table 4.2 and Figure 4.4). Figure 4.4 (and maps of other labels) have been described

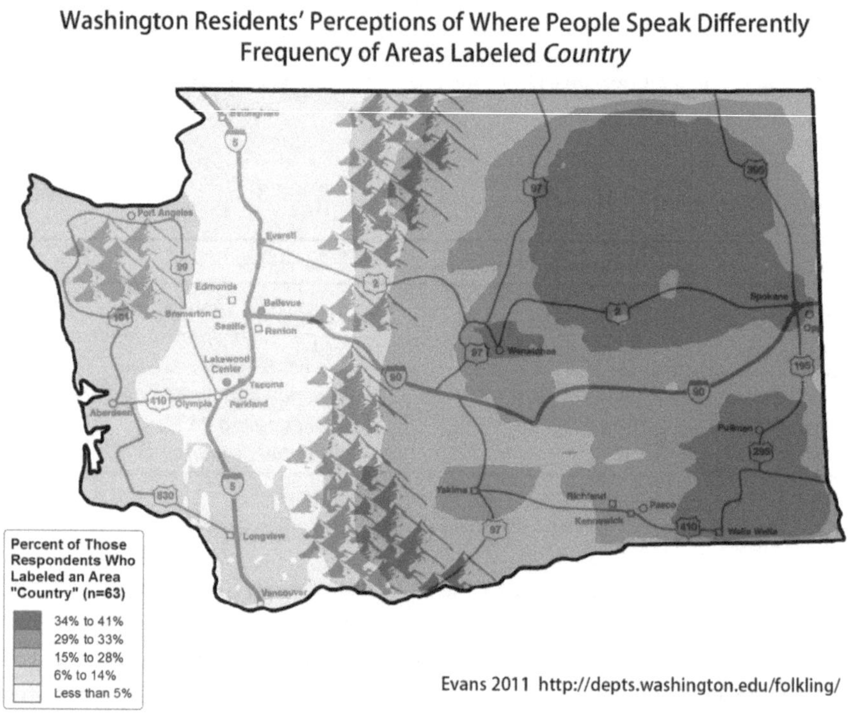

Fig. 4.4: Composite map of areas labeled in the Country category.

Tab. 4.2: Labels in the Country category[3].

Sample labels in the Country category (total number of labels in this category = 86)
– *In eastern Washington it sounds a bit country*
– *Boondocks*
– *Sounds like farmers*
– *more rural +southern-y*
– *Countryfied more southern drawl*
– *sounds more redneck country talk Get er done!*
– *Hillbillish Oroville in particular*
– *Mountain Man Draw*
– *cowboy*
– *Hick, country talk*

elsewhere (Evans 2011, 2013a, 2013b) but here I consider its importance for highlighting the significance of rural communities in sociolinguistics and with specific reference to the notion of the urban/rural dichotomy. Figure 4.4 shows the results of mapping polygons associated with the Country label. Firstly I note that a large part of eastern WA was marked as *different* and given a Country label by at least 29 % of respondents. The largest amount of overlap of polygons is in the northeastern area of the map with another large amount of overlap in the southeast corner of the state. These areas on the RUCA map (Figure 4.2) are largely "isolated" excepting the southeast corner of the state which is considered "urban". This southeastern area is a US Census Metropolitan Statistical Area consisting of two counties; Asotin County in WA and Nez Perce County in the Idaho Panhandle. Two principle cities, Clarkston, WA and Lewiston, Idaho, are population hubs for the region. Respondents seem to perceive this region as *country* in spite of the US Census Bureau designation as an urban area.

The considerable overlap of areas means that there is a certain amount of agreement among the respondents about these areas. Namely, eastern and western WA are perceived as "different". If we consider the labels in the Country category, we obtain a picture of what is "different" about that region according to the respondents. The Country category is combined with other similar categories like *hick* and *southern* which were also frequently associated with eastern WA, making this grouping 86/336 labels or 25 % of the labels (Table 4.2). This category of labels reflects typical negative associations with rural regions such as insularity and backwardness in addition to arguably neutral

3 All labels are quoted exactly as found on the maps.

associations with farming and agriculture (Woods 2011). These negative labels suggest a belief that English in eastern WA is at the very least non-standard but possibly even "worse" such as *hillbillish, southern* (Preston 1996; Niedzielski and Preston 2003). The positive associations with the rural such as tranquility and the "rural idyll" (e.g., Woods 2011; Williams 1973) are not present.

The label *rural* appeared 15 times in the overall data set and was included in the Country category. 12 of the 15 rural labels were associated with eastern WA. This label must be considered with caution as Jones (1995: 42) found that the term rural was not frequently used in lay discourses and that terms like country/countryside were more likely to be found.

The western coast of WA is also indicated as Country by some respondents (at least 6 % of polygons). Further inland however, the Interstate 5 corridor from Bellingham to Olympia contained less than 5 % of polygons with the Country label. This region is the most highly populated in WA according to RUCA data (Figure 4.2). This pattern of marking on the maps suggests, again, that a dichotomous perception of urban/rural difference is strong for these respondents.

2.3.2 Gangster

An interesting category related to urban/rural perceptions is Gangster (Table 4.1 and Table 4.3). Fifteen labels in the data set related to gangs (Table 4.3). Figure 4.5 shows the areas labeled as Gangster include urban areas but also include a large part of southeastern WA. Rural areas in WA do experience gang activity. The National Youth Gang Center (2009) reported that while only 5.9 % of gang members live in rural counties nationally, the percent change in the number of gangs between 2002 and 2007 was greatest for rural counties (64 %). These respondents seem to have noticed this. In fact, Puget Sound Public Radio presented a weeklong series in April 2011 titled *Living in Gangland* that documented the problem of rural gang activity in the Pacific Northwest. In addition, Yakima County (south central WA) established a Gang Commission in 2010 to solve "the problem of gangs" there (Board of Yakima County Commissioners 2010).

Although Gangster labels have not turned up in previous national-scale PD studies (e.g., Hartley and Preston 2001; Niedzielski and Preston 2003), *gangster* was a category also attributed to urban areas in the study of perceptions in California by Bucholtz et al. (2007), St. Louis, Missouri (Gordon 2010) and Houston, Texas (Cukor-Avila, forthcoming). These PD studies, like the present one, surveyed respondents on a smaller scale map that focused respondents' attention on more local space (their own state/region) rather than a full US

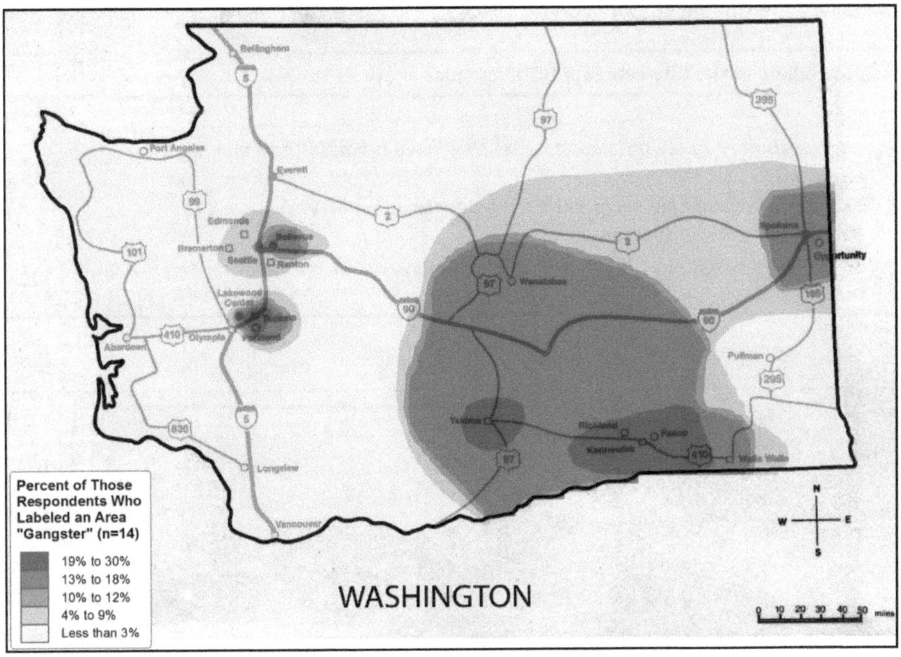

Fig. 4.5: Composite map of areas labeled in the Gangster category.

Tab. 4.3: Labels in the Gangster category.

Sample labels in the Gangster category (total number of labels in this category = 14)
– *Ghetto*
– *gangsta slang*
– *Gangster*
– *gangster -yo! homie*
– *gang slang*

map. A smaller-scale map may have enabled respondents to give more detail that would be unnoticed by outsiders.

2.3.3 City

Labels such as *city* (see Table 4.1 and Table 4.4) also suggest that the socio-cultural and linguistic distinctions of urban and rural are salient for non-linguists, especially when compared to the Country category and maps described above (Table 4.1 and Table 4.3, Figure 4.6). *City*, for example was used largely

Tab. 4.4: Labels in the City category.

Sample labels in the City category (total number of labels in this category = 14)
– *City*
– *City vocabulary valley girls sounds like they have a ball in their mouth*
– *City English*
– *City talk. These people seem to talk like people do on MTV*
– *Seattle-city slickers "yo, man"*
– *Speak like city folk, use language commonly associated with popular music e.g., hip-hop*
– *urban*

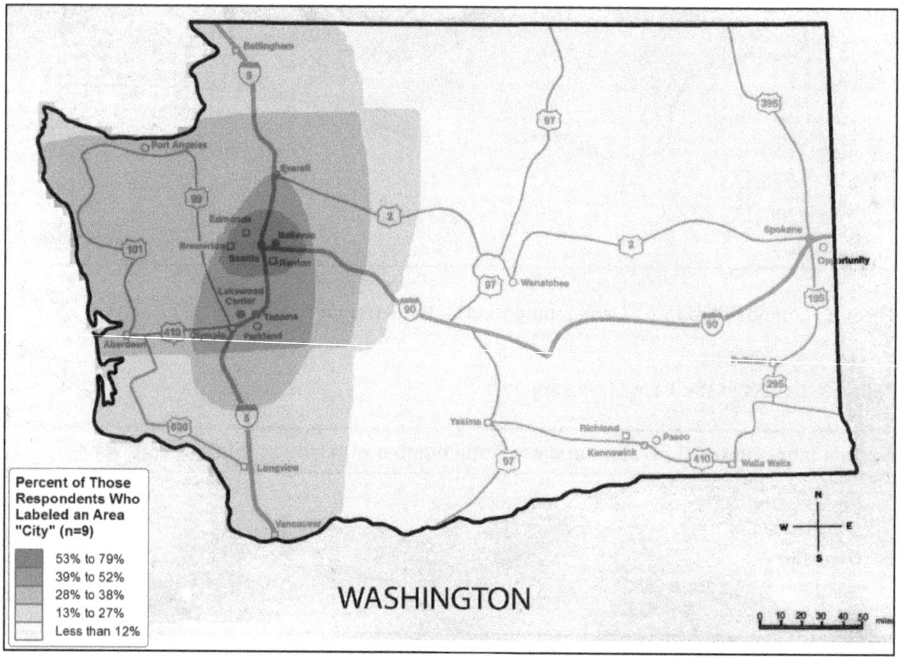

Fig. 4.6: Composite map of areas labeled in the City category.

to describe western WA. In fact, half of the *city* labels[4] were associated with the Seattle area. Figure 4.6 is a composite map of the polygons drawn on maps and given a *city* label. This map shows the strong perception that western WA,

4 The category City includes the label "urban". Although urban has various related meanings, its primary association of 'city' was assumed in the absence of verification that other meanings were intended (e.g., through follow-up with respondents).

especially Seattle, is *city*. It is difficult to determine from these labels the exact meaning of *city talk* because many of the labels are neutral. The affective values of *city talk* become clearer when considered in conjunction with labels associated with Seattle, discussed below.

While inhabitants from urban and rural areas seemed to agree on an urban/rural dichotomy, there were a few eastern respondents who attributed labels like *normal* to their own regions, suggesting they are unaware of the perceptions of their own region as non-standard. This highlights the complexity and the importance of locality with regard to spatial perceptions and linguistic variation in urban and rural contexts.

2.3.4 Labels associated with a specific place

One advantage of the GIS mapping technique for PD is that data can be explored in multiple ways. For example, in addition to identifying patterns drawn by respondents, specific regions can be explored for patterns of perceptions. In this case an area on the map is selected, using a qualitative designation such as a circle encompassing the majority of polygons in the location of Seattle (e.g., Figure 4.7), and the labels associated with that region can be obtained. A content analysis of the labels can then be examined for consistent themes. Exploring labels for major cities in eastern and western WA may help refine interpretations of respondents' associations with urban/rural. Therefore Seattle and Wenatchee were selected for this analysis, and the labels associated with those cities are described below.

2.3.4.1 Seattle

In 2010, the population of the city of Seattle was 608,660 (US Census Bureau 2010a). In terms of race and ethnicity, approximately 70 % of the population is white, 16 % East Asian, 9 % Black or African American and 6 % Latino (ibid.). King County, where Seattle is located, depends largely on non-farm employment such as professional, government, education and health services (Vance-Sherman 2013). Figure 4.7 shows the selected area around Seattle, and the labels associated with polygons that fall in the selected area. In this case, 72 labels were associated with this area. These labels can be explored for patterns of consistent themes and provide insight on the respondents' perceptions of Seattle English. The labels associated with Seattle represent three primary categories (after a content analysis): High Status, Ethnic Diversity, and Metalinguistic Comments. Table 4.5 provides examples from each category. These categories present main themes although they do not account for all labels as some were too diverse to fit

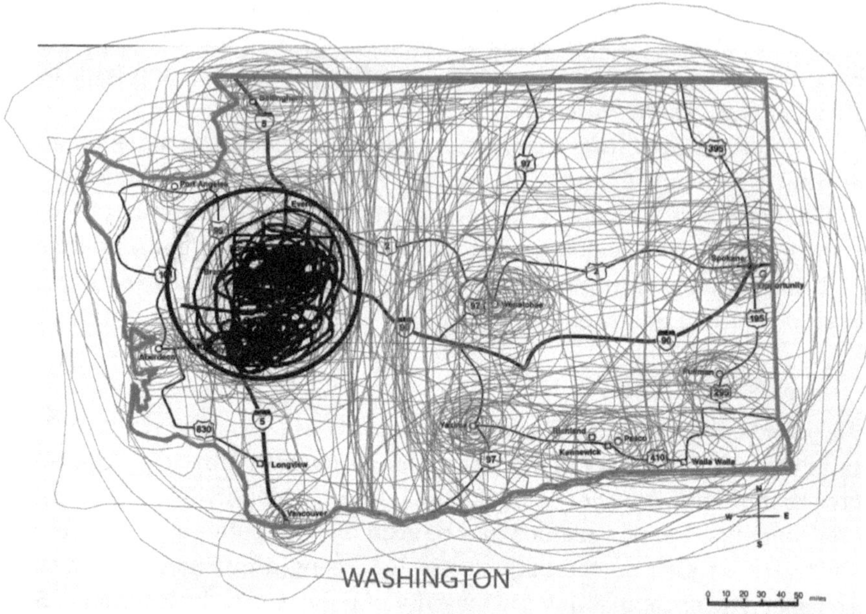

Fig. 4.7: Overlapping polygons in the Seattle area.

into these or any other categories. This collection of labels for the Seattle area suggests that while *city talk* is "high status talk", it is also "varied talk" because of the perceived diversity of the population in cities.

The smallest category was *High Status*, representing labels such as *high society* or *more pure*. Metalinguistic commentary relating to high status often reflects perceptions of correctness with regard to language (Niedzielski and Preston 2003: 147). It seems that, for some of the respondents in the present data, the city is a place where educated wealthy people live, supporting Fridland and Bartlett's (2006) findings relating to a self-perceived higher status associated with urban centers. In the present study, comments about high status for Seattle came from both western (in-group) and eastern (out-group) WA inhabitants. Of the 11 *high status* comments, 6 were from eastern and 5 were from western respondents.

The categories of Ethnic Diversity and Metalinguistic Comments both suggest there is an important amount of perceived linguistic and cultural variety in the city. For example, comments such as "Ethnic groups diverse" and "different styles of speaking" reflect a perception that the city is a place of contact and mixing of cultures. Some of the Metalinguistic labels contradict the High status category (discussed above) in that they point to non-standardness (e.g.,

Tab. 4.5: Labels associated with Seattle metro area (n = 72).

High Status (11/72)	**Ethnic diversity (15/72)**	**Metalinguistic Comments (16/72)**
– *High society proper English* – *Educated words wealthy* – *More "pure"* – *Ivy league* – *Rich sounding*	– *Ethnic groups diverse* – *A lot of people live here that are from different places* – *All Seattle area or town on the side of Seattle might sound a little different because of the different ethnic groups mixed together*	– *Slang* – *Faster shorter speech* – *They use different styles of speaking* – *Different word choice mainly in the city* – *iSpeak like city folk, use language commonly associated with popular music e.g., hip-hop*

slang, language associated with hip-hop). While some labels seem neutral about the effects of ethnic diversity on the language of the city, others suggest overtly that language in the city is non-standard. It is possible that this perception of ethnic diversity is directly related to the perception of non-standardness. Niedzielski and Preston (2003: 140) found that ethnicity, and especially African American English, were salient features in metalinguistic commentary and that respondents equated such varieties with non-standard English. It seems that, where there is ethnic diversity, there is non-standard English. These kinds of perceptions of the city as ethnically diverse are unsurprising as they are consistent with popular discourse about cities.

2.3.4.2 Wenatchee

Wenatchee is significantly smaller than Seattle. US Census Bureau indicates that the 2010 population of Wenatchee was 31,925, approximately 29 % of which is Latino and 70 % of which non-Latino (US Census Bureau 2010b). The region's economy relies heavily on agriculture (largely fruit trees) and plays an important role in employment (Meseck 2013).

Figure 4.8 shows the polygons drawn around the city of Wenatchee, a city in the center of the eastern side of the Cascades. English in Wenatchee was most frequently labeled as influenced by Spanish (see Table 4.6). Thus it appears that the presence of Latino/as is more salient to the respondents than aspects of the Country category. This may be due to the large population of Mexican and Mexican-American residents in Wenatchee and surrounding cities. Latino/as have been present in the WA region since before WA was a US

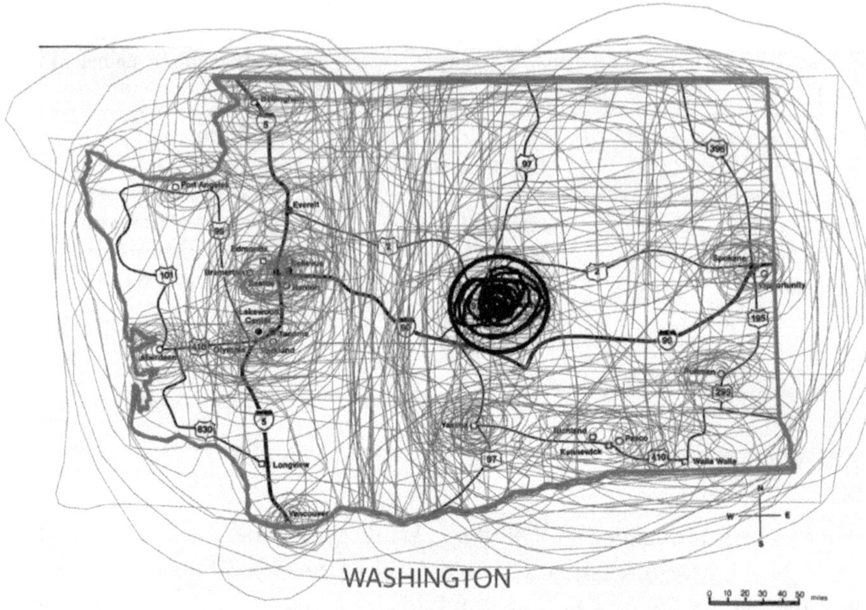

Fig. 4.8: Overalapping polygons in the Wenatchee area.

Tab. 4.6: Labels associated with Wenatchee (n = 15).

Spanish (9/15)	Country (3/15)
– *Spanglish*	– *hicks*
– *Spanish English mix*	– *Sounds like farmers*

state; the Yakima Valley (southeastern WA) specifically experienced a large settlement of people from Mexico after WWII for agricultural work (Gamboa 1981). This region continues today to have a high percentage of Latino/as (WA Office of Financial Management 2010).

3 Discussion

The data presented here sheds light on Washingtonians' perceptions of English in urban and rural areas. The data shows that an urban/rural dichotomy is salient for most respondents. They also expressed perceptions of rural English that related to Country, Southern, Gangsta, and Hispanic identities. Urban areas were associated with high status, ethnic diversity, and nonstandard language.

The fact that respondents used *city* and *country* as explicit labels suggests these categories are meaningful and productive. It may seem difficult to reconcile this with Lichter and Brown's (2011) assertion of a new urban/rural interface in which the interdependencies of the urban and rural render sharp urban/rural distinctions "obsolete" (566). Nevertheless, by considering the nature of academic and lay discourses and by accepting that they can rationally coexist, these two seemingly contradictory points can be brought together. Sayer (1989) and Jones (1995) distinguish two fields of interpretation of urban/rural: "academic discourses" and "lay narratives/discourses".

These two perspectives have different purposes in that the aim of the "academic discourse" is to objectively and rationally describe the urban/rural while the aim of "lay narratives" is to make sense of urban/rural in everyday contexts. That is, for lay narratives, urban/rural is necessary folk shorthand for population distribution that is admittedly complex. In fact, while social geographers have criticized the oversimplification of the urban/rural binary in academia (e.g., Vanderbeck and Dunkley 2003: 244), this property of oversimplification, or shorthand in lay narratives, should not render them useless. Indeed, Halfacree (1993) argues that this "shorthand" is a type of "social representation" (Moscovici 1973) of space that serves the purpose of understanding and communicating about one's social world. Thus, it seems reasonable to expect that social representations of rural/urban will continue to be productive in lay discussions of space. In fact, Halfacree (2006: 44) argues that rural space is "far from disappearing as a significant conceptual category". As Moscovici (1998: 210) explains, "most individuals prefer popular ideas to scientific ideas, making illusory correlations which objective facts are incapable of correcting".

As for the rational coexistence of the academic and lay perspectives, Jones (1995) describes how lay discourses can be incorporated into rural studies and argues strongly for this incorporation: "[...] academic definitions of the rural must incorporate bottom-up, descriptive, narrative approaches, which draw on lay discourses, as well as the top down, theoretical, explanatory approaches which have normally dominated" (1995: 47). Preston has also argued for some time that lay descriptions of language are a crucial part of linguistic research (e.g., 1986b, 2011). Thus scholars across disciplines have recognized the value of incorporating lay perceptions of socio-cultural objects in their research for revealing emic categories hitherto hidden from view.

While a strong urban/rural dichotomy has not emerged in many PD studies, even in conversational data (Niedzielski and Preston 2003: 121), this may be the result of the geography of the map used in previous surveys. PD research involving maps has often used a map of an entire country while, here, respondents were only asked about their own state (see also Cramer's category Louisville/Lexington," this volume). A map of the whole US does not lend itself well to an urban/rural breakdown because rural and urban communities exist all over the country and therefore may be salient only at a more local level. Limiting the space surveyed to the respondents' own state may allow them to indicate finer constructs of difference in their community (see also Lonergan, this volume). For example, according to Cloke and Milbourne (1992: 364), "localised constructs of the rural are [...] likely to be complex and multifaceted" and one facet of that complexity of perceptions of rural arise from context. That is, "important meanings associated with the rural are circulated and negotiated also at the [national], regional and local scales and indeed may be crucially mediated by the individual at all scales" (361). Bucholtz et al.'s (2007) data from California (described above) suggests that perceptions of residents in other states may be similarly local and dichotomous. Data from other communities should be collected to confirm the salience of urban/rural dichotomies on the local level in other states (see Cukor-Avila, forthcoming).

Halfacree (1993: 32) emphasizes the importance of research on perceptions of space: "Lay discourses are [...] not to be regarded as being rooted in a probable myth but should be seen as interpretive repertoires derived from a disembodied but none the less real social representation of the rural". Jones (1995) echoes the notion that lay perceptions of rurality are crucial for understanding the nature and importance of rurality in society. In order to better understand folk perceptions of urban/rural, further studies involving interviews with respondents could clarify what that distinction means to them and whether the blurring of urban/rural in modern society is evident to them (see Mann, this volume). In addition, the role of urban/rural identity for Washingtonians, especially those in eastern WA, should be explored to see if *country talk* is leveraged for identity purposes such as Hall-Lew and Stephens (2012) found for Texoma respondents. Understanding perceptions of rurality has a critical role in the overall picture of perception of and production of language.

Stephen L. Mann

Chapter 5
Rural "rednecks" and urban "bluebloods": The (in)compatibility of sounding gay and sounding southern

1 Introduction

One of the aims of the current volume is to consider ways in which linguistic notions of place are realized by "real people" (Preston 1998: 255), i.e., non-linguists, residing in both urban and rural locations. With that in mind, while preparing this chapter I kept returning to the conversation in (1).

(1) "Southern by the grace of God"[1]

SLM:	Do you consider yourself a southerner?
Andrew:	*[Jokingly]* Not with an accent like this. Of course not. Are you kidding? *[Now being serious]* Absolutely.
SLM:	Okay. What does that mean to you?
Andrew:	Um, (I don't know, like) most of all you're southern by birth and, uh, how does it go, um, no, um, I- sh- I can't remember how the thing went. Uh, something about the grace of G- oh, southern by the grace of God. American by birth and southern by the grace of God. That's the way it goes. But, uh, anyway, um, yeah, born in the South. But it's, um, I don't know- when I- um, I don't know, we've got Atlanta and places like that, but, uh, much faster pace places.

The example in (1) comes from an interview I conducted with Andrew, a 50-year old gay man from South Carolina, in the southeastern United States. This short clip provides insight into several key components of what it means to be a "southerner," or to be affiliated with the U.S. South. First, there is a direct association with being a southerner and speaking in a certain way. Andrew implies that it is impossible to have an accent such as his and not identify as

1 "SLM" is the interviewer/author. All other names provided in transcripts and discussions are pseudonyms.

a southerner. There is also an association between a southern identity and a certain pace of life, namely a pace of life not expected in a large city such as Atlanta, even if that city is, like Atlanta, situated within the geographic boundaries of the U.S. South. Finally, there is an implied link between southern identity and Christian religious beliefs.

Another conversation to which I kept returning took place during an interview I conducted with Mark, a 49-year old gay man, also from South Carolina.

(2) "The person I feel sorriest for in the entire world"

Mark: Okay, I- I'll put it this way. Th- the person I feel sorriest for in the entire world is an effeminate-speaking straight man who is actually straight, because there are certain people who speak in a way that the presumption of you're being gay is irrebuttable[2], no matter, I don't care if you bring a woman in and have sex with her right there in front of these people, uh, uh, uh, people are gonna still presume you're gay.

This example highlights a folk assumption, which is consistently present in interviews and focus groups I have conducted, namely that "sounding gay" is equivalent to "being gay".

The questions I address in this chapter relate to the folk beliefs about language highlighted in (1) and (2). If, on the one hand, identifying as a southerner requires adherence to a specific phonological system, and if, on the other hand, "sounding gay" directly implicates a speaker's sexuality, what does that mean for gay men from the U.S. South? Andrew, the man who identifies as southern in (1), openly identifies as a gay man. Is it possible for him to be perceived by listeners as both southern *and* gay?

The folk belief that there is an inherent incompatibility between "sounding southern" and "sounding gay" is a common theme that runs throughout interviews and focus groups I have conducted. For example, in a focus group interview with five self-identified gay men, the participants had been discussing whether a speech sample they heard was from a gay man or a straight man. In the sample, the speaker lengthened the vowel in the word *there*. Scott thought the lengthened vowel made the speaker sound gay, but Oscar, who was born and raised in the South, said that vowel lengthening was also a common feature of Southern U.S. English. But rather than judge the speaker to be both gay and southern, he only chose southern, providing the explanation in (3).

---

2 I have opted not to use "[*sic*]" in transcripts so as not to assign negative value to any linguistic form.

(3) “It throws your gaydar all completely out of whack”

Oscar: But the southern accent for me-

Brian: Right.

Oscar: It’s like- it’s like being in Europe when you see all these men dressed in these very form-fitting shirts, and the pants are very form-fitting, and in the U.S. you automatically think “gay,” but, you know, in Germany or England it’s like, “Oh, that’s how they dress”. I mean, it throws your gaydar all completely out of whack.

Brian: Mm-hm.

Oscar: The southern accent (will do that).

Scott: Right. But that’s- yeah, that’s visual. But sound-wise, it’s <unintelligible>.

Oscar: [*overlapping Scott*] But- but that’s the- but that- the southern accent to me is the equivalent of that visual in Europe.

In another focus group interview I conducted, one of the participants talked about his own speech and the fact that he is often mistaken as a northerner because of the fact that people think he sounds gay. A transcript of that discussion is provided in (4).

(4) “He’s just a faggot”

Juan: When I was younger, and it- it’s something that I was- have been very conscious of, especially probably the last ten years of my life, say, like, from sixteen to, like, twenty-five, I always had people here, and I grew up, you know, rural South Carolina, but I have people here ask me where I was from. And it was just constantly, and my friend [name omitted] used to think it was hilarious, ‘cause she’d be like, “Oh, where are you from? Are you from the north?” It’s like, “Oh, no. He’s just a faggot”.

In this chapter, I address the question of whether speakers can be heard to be both southern *and* gay. I focus my analysis on two men I interviewed to show that sounding gay and sounding southern are not necessarily incompatible. Prevailing ideologies of gender, sexuality, class, and rurality constrain listener perceptions, however, causing some listeners to perceive a speaker as being either southern or gay but not both. Before addressing this question, I will first provide some background on language attitudes and perceptions, an overview of my methodology, and a discussion of folk notions of southern identity.

2 Language attitudes and perceptions

Language attitude studies can be divided broadly into two areas: the social psychology of language and folk/perceptual dialectology[3]. Social psychologists use techniques such as matched-guise tests to elicit listeners' attitudes toward specific speech samples. Perceptual dialectologists have created or adopted techniques such as imitation studies and folk dialect maps to garner the individual and shared beliefs that people have about different language varieties. The current chapter builds on the foundation provided by both the social psychology of language and perceptual dialectology. To determine listener perceptions of speakers as "gay" and/or "southern," I rely on variations of the matched guise technique introduced by Lambert et al. (1960), but the ensuing discussion is based primarily on an analysis of interview and focus group data to determine what "real [gay] people" (to borrow from Preston [1998]) have to say about their perceptions of and attitudes toward southern and/or gay male varieties of U.S. English.

A major breakthrough in language attitude studies was seen with the advent of the matched-guise technique. This technique, first introduced in Lambert et al. (1960), "makes use of language and dialect variations to elicit the stereotyped impressions or biased views which members of one social group hold of representative members of a contrasting group" (Lambert 1967: 93). A single bilingual or bidialectal speaker is recorded reading the same passage in each language or dialect. By changing the language variety but not the speaker, researchers are better able to attribute listeners' reactions directly to language varieties rather than speaker-specific attributes (Giles and Bourhis 1976: 294).

Over the past twenty years, several researchers have adopted and adapted the matched guise technique to gain insight into listener attitudes toward "sounding gay" in U.S. English. Gaudio (1994), for example, shows judges to be consistently correct in identifying speakers as gay or non-gay. In addition, judges rated gay-sounding voices based on common stereotypes of gay men, i.e., "effeminate," "emotional," and "affected". More recently, new technolo-

---

3 In clarifying their use of the word "folk," Niedzielski & Preston (2003) state that:

> we use *folk* to refer to those who are not trained professionals in the area under investigation (although we would not for one moment deny the fact that professional linguists themselves are also a folk group, with their own rich set of beliefs). We definitely do not use *folk* to refer to rustic, ignorant, uneducated, backward, primitive, minority, isolated, marginalized, or lower status groups or individuals. That is an out-dated use in folklore and an absolutely useless one for our purposes. (xviii; emphasis in the original)

gies have given researchers the ability to do more sophisticated manipulation of speech samples to pinpoint reactions to specific linguistic variables more accurately. Levon (2006), for example, suggests that speakers are less likely to be perceived to be gay and more likely to be rated masculine if the speech sample contains shorter sibilants and a narrower average pitch range across the sample. Campbell-Kibler (2007) uses an adaptation of the matched guise technique to study the "ideological baggage" (32) carried by the velar [ŋ] and alveolar [n] pronunciations of the English progressive *-ing* suffix. She concludes that the phonetic realization of the (*-ing*) variable by itself does not cause a listener to perceive a speaker as "southern" or "gay" but can increase the level of perception and contribute to the attitudes toward it.

Perceptual dialectologists strive to differentiate between the etic views of language studied by linguists and the emic views of language held by non-linguists (Preston 1986b). Preston (1998) describes two taxonomies of language: the folk taxonomy, which represents the emic view, and the linguistic taxonomy, which represents the etic view. Focusing on emic views of language, Preston (2004) discusses three types of metalanguage available to non-linguists, two of which are relevant to the current discussion. The first type, Metalanguage 1, includes all conscious, overt comments about language. An example is provided of a woman in southeastern Michigan commenting on how her new coworker from Cleveland pronounces the name of the town [mən'ro] as ['manro] (75). Folk taxonomies of language fall into this category. Metalanguage 3, however, refers to the usually unspoken shared community beliefs that drive Metalanguage 1. Because Metalanguage 3 tends to remain below the surface, perceptual dialectologists use research on Metalanguage 1 to gain insight into Metalanguage 3. For example, a reference to African American English as incorrect Standard U.S. English (an example of Metalanguage 1) may be drawn from racial stereotypes shared by members of the community. This shared belief is an example of Metalanguage 3. (See Preston 2004: 86–87). Similarly, associations between sounding gay and sounding effeminate and/or urban on the one hand and sounding southern and sounding masculine and/or rural on the other hand may stem from broader ideologies of sexuality and gender rather than more specific language ideologies.

3 Methodology

My analysis draws on data from three sources: interviews, an online attitude and perception study, and two focus group sessions.

3.1 Interviews

In 2009, I individually interviewed eight self-identified gay men, whom I recruited through networks in the local community – a midsized city in the southeastern U.S. – where I was conducting my research. Each interview was approximately 90 minutes long. To reduce observer's paradox effects (Labov 1972), interview participants were initially informed only that they would be taking part in a study on gay men's social networks. All men were asked questions about their upbringing, their relationships with their families, their primary social networks, their social activities, and the coming out process[4]. Approximately two-thirds of the way through each interview, I debriefed the participants regarding the linguistic aspect of the research and then continued with a series of questions about gay men's language and what it means to "sound gay". Details on each of the interview participants are provided in Table 5.1.

Five of the eight men I interviewed were born and raised in South Carolina, which is geographically located within the South. Chris was born and raised in the North in Massachusetts. Mickael and Tommy are from Kentucky and Maryland, respectively, both states which lie on the border between the North and the South. (A discussion of the relevance of border states is provided in Section 4.3 below; see also Cramer, this volume.) Six of the men self-identified

Tab. 5.1: Interview participants.

Pseudonym	Age	Occupational category	Raised	North or South	Self-identified southerner?
Andrew	50	Retail	South Carolina	South	yes
Chris	46	Corporate	Massachusetts	North	no
David	67	Healthcare	South Carolina	South	yes
Len	35	Corporate	South Carolina	South	yes
Mark	49	Law	South Carolina	South	yes
Mickael	39	Corporate	Kentucky	North/South border	yes
Thomas	31	Corporate	South Carolina	South	yes
Tommy	39	Performing arts	Maryland	North/South border	no

4 Leap (1996) argues for the use of "life-story narratives" to gain insight into gay men's early socialization as gay-identified men (125–139). See Liang (1997) and Chirrey (2003) for a discussion of coming out as a speech act.

as southerners, and two of the men self-identified as northerners. All of the interview participants were residing in South Carolina at the time of the interview. Tommy had relocated to South Carolina three years earlier; Mickael and Chris had each been living in South Carolina for over ten years.

Although I did not make use of long-term participant observation, I was still able to build a rapport with each interview participant rather quickly. The men I interviewed seemed very comfortable talking to me, at times joking with me, and, most importantly, sharing very personal information with me. I attribute the ease with which we were able to develop a rapport to the fact that I am also a gay man who – at the time of the interview – had been a member of the local community for approximately five years. These men knew or recognized me as a participant in local activities. Also, because I am gay, they assumed that we shared similar past experiences regarding coming out and living as a gay man in a heteronormative society. Andrew, for example, when asked to define "being out," stated that "it can be tough. God knows what I- Lord knows we all lived through it. I'm satisfied that you had to deal with whatever". I was also able to build a rapport with the focus group participants described below quickly for similar reasons.

3.2 Attitude and perception study

I also conducted an online attitude and perception (A&P) study in which 50 respondents listened to short speech samples – approximately 10 to 13 seconds each – which I drew from each of the interviews. After listening to each sample, A&P study participants indicated on a survey response form whether they thought the speaker was gay or straight and provided their personal attitudes toward the speaker, including ratings of masculinity, femininity, intelligence, and friendliness. The survey instrument was adapted from Campbell-Kibler (2007).

A&P study participants were recruited in several ways. First, a recruitment letter was sent through Facebook, a popular online social networking site with a diverse user demographic and wide geographic reach. The letter was initially sent to my personal network but with a request to forward it to others, thereby establishing a snowball sample. Additionally, A&P study participants were recruited through the gay-straight student alliance at a local university. Wide distribution of the survey link was critical, because the 20–30 minute length of the survey – while necessary to answer the research questions – introduced the potential for survey fatigue. The probability of a low response rate required a wide reach in order to meet the response quota.

Tab. 5.2: Attitude and perception study participants by sex and sexual orientation.

Sex	Sexual Orientation			
	LGBQ	Straight	Unknown[5]	Total
Female	0	13	0	13 (26 %)
Male	13	11	0	24 (48 %)
Other	1	0	0	1 (2 %)
Unknown	0	0	12	12 (24 %)
Total	14 (28 %)	24 (48 %)	12 (24 %)	50 (100 %)

Tab. 5.3: Attitude and perception study participants by identification as southerner.

Self-identified southerner?	Number of participants	Percent of total participants
Yes	17	34 %
No	21	42 %
Unknown	12	24 %
Total	50	100 %

Because a primary focus of the larger project for which these data were collected is gay men's attitudes toward gay male varieties of U.S. English, I needed to ensure a sufficient number of gay male participants. A cross-tabulated breakdown of A&P study participants by sexual orientation and sex is provided in Table 5.2. Of the A&P study participants who provided demographic information, 14 identified as lesbian, gay, bisexual, and/or queer (LGBQ) and 24 identified as straight. 13 participants identified as female and 24 participants identified as male. This balance, however, is not maintained when comparing sex and sexual orientation combined. All but one of the LGBQ-identified A&P study participants were men[6], and none of the women identified as bisexual, lesbian, or queer. One of the LGBQ-identified men identified as queer, another identified as bisexual, and eleven identified as gay[7].

5 The large number of "unknown" responses is due to a flaw in the survey design. The demographic questionnaire was the last page of the survey. Participants who only partially completed the survey (most likely due to survey fatigue) provided useable information about their attitudes toward and perceptions of a portion of the speech samples. Their demographic information, however, was never captured.

6 Of the participants who provided demographic information, the only one who did not identify as either "male" or "female" identified as "gender queer".

7 One of the male participants who identified as "gay" also identified as "bisexual".

A summary of A&P study participants based on whether they identified as southerners is provided in Table 5.3. Of the participants who provided demographic information, 17 identified as southerners and 21 did not identify as southerners. The participant population, therefore, is not skewed in terms of southern identification.

3.3 Focus group sessions

Following Campbell-Kibler (2007), I also conducted two focus group sessions in conjunction with the A&P study. During each focus group session, 3–5 self-identified gay men individually performed the tasks required of A&P study participants. Then, as a group, we discussed the samples and the ratings that each participant had provided on his survey sheet. Six of the focus group participants self-identified as southerners and two of the participants self-identified as northerners. Details on each of the interview participants are provided in Table 5.4.

Tab. 5.4: Focus group participants.

Pseudonym	Age	Raised	Currently living	Self-identified southerner?	Focus group
Alan	32	South Carolina	South Carolina	No	1
Brian	40	Arkansas	Pennsylvania	Yes	2
Chuck	45	Pennsylvania	Pennsylvania	No	2
James	26	South Carolina	South Carolina	Yes	1
Juan Carlos	37	South Carolina	South Carolina	Yes	1
Kenneth	37	Louisiana	Pennsylvania	Yes	2
Oscar	37	North Carolina	Pennsylvania	Yes	2
Scott	40	Massachusetts	Pennsylvania	No	2

4 Connectedness to the U.S. South

The introduction to this chapter briefly considered what it means to be a "southerner" from the perspective of Andrew, one of the men I interviewed. At the beginning of each interview, I asked participants where they were from, where they had lived, and whether they identified with the regional label generally associated with their state of origin (e.g., "southern" for South Carolina).

Because not all of the men I interviewed are originally from the South, I separate the discussion in the sections that follow based on where the men were raised: in the South, in the North, or in so-called “border states” (Kentucky and Maryland).

4.1 Raised in the South

The five men from South Carolina gave two primary responses when asked whether they considered themselves to be southerners. Andrew, David, and Mark gave definitive “yes” or “absolutely” answers to the question. Len and Thomas also answered affirmatively, but they were more hesitant when doing so[8]. Len’s hesitancy is shown in (5).

(5) “They tell me I don’t have much of a southern accent”

SLM: Um, so would you consider yourself a southerner?

Len: I would s- yeah, sure.

SLM: Okay. Well, you seem hesitant. I mean-

Len: Um, no. It’s just funny you say that, because of- I travel around a good bit, and I- I get a lot of times, uh, when I tell people I’m from South Carolina, they’re like, “Oh, you weren’t born there obviously”. Or, “You just moved there”. I’m like, no, been there all my life. An- they tell me I don’t have much of a southern accent, so I gu- it comes and goes. But, um, they’ll tell me- so, I me- you just say that, it kind of made me laugh a little bit, ‘cause I get that often when I’m out west or up north, and, uh, “You don’t sound like a southerner”. Um, so, I- I didn’t pick up the drawl as bad as my sister did, but, um, yeah, I would say I’m a- I’m a southerner. Didn’t grow up on a farm or anything like that, but, um, uh, I did, uh, grow up in an area, you know, you know, about an acre or so of land, you know, so you weren’t right up on top of your neighbors, and you’re in a neighborhood, uh, where you’d leave your door unlocked. Um, leave your car unlocked at night. You just walked over to people’s houses and everybody knew everybody and, so, I would say from that standpoint absolutely.

8 It is possible that there are generational differences at work here. Len and Thomas were both in their thirties at the time of the interview; Mark, Andrew, and David were 49, 50, and 67, respectively. Unfortunately, I do not have the data to test this hypothesis.

Len's hesitancy here is initially based on the fact that he is not perceived by others to be a southerner because he does not sound like a southerner is expected to sound[9]. He continued to add to that hesitancy when trying to discern whether his childhood home met certain criteria for "southernness": a rural location, sufficient land, close relationships with neighbors, and a feeling of security. He eventually agreed that he is "absolutely" a southerner, using the same modifier that Andrew uses in his response to the question, but it took him much longer to get there.

Thomas, on the other hand, provided a different type of hesitancy, as seen in (6).

(6) "You can't get much more southern than that"

SLM: Um, do you consider yourself a southerner?

Thomas: Yes.

SLM: What does that mean to you?

Thomas: To me it's kind of- I mean, it's more im- it's like, I was born in the South, you know. I was born in Columbia, South Carolina. You can't- outside of Alabama, you can't get much more southern than that. Um, you know, so I think it's, you know, I just- I don't- I don't play into the stereotypical what most people think of a southerner. Um, the whole redneck, backwoods, uneducated, you know, portrayal that a lot of people seem to t- think about it. You know, I think it's any- you know, it's really- it's just the area where you're from.

Thomas initially answered the question of whether he considers himself to be a southerner with a concrete "yes". But when asked what it means to be a southerner, instead of answering the question, he started to question the original affirmative answer he gave. His hesitancy stems from the fact that he feels that being born and raised in South Carolina makes him a southerner by default, but he also wants to distance himself from the negative associations of the label. It appears that Thomas has two notions of "southerner". One is a label used for people who are born and raised in the South and the other refers to stereotypical behaviors associated with them.

For all of the men from South Carolina, being born and raised in the South is a prerequisite for being able to identify as a southerner. But for Andrew, David, and Mark – the three men from South Carolina who identify as south-

9 Len's statement is confirmed by the A&P study results. He was identified as a southerner with only 13.2 % accuracy.

erners without hesitation – being born in the South is not sufficient. When asked what being a southerner means to him, Andrew quoted the saying "American by birth and southern by the grace of God," as seen above in (1). He continued with the explanation in (7).

(7) "You got some Yankee blood in here somewhere"

Andrew: But, uh, anyway, um, yeah, born in the South. But it's, um, I don't know- when I- um, I don't know, we've got Atlanta and places like that, but, uh, much faster pace places, a little easier going, then, yet, I've since not r- growing up north, I- or living there, I don't really, you know, that's- it's the, um, I- well, it is, it's how, uh, yes, it is a preconceived notion that we're all slow and stupid down here, but, uh, we know better. I guess, the- the- you know, um, I've got my, uh, granddaddy on my mother's side's from Michigan, so, um, yeah, you got some Yankee blood in here somewhere, but, uh, it- it's just a- a slower pace, take it easy sort of thing. Uh, and, uh, you enjoy the hot weather better than you do the cold. And especially me.

Andrew started by saying that being born in the South is part of the definition of being a southerner, but then he hesitates. Apparently he does not want to consider some people born in Atlanta to be southerners even though they are born in the South, because being a southerner is not just about where someone lives; it is also about how a person lives (e.g., "a slower pace") and the values s/he holds, such as a value placed on knowledge gained from life experiences to which Andrew alludes in his claim that "we know better". Andrew's description of southern identity supports Lippi-Green's (2012) argument about regional stereotypes in the U.S.: "Northern construction of intelligence [...] is closely linked to a high level of education, [...] [while] a construction of Southern intelligence [...] has more to do with common sense and life experience" (225).

David was much more specific in his details, as seen in (8).

(8) "The old basic traditions and cultures survive"

David: Being a southerner. Okay, if you're a southerner, you are raised in a culture that is very Christian-oriented, very religious faith-based. Um, you are schooled in the history of the old South. Every little kid has to see the movie *Gone with the Wind*. I mean, that is just one thing that you have to see, and, uh, even though today I can watch it, and I think, "Oh, what a distorted picture that makes". But, you know, it really impressed me when I was a little kid. And, um, I don't know, you know. I've, um, kind of lived out and away from South Carolina, but I enjoy living here, but still I find that, uh, the old basic traditions and cultures survive, especially here in this particular part of the southeast.

For David, being a southerner requires adhering to "the old basic traditions and cultures," which includes knowledge of southern history and a belief in a Christian value system.

Mark's definition of "southerner," provided in (9), is even stricter.

(9) "Many of them are northerners [...] living in the South"

Mark: When I think of someone being a southerner, I think of someone above Florida and, um, someone who lives here, I- and- and has lived here at least generationally if not multi-generationally. Um, with the sunbelt being where everybody's moving, I guess there's lots of southerners, but many of them are northerners who just are living in the South. Um, so, for me, uh, it- it means, really someone who at least for a few generations has- has grown up in the South.

According to Mark, a person can be born in the South but not be a southerner if that person's family has not been present in the South for at least a generation, because it takes time for the southern value system to be instilled. Mark explained later in the interview that "when you meet people, you can usually tell if someone's [...] been raised in the South, as opposed to whether or not they've come to the South," because they have different notions of politeness, different attitudes toward the land, and little or no understanding of and appreciation for southern history.

4.2 Raised in the North

Chris is originally from New England in the northeastern U.S. He was born and raised in the North, but at the time of the interview had been living in South Carolina for approximately eleven years. When asked whether he identifies as a northerner, Chris provided the response in (10).

(10) "I guess so"

SLM: Would you consider yourself a northerner? Is that a label you use to describe yourself?

Chris: I- yes. I- I guess so.

SLM: You guess so. I mean, you don't have to.

Chris: I mean-

SLM: I'm just, uh, I don't want-

Chris: Yeah.

SLM: To put words in- yeah. Okay. Um, living down south, is that- how has that been for you, as a northerner?

Chris: Um, I'm trying to, uh, I'm trying to think o- how to answer that question. Um, actually I really enjoy living in the- living in the South. Everything, uh, the eleven years or so that I've spent here so far have really been atypical in most respects. It really hasn't been any- any different than living up north, except for the climate, obviously. Um, I have noticed that there are some communities or some areas- it's a- it's- it's more difficult to live as a gay man out in the open. Um, but the people that I've surrounded myself with, and when I- I- I guess I should preface all this by saying that when I moved to South Carolina, I already had a network of friends, um, and family that were extremely supportive of me as a person and my identity as a gay man, so when I came down here, I already had this network and a family and a family of friends that I- that I could rely- rely upon, um, for, you know, guidance, advice, and just for friendship and good times. So I really didn't have- I didn't feel like I had a lot of obstacles or a lot of difficulties that others may have had not having a network and being- uh, being gay, moving to the South, as far as being open, it was just a very seamless transition for the most part. [...] And I think another aspect of that, too, is that in [this city] you do have the, um, a more liberal and progressive environment because of the concentration of the universities, the colleges, and there is more youth, too.

Like Thomas and Len, Chris was hesitant to apply a regional label to himself because he has lived in the South for so long and does not necessarily want to continue to refer to himself as a "northerner". He finds himself lucky to have transitioned so easily to living in the South and enjoys living there. His network of friends for the past eleven years consists primarily of people born and raised in the South, which may also be a factor in his hesitancy toward referring to himself as a "northerner". At no point, however, does Chris adopt the label "southerner" for himself, because one of the one major differences he sees between living in the North and living in the South is the extent to which LGBQ people can be open about their sexual identities. Chris admitted that he probably would not be as content in the South if he were not living in a city, which has "a more liberal and progressive environment".

4.3 Raised on the border

The final two interview participants, Mickael and Tommy, are originally from states that lie on the border between the North and the South (see also Cramer, this volume). For these two men, therefore, unlike the five men from South Carolina, identification as “northerner,” “southerner,” “midwesterner,” or any other regional label is much less about where they were born and raised and more about their values, practices, and beliefs. When asked if he identifies as a “northerner,” Tommy responded, “Definitely. Yeah”. Earlier in the interview, he described the neighborhood in which he was raised. His description is reproduced in (11).

(11) “It was kind of soulless [...] out there”

Tommy: Um, rural Maryland is- is more- the part that I lived in was the- the subd- community I lived in was sort of blue collar Baltimore white flight. These are people that wanted to send their kids to quote unquote good schools away from black families that had sort of encroached on their, um, blue collar white neighborhoods in the late sixties and seventies. So, they, you know, my parents and some of their other, um, contemporaries moved out to these soulless neighborhoods of new, um, ranch houses, and there I was with these other, um, children of, you know, blue collar parents, and, um, it was kind of soulless, um, out there. I’d say it was definitely more northeastern than, um, our neighbors, who had been in Harford County, Maryland, for generations, were more farm folk, and, um, I’d say that they maybe had a little bit more southern in them, certainly than the blue collar, um, immigrants that had moved in from Baltimore out- out to the, um, country.

Tommy described rural Maryland as a border between the North and the South because recent migration into the area had brought people from Baltimore who held the blue collar, industrial, big city values of the North. They brought these values into a community that had historically upheld the traditional values of the South described by Mark, David, and the other South Carolinians I interviewed. Because Tommy’s family was one of the families that relocated from Baltimore, he was not raised with traditional southern values, and, therefore, “definitely” considers himself to be a northerner.

Mickael is originally from Kentucky and identifies as a southerner. His explanation for this self-labeling is provided in (12).

(12) "Kentucky was a border state"

SLM: Um, so you say you're from Kentucky. You've lived [in South Carolina] now fifteen years. Do you consider yourself a southerner?

Mickael: Yeah, actually I do. Uh, Kentucky was a border state.

SLM: Mm-hm.

Mickael: And, um, we went both ways. But I've always kind of, um, for the most part associated myself with more of the southern lifestyle then I would with the northern lifestyle. Uh, I- to me they're- they're different. Things are just a little more, um, um, they're a little slower paced in the South. Um, I think people tend to be a little more friendly, a little more outgoing than what I would normally see from- in the North. Uh, so, yes, and my dad is born and, uh, raised Kentucky. South part of Kentuck- Kentucky. Um, backwoodsish. So that kind of also was part of- of that.

SLM: What about your mother?

Mickael: My mom's from California.

SLM: Okay.

Mickael: Uh, so she was more of the city girl, but I think, um, she lived in Kentucky long enough to- to just kind of mellow a little bit. She was a big city girl come into, um, small Kentucky and backwards Kentucky, but, um, I think that she ma- (made) do. Um, I- I think my dad, though, had more of an influence on whether I would consider myself a sor- southerner or a northerner.

Mickael used the actual term "border state" to refer to Kentucky and argued that as a result the state goes "both ways," i.e., has both northern and southern aspects to it. But the specific home environment in which Mickael was raised was also on the border of the South and the non-South, because his father was from the rural, southern-identifying part of Kentucky while his mother was originally from California. As with Tommy, Mickael's choice of either "southerner" or "northerner" as a label is not based on where he was born and raised, because being on the border gives an individual flexibility. Instead it is based on the core set of values, practices, and beliefs to which an individual adheres. For Tommy, the decision was a simple one; both of his parents shared a similar set of values. Mickael, on the other hand, had to choose. He was more influenced by his father's and his own rural Kentucky upbringing and, therefore, adopted the southern values associated with that upbringing.

Based on these interviews, strong connectedness to a southern identity includes being born and raised in the South, but that is not sufficient, because it also includes a definition of “southerner” based on values, practices, and beliefs. What is most relevant to this chapter specifically, and to the volume more generally, is that these values, practices, and beliefs are primarily ones associated with rural rather than urban areas.

5 Intersections of sexuality, gender, rurality, and class

The remainder of this chapter focuses on two of the men I interviewed: Andrew and David. Andrew was 50 at the time of the interview. He has lived his entire life in the South, has a high school education, and has held a variety of working-class jobs. He was trained in high school as a welder but ended up working in printing for many years. He then briefly ran a trucking company before moving into retail grocery. He is extremely proud to be a southerner, which for him includes devotion to his Christian faith and strong biological kinship ties.

David was 67 at the time of the interview. He was born and raised in the South. After serving a few years in the Army – stationed in the South – he went to college in his hometown and studied to become a nurse. He soon thereafter earned a master’s degree to work as a nurse anesthetist. He lived for many years in different parts of the southwest and west, but he returned to the South about 10 years prior to the interview for family obligations. David considers himself to be a southerner and has pride in that label, but he often has trouble reconciling the hypocrisy he sees in southern religious beliefs and attitudes toward homosexuality. Both David and Andrew have strong ties to the same local gay bar, around which they have built their primary friendship networks.

Andrew and David both consider themselves to be speakers of a southern variety of U.S. English, but their linguistic practices are perceived quite differently. All focus group and A&P study participants correctly identified Andrew and David as southerners; however, David was almost always identified as gay, while Andrew was almost always identified as straight. Overall attitudes toward the two speech samples were also vastly different. Juan Carlos’ attitudes toward David’s and Andrew’s speech samples are representative of the other focus group participants’ reactions to the two clips. Talking about David, Juan Carlos made the comment in (13).

(13) “I loved his voice”

Juan Carlos: He seemed really warm, compassionate. I loved his voice. That’s a voice I could sit and listen to a lot.

But after hearing Andrew's sample, Juan Carlos made the comment in (14).

(14) "An overly confident dumb ass"

Juan Carlos: I was thinking some big-ass, dumb redneck. [...] The image I got was the greasy, redneck, mullet thing. Some guy that was not very confident. Well, he's confident but not confident. An overly confident dumb ass who really shouldn't be confident.

I argue that differences in perception of sexual orientation and overall assessment of each speech sample in positive or negative terms stem from prevailing ideologies of sexuality, gender, class, and rurality.

5.1 Gender

When asked about what it means to sound gay, David and Andrew provided very different answers. For Andrew, sounding gay means sounding "effeminate". David, on the other hand, does not associate sounding gay with sounding effeminate. And he does not just avoid the semantically weighted term "effeminate"; he avoids assigning any gendered label to it at all.

Focus group participants frequently made references to effeminacy, femininity, or masculinity when explaining why they judged a speech sample to sound gay. Juan Carlos' argument about why we make these associations is provided in (15).

(15) "We also put these signifiers with it"

Juan Carlos: We- we also put these signifiers with it. "Gay" means "effeminate," you know, "straight" means "masculine". And so that's really difficult, because are you- it- it seems to be so wrapped up in these ideas of homophobia and this heterosexist ideal that we're fed that if you sound this way, then you are gay. Or if you are gay you have to sound this way. So it's- it's a real difficult thing for me to separate that.

The signifying relationship highlighted in this example is not surprising based on mainstream ideologies of the relationship among sex, gender, and sexuality (as reported, for example, in Gaudio 1994: 48), and it is supported, to some extent, by an analysis of A&P survey participant's ratings of masculinity and femininity[10]. Juan Carlos' comment in (15) lead to the discussion in (16).

10 Due to space limitations, I am not able to provide that analysis in this chapter.

(16) "[It] might mean that you're not straight"

Alan:	And it's not just the- the so-called gay lisp. I- I think it's, uh, the way that, uh, your inflection changes.
Juan Carlos:	Uh-huh.
Alan:	Like, you s-, uh, sound particularly – I mean, the word is "gay" – happy.
Juan Carlos:	Yeah.
Alan:	Uh, like, I- I guess the idea of masculinity is, uh, a certain stoicism or something.
Juan Carlos:	Yeah.
Alan:	And, um, if you sound too open and friendly, uh, then that, uh, might mean that you're not straight.
James:	Yeah. Or make people assume that you're not straight.
Juan Carlos:	Yeah.
James:	Um-
Alan:	If you're, uh- I- I guess that goes with the idea of, uh, traits that aren't, uh, masculine. Traits that are more feminine, you know, being, uh, engaging or empathetic to other people and trying to, uh, be open to them.

Alan discussed a complex relationship among language, gender, and cultural expectations. This relationship becomes even more complicated when considered in conjunction with sounding southern, as seen in a comment from James after he heard a clip that he thought sounded very masculine. In (17), the "perfect example of a country queen" to whom James referred is David.

(17) "'That's a perfect example of a [...] country queen'"

James:	[...] if there's a- a- a level of masculinity about a southern voice. And there's something about, like, the southern accent that, um, it amplifies masculinity or it amplifies femininity, because going to back to the one that I was like, "That's a perfect example of a country- country queen". If he was probably from any other region, it wouldn't have sounded as feminine. And there's something about the southern accent that in- in my mind, especially with all of these, um, the more southern they were, the more they seemed to come across- 'cause this was another very masculine- [...]

Because of this apparent "amplification" of masculinity and femininity, therefore, listener judgments of sexual orientation for southern-sounding voices may rely more heavily on cultural assumptions regarding the relationship between language and gender in addition to assumptions regarding the relationship between language and sexuality.

5.2 Rurality

One discussion that arose in each of the focus groups concerned what it means to be a "redneck". For most participants, the term "redneck" implies a rural upbringing, overt displays of masculinity and bravado, and ignorance. After listening to Andrew, Scott gave the description in (18).

(18) "Beer-drinking good ol' boy"

Scott: I put, uh, southerner, farmer, redneck, and beer-drinking good ol' boy. Has a beard, s- scraggly beard, ball cap, and wears plaid.

But focus group participants were often undecided regarding whether to describe David as a redneck. This indecision led one focus group to start teasing apart what it means to be a rural southerner and whether the rural/urban distinction is sufficient. In the conversation reproduced in (19), Juan, Alan, and James worked through this question in order to come up with a better assessment of David, because they did not necessarily want to label him a redneck as they did Andrew.

(19) "He sounded to me very provincial"[11]

Juan: I mean, there's- I mean, there's also, like, small-town rural, you know, and that brings a whole lot of- with it, too, especially linguistically, I mean, in the way that people carry themselves, and-

Alan: There's a lot between southerner and redneck.

Juan: Yeah.

Alan: I- I mean, a "redneck" has a connotation of somebody who-

James: Of ignorance.

11 Because one of the towns mentioned in (18) is very small, I omitted town names from this example to reduce the risk that any of the focus group participants could be identified. According to the U.S. Census Bureau, Town A has a population of approximately 5,500 with a population density of approximately 1,200 per square mile. Town B has a population of approximately 18,000 with a population density of approximately 2,000 per square mile. Town C has a population of 500 with a population density of approximately 150 per square mile.

Alan: You know, uh, is aggressively ignorant.
Juan: Yeah. And proudly so.
Alan: Yeah. As opposed to just “provincial”.
James: “Provincial” sounds <unintelligible>. He- he- he sounded to me very provincial. I mean, I could totally- I could totally see him running the florist- the florist shop in [Town A].
Juan: That’s what I was about to say. He sounds like he’s from [Town A].
???: <unintelligible>
James: He completely sounds like somebody from [Town A].
Juan: [*overlapping James*] And there’s a big difference if you think about [Town B], which is, you know, provincial enough, and [Town A], and then where I’m from, [Town C]. I mean, there’s a whole- there’s a ch- there’s a change. Even those of us who come from there who are queer, you know, and the people who live there never really even identified, but everybody knew. I mean, ‘cause [Town A] probably had the florist, and everybody knew and loved, whereas in, you know, [Town C], you didn’t do that. You went to the trailer park and you hid. And so it changed the way their whole perceptions and the way they presented themselves were.

According to the focus group participants, Andrew sounds like a redneck, which – according to these discussions – is incompatible with sounding gay. But David was judged differently. It is compatible for him to sound both southern and gay, because his speech patterns did not evoke an image of a redneck. David was instead heard to be provincial, which does not share the same connotations of ignorance and overt masculinity.

5.3 Class

The provincial/redneck distinction is also connected to ideologies of social class. Consider the two judgments of Andrew and David in (20) and (21), respectively. Scott assumed Andrew to be a member of the working class.

(20) “He sounds [...] poor”

Scott: He sounds, uh, poor. He just reminds me of the family- the type of, uh, of income level that I would see a lot of when I would go visit my little brother in the Big Brother/Big Sisters program. And, uh, he talked much the same way that those people do.

Conversely, when talking about David, Juan on multiple occasions used the word "blue blood".

(21) "He has that inflection in his voice"

Juan: I was thinking more of a- that generation, like, right above me, uh, the guys who don't quite come out but maybe should, maybe shouldn't. Definitely a class thing going on there, you know, a little blue blood. Uh, 'cause he talks like the straight guys that you kind of are always confused about that are in their fifties- you know, mid- to late-fifties, maybe sixties. He has that inflection in his voice.

I argue that listeners are basing their judgments of sexual orientation in part on preconceived notions of mainstream metropolitan gay male communities of practice (discussed in Leap 1996: 19–20) in which gay men are seen as having plenty of disposable income, which they spend on the latest designer fashions, expensive cars and homes, gourmet food, top-shelf liquor, and world travel. Gay men in the rural U.S. South – whether redneck rural or provincial – are expected then to hold similar aspirations. If a speaker like Andrew speaks in a way that indexes his working-class, southern identity, he is considered not to share these aspirations and, therefore, must not be gay.

6 Conclusion

Throughout the interviews and focus group sessions, participants did make reference to specific linguistic features, including the features that have been the focus of the sounding gay literature, such as lengthened [s] and dynamic pitch (c.f. Gaudio 1994; Levon 2006). But as I have shown in this chapter, judgments of sexual orientation are complicated by ideologies of sexuality, gender, class, and rurality – ideologies, which are also linked together in complex ways. It is this interplay of class, gender, sexuality, and rurality, which may lead at times to the incorrect assumption of the complete incompatibility of sounding gay and sounding southern. Male speakers who sound both southern and masculine cannot also sound gay, because ideologies of masculinity in the South rely on images of hypermasculine, rural, working-class men, which conflict with local attitudes toward and beliefs about homosexuality. Male speakers who sound southern and either feminine or effeminate, however, can also be heard to sound gay, because listeners do not associate them with the rural, working class South or label them "rednecks". Instead, they are labeled middle-class "provincial," i.e., associated with a small town in the South, or they are labeled "blue blood," i.e., associated with the upper class in a prominent southern city. Attitudes toward and perceptions of southern gay male speakers in the U.S. South, therefore, rely on associations between rural, "provincial," and urban, on one hand, and working-class, middle class, and "blue blood," on the other.

Lisa Jeon and Patricia Cukor-Avila

Chapter 6
Urbanicity and language variation and change: Mapping dialect perceptions in and of Seoul

1 Introduction[1]

On January 21, 1991, the president of South Korea created the Seoul-based National Institute of the Korean Language (NIKL) to act as a regulatory body to promote and maintain a standard variety of Korean based on the speech of "well-cultivated" middle class Koreans from Seoul (NLRI 1992). Since its creation, NIKL, with support from the South Korean government and the National Language Research Institute (NLRI), has built a successful campaign to reinforce the notion that there is only one correct and prestigious dialect of Korean – the dialect spoken in Seoul and the surrounding metropolitan area within the Gyeonggi province (see Figure 6.1). This is not surprising considering that Seoul is the largest urban area in South Korea that more than half of the country's 50 million inhabitants call home. However, dialect variation exists in Korea despite the language standardization efforts of the Korean government that disseminates the conceptual myth of "one country, one language" central to a "Korean identity".

First, there is variation in the number of vowel phonemes that corresponds to geographic region and speaker age (King 2006). The phonemic contrast between long and short vowels in the standardized speech of North and South Korea is disappearing in the language of younger speakers from Seoul (Nakamura et al. 1991), and this distinction has disappeared altogether in some dialects like Hamgyeong in North Korea and North Jeolla in South Korea. Second,

1 This research was supported in part by funding from the Greater Dallas Korean American Chamber of Commerce. We would also like to thank the following people, without whom this study could not have been possible: Dr. Dennis R. Preston (Oklahoma State University), Dr. Chetan Tiwari and Patricia C. Rector (University of North Texas), Dr. Kigap Yi and Dr. Edmundo Luna (Mokpo National University), Dr. Young-lim Ko and Dr. Young-taek Yoon (Jeju National University), Dr. Bae-kyun Yoo, Dr. Kwang Ho Baek, and Hyunsook Park (Baekseok University), Dr. Sang-yong Ohm (Woosong University), and Brenda Paik Sunoo, Yong-su Ahn, and Shine Kang.

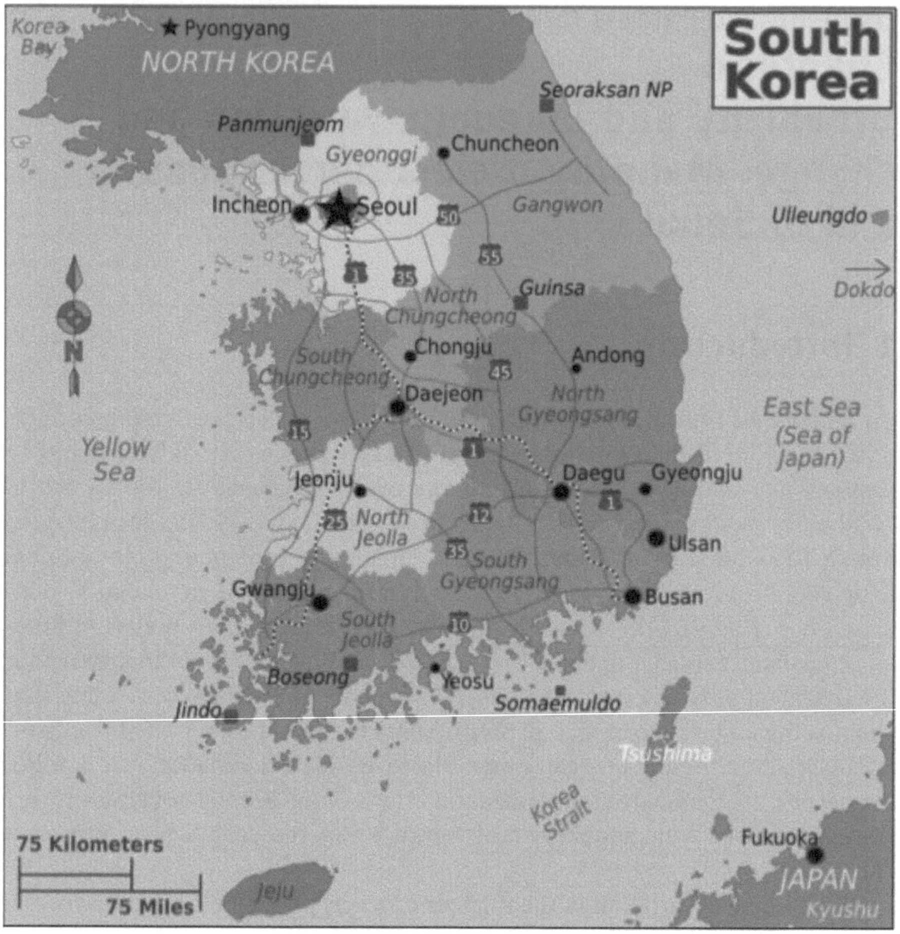

Fig. 6.1: South Korean provinces and major cities (Source: Fitzgerald 2009).

variation in pitch accent occurs in several regions, like Geyongsang, Hamgyeong, South Jeolla, and some areas of Gangwon, but not in Seoul or in the surrounding Gyeonggi area (Hayata 1976; W.-J. Kim 1983; Lee 1983). Finally, there is considerable lexical and grammatical variation, especially in provinces in the southern dialect regions like Gyeongsang, Jeolla (S. Yi 1998; K. Yi 1998), and the dialect spoken on Jeju Island. In fact, the Jeju dialect is often referred to as the most divergent dialect in Korea, and many people from the mainland claim that it is unintelligible (Jeon 2011).

Social and cultural stereotypes associated with provincial regions are also prevalent in Korean society (Ask a Korean 2013). For example, the people from

Fig. 6.2: Korean dialect divisions (Source: Korean Wiki Project 2009).

Gyeonggi, the province that surrounds Seoul, are typically described as being cultured. Residents of Chungcheong, the region just south of Gyeonggi, are considered to be mild-mannered, manifesting true *yangban* 'scholarly' virtues. The people of Gyeongsang are viewed as loud and crude, those in Gangwon are commonly thought to be poor and stolid, and the women of Jeju Island are famous for being strong-minded and independent.

Previous dialect studies have also mapped dialect variation in Korea. Figure 6.2 is a map of five dialect areas on the peninsula first identified by Ogura (1940) and substantiated by more recent work (P. Kim 1988; S. Yi 1998; King 2006). Interestingly enough, the dialect areas roughly approximate the present-day province boundaries[2]: (1) Central dialects (Seoul and Gyeonggi province, Yeongseo region west of the Taebaek Mountains neighboring Gangwon province in North Korea, and Chungcheong province); (2) Yeongdong dialect (Yeongdong region east of the Taebaek Mountains neighboring Gangwon province in North Korea); (3) Southwest dialect (Jeolla province); (4) Southeast dialect (Gyeongsang province); and (5) Jeju dialect (Jeju province).

In this paper, we investigate the differences between traditional and perceptual dialect boundaries in South Korea, stratifying the data by perceptions of standard and non-standard Korean. More specifically, we focus on dialect perceptions in and of Seoul, South Korea's most densely populated and urbanized city, and compare them to perceptions of other urban and rural areas of the peninsula to gain a better understanding of the relationship between urbanicity and perceptions of linguistic variation. Finally, we also examine the data stratified by respondents' identification as urban, suburban, and rural to determine whether urbanicity has also become an important catalyst for changing the way people perceive dialect areas within the country.

2 Perceptual Dialectology in South Korea

There has been relatively little perceptual dialectology research conducted in Korea, with early studies using only short language attitude surveys (Yim 1993; Sanada and Yim 1993). The most extensive study to date that incorporates the methods of perceptual dialectology and language attitude research is Long and Yim (2002). Using the draw-a-map task (see Chapter 2, this volume, and section 3), Long and Yim examined the dialect perceptions, including perceptions of the most pleasant and most standard speech, of 471 Korean university students in Seoul. They report that the overwhelming majority of respondents perceived dialect areas to fall within province boundaries (2002: 251), and almost two-thirds of the students also indicated a separate dialect area around Seoul. A few others also singled out Busan as having a separate dialect; however, no other major cities were identified or labeled by respondents. Long and Yim note that the salience of Seoul as a distinct speech region is remarkable considering the "metropolitan area" status accorded to the five other major cities in

2 In this study, we romanize the names of places in South Korea following the romanization given to that place by the South Korean government. Therefore, all names of places are romanized according to the Revised Romanization of Korean, the official Korean language romanization used by the South Korean government.

South Korea (Busan, Daegu, Incheon, Gwangji, and Daejeon) (2002: 251). Results from the "most pleasant/most standard" task reflect the success of the NIKL and NLRI campaigns as nearly all respondents in the study perceived that standard Korean was spoken in the province of Gyeonggi, the area surrounding Seoul.

The present study is an extension of Long and Yim (2002), but differs in its method of investigation in the following key ways. First, this study expands the respondent pool with data from 436 South Koreans (aged 18–84) from a variety of social and economic backgrounds living in all six provinces of South Korea. Second, it incorporates Geographic Information System (GIS) technology to digitize and aggregate hand-drawn maps from the respondents (see Montgomery and Cramer, this volume). During the aggregation process, dialect categories for subjects' perceptions were identified and mapped and the most salient perceptual categories were examined to get a better understanding of the dialect image and ideology associated with Seoul as well as with other urban and rural areas within the country. Finally, demographic information from the respondents has been correlated with their perceptual data to investigate how dialect perceptions are stratified by age, sex, and identification as rural, suburban, or urban. By expanding on the methods used in Long and Yim (2002) in the ways outlined above, this study provides detailed quantitative and qualitative analyses of perceptions of linguistic variation in South Korea.

3 Methods

3.1 Survey instrument

This study follows the primary approach for perceptual dialectology research outlined in Preston (1999b: xxxiv): the draw-a-map task. Respondents were randomly approached and asked to draw lines on minimally-detailed maps of Korea indicating places where they believed people speak differently[3]. We then asked respondents to provide names or labels for the areas they indicated. In

3 We used two different map types with varied geographic information in this study to test for the effect that geospatial reference information provided on survey instruments might have on respondents' answers. With the exception of Lamelli et al. (2008) and Jeon (2011), map type differences have not been quantitatively analyzed in perceptual dialectology research. In this paper, we have grouped the data from both map types because our initial findings are that map type did not significantly affect the type or amount of information these respondents put on the maps. We are currently investigating this question in more detail.

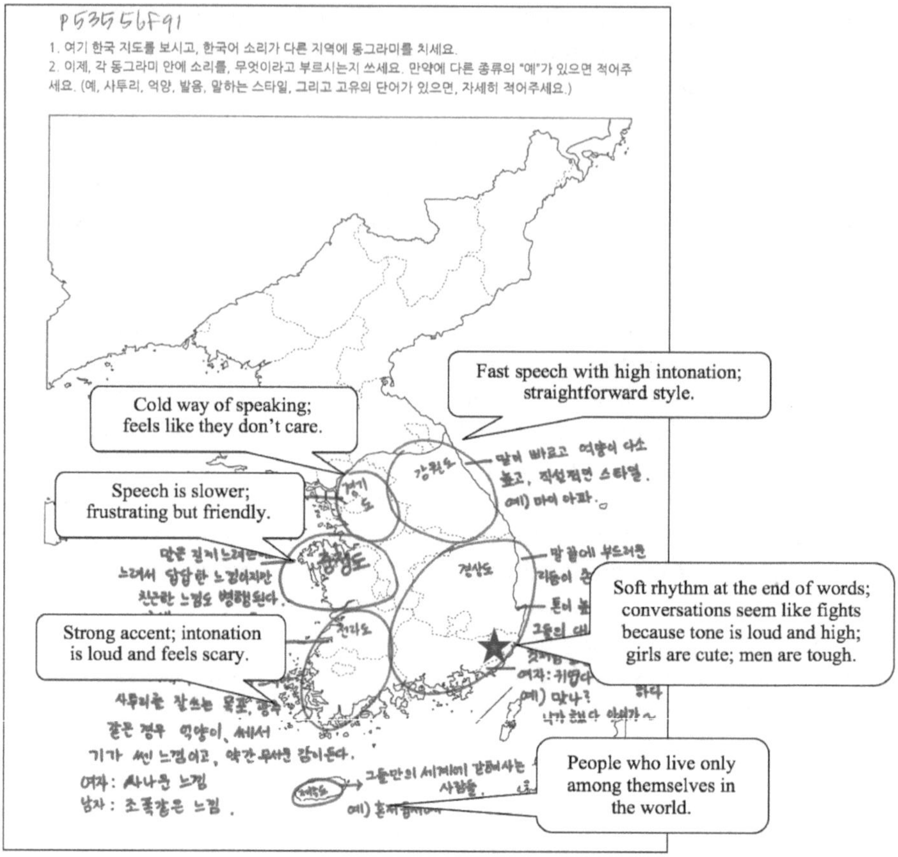

Fig. 6.3: Hand-drawn map from a female born in 1991 in Busan.

addition, we collected qualitative data from open-ended conversational interviews with respondents about the information they provided on the maps and any other comments they had about linguistic variation in South Korea.

Respondents also provided answers to nine demographic questions listed on the back of the map including year of birth, sex, ethnicity, educational background, ability to speak a language other than their first, time lived in Korea, place lived in the longest, self-identification as urban, rural, or suburban, and self-identification with a place. Identification with a particular province or city is an important cultural concept for South Koreans often discussed in terms of designating oneself as being an "X person" (e.g., 서울 사람 *Seoul saram* 'Seoul person'). This demographic feature had an impact on our results, as will be further discussed in section 4.2.8.

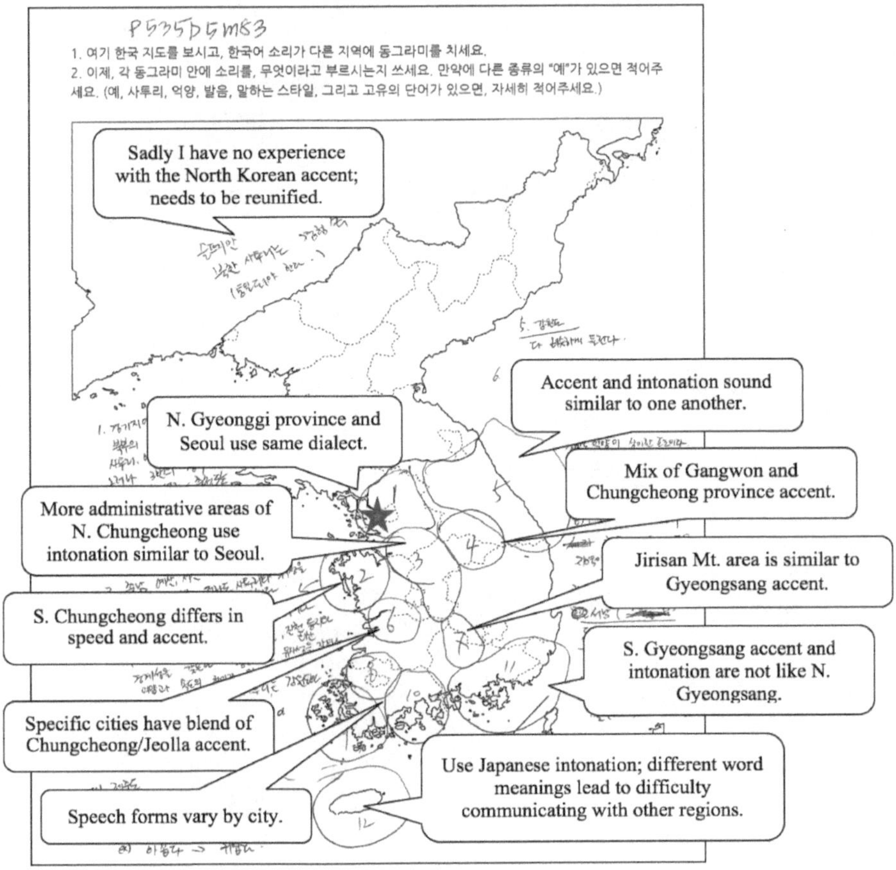

Fig. 6.4: Example hand-drawn map collected from a male born in 1983 in Seoul.

The hand-drawn maps in Figure 6.3 and Figure 6.4 illustrate the kinds of varied responses we collected. (The blue star on each map indicates the place where the map was collected). Some respondents drew perceived dialect areas that roughly approximated the South Korean province boundaries on their maps, such as the map in Figure 6.3. Others, as in Figure 6.4, wrote very detailed information about linguistic variation throughout the peninsula and indicated several distinct perceived dialect areas that included cities (both large and small), province regions, and geographic landmarks.

3.2 Data collection

Our sample includes native Koreans of various ages and backgrounds from major urban areas and their surrounding rural communities. As shown in Fig-

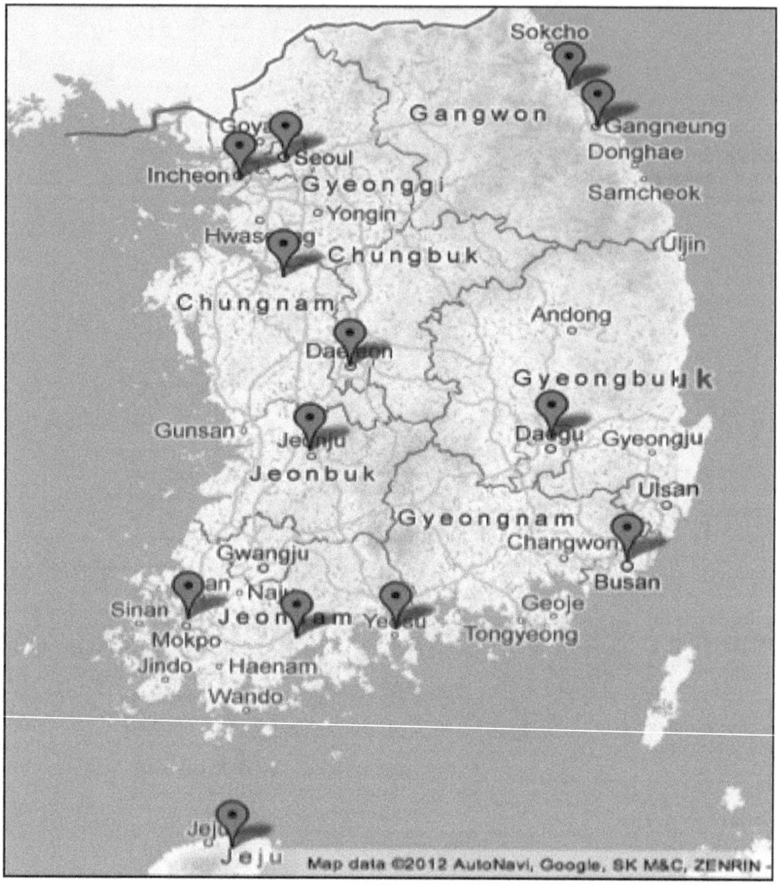

Fig. 6.5: Data collection sites (Source: Google.com).

ure 6.5, we surveyed people from 13 cities (Seoul, Cheonan, Daejeon, Jeonju, Mokpo, Yeosu, Boseong, Busan, Daegu, Yangyang, and Jeju City) located in all six provinces (Gyeonggi, Chungcheong, Jeolla, Gyeongsang, Gangwon, and Jeju). Fieldwork sites were diverse and included university campuses, bars and restaurants, coffee shops, malls, museums, hotels, taxis, buses and bus terminals, trains and train stations, airplanes and airports, and the 2012 World Expo in Yeosu. We also collected maps from respondents we randomly approached on the street.

During three weeks of fieldwork we collected a total of 488 maps; however, 52 maps were discarded from the final analysis because they either had ambiguous information or were left blank. This left a total of 436 maps from 197 male (45 %) and 239 female (55 %) respondents. As Table 6.1 and Table 6.2

Tab. 6.1: Total respondents by year of birth adapted from Jeon & Cukor-Avila (2015).

Year of Birth	# of Respondents	% of Total
1930–1962	47	10.8 %
1963–1982	92	21.1 %
1983–1994	186	68.1 %
Total	436	

Tab. 6.2: Total respondents by education level adapted from Jeon & Cukor-Avila (2015).

Education Level	# of Respondents	% of Total
Some high school	15	3.4 %
Completed high school	35	8.0 %
Some college	246	56.4 %
Bachelor's degree	96	22.0 %
Graduate degree	43	9.9 %
Unknown	1	0.2 %
Total	436	

show, the majority of respondents (68.1 %) were between 18–28 years old and were well educated (78.3 % reported some college or higher)[4].

In addition, most respondents (83.7 %) were born in Korea and had lived there all their lives, self-reported as urban (63.3 %)[5], and reported that they could speak a second language (86.2 %).

3.3 Data analysis

The nature of the draw-a-map task presents challenges in uncovering relationships and trends from the hand-drawn map data because respondents are able to

4 This is representative of the Korean population. According to annual reports by the Organization for Economic Cooperation and Development (OECD) since 2005, South Korean people are the most likely among industrialized countries to be literate (97.9 %) and have university degrees.

5 This is also representative of the Korean population. According to the World Bank, most of the population of South Korea has been concentrated in urban areas since the country underwent industrialization in the 1960s. In 2002, over 77 % of the population was concentrated in major urban areas, and the country continues to see a growing number of people migrating from rural to urban areas.

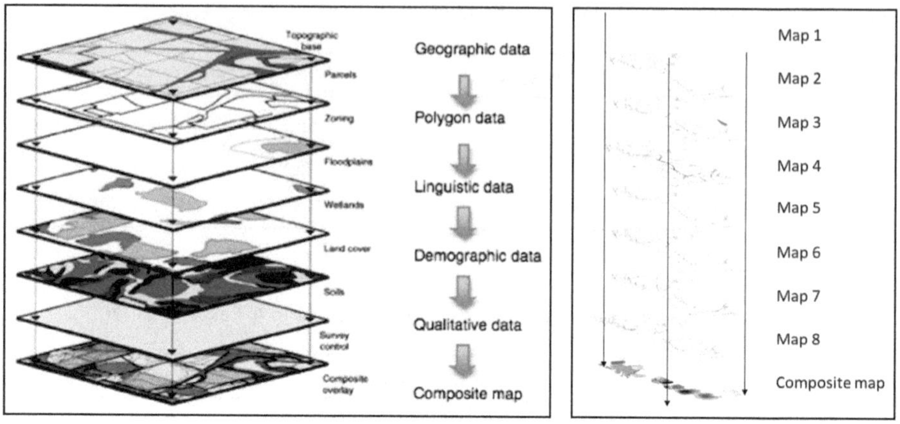

Fig. 6.6: Multi-layering of data types in a GIS (adapted from Montgomery 2011b and Jeon & Cukor-Avila 2015).

draw lines (representing perceived dialect areas) and make annotations as they see fit (cf. Evans 2011). There is also considerable variation in the size and number of the areas that they draw. While this free choice allows for respondents to provide a wealth of information about their dialect perceptions, it does not easily lend itself to data aggregation (cf. Montgomery and Stoeckle 2013). For example, the hand-drawn maps collected in this study yielded the following "layers" or categories of data: (1) geographical data that included the extent, placement, and names of dialect areas; (2) perceptual data that consisted of qualitative comments; and (3) linguistic data that consisted of lexical, grammatical, and phonological features that respondents associated with different regions of Korea.

An ideal solution for digitizing and analyzing multi-layered data from draw-a-map tasks is the use of a GIS, as discussed further in Chapter 2. The primary function of GIS for perceptual dialectology is "geo-referencing" or linking different layers of data to the earth's surface, allowing for the aggregation of both "semantic and geometrical information" (Gomarasca, 2009: 481). Using a GIS, perceptual data can be aggregated, analyzed, and displayed according to the different layers of data collected from respondent maps, as is illustrated in Figure 6.6. Following this approach, we used two open-source programs that interface with GIS (QGIS and PostGIS) to stratify and analyze the data by perceptual categories and demographic factors. We then used ArcGIS 10.0 software to visually represent the hand-drawn map data and to identify trends and relationships among the data.

4 Results

4.1 ArcGIS analysis

Figure 6.7 compares a composite map generated using ArcGIS (on the left) with a map of traditional Korean dialect boundaries (on the right) of all the draw-a-map task responses. The color intensity on the ArcGIS map indicates the level of agreement about perceived dialect areas. The degree of overlap, shown in the map's legend, is calculated as the percent of respondents who identified an area out of all 436 respondents. Thus, the darkest areas on the map indicate where the most overlap of perceived dialect areas occurs (60.2–75 % of people surveyed), while the lighter areas indicate where the least overlap occurs (less than 15 % of respondents). The composite map confirms that South Koreans perceive language variation despite the conceptual myth of one "standard" dialect spoken throughout the peninsula.

In addition, the analysis suggests that the dialect regions perceived by our respondents are not contained within province boundaries, as previously suggested by Long and Yim (2002). A notable difference between the two maps in Figure 6.7 is that the Central dialects spoken in Seoul/Gyeonggi and Chungcheong, which are distinct on the traditional map, are merged into one large

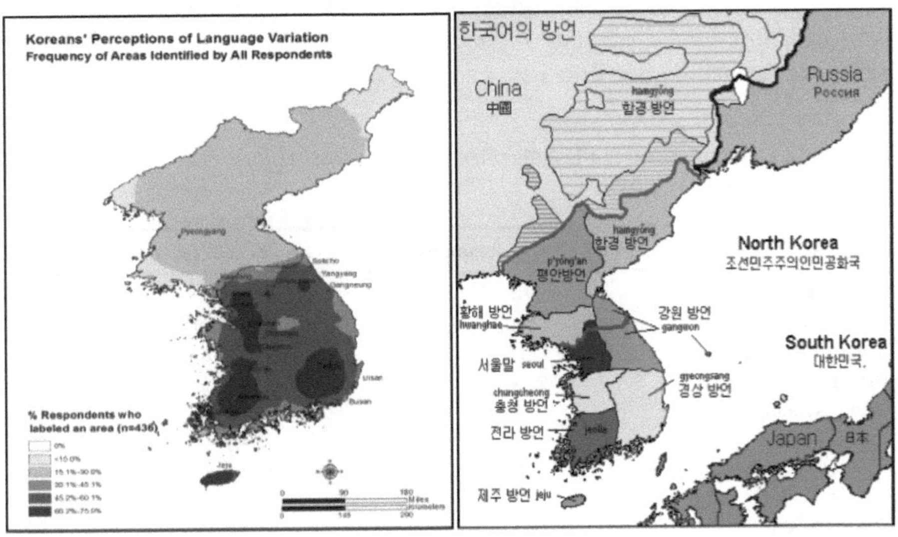

Fig. 6.7: Composite map showing the most salient perceived dialect areas for all respondents (left; adapted from Jeon & Cukor-Avila 2015); map of traditional dialect boundaries in South Korea (right; Korea Wiki Project 2009).

dialect region on the perceptual map. The perceptual map also shows that respondents perceive dialect leveling across North Korea, whereas the traditional map outlines separate dialect areas in the Northwest and Northeast (Pyeongyang and Hamgyeong) and in the Central region (Hwanghae). And not surprisingly, our respondents also perceived a distinct dialect boundary at the 38th parallel north, a line of latitude used as the pre-Korean War division line and current Military Demarcation Line separating North and South Korea. This boundary is not represented on the map of traditional dialect divisions.

4.2 Content analysis

We performed a content analysis using the "keywords" technique outlined in Garrett, Williams and Evans (2005) and Evans (2011) to analyze the comments and qualitative labels that respondents provided on the maps. During this process, we combined words and phrases in the same semantic field to find emerging themes. For example, we combined "correct" and "official language" to form one category called *Standard*. The keyword analysis resulted in eight categories. We have omitted three of the categories, *Lexical Items*, *Sentence Final Endings*, and *Manner/Personality*, from the present analysis because they were identified in every region and thus were not as relevant perceptually. The remaining five categories are listed in Table 6.3 by order of frequency from most to least identified: (1) *Standardness*, which we further divided into *Non-Standard* and *Standard*; (2) *Strong Intonation/Tone*; (3) *Strong Accent*; (4) *Speed*, which we further divided into *Slow* and *Fast*; and (5) *Gender Association*, which

Tab. 6.3: Five perceptual categories identified by respondents at least once per map adapted from Jeon & Cukor-Avila (2015)

Category	# of Respondents	% of Total (n = 436)
Standardness	420	96.3 %
Standard	(202)	(46.3 %)
Non-Standard	(218)	(50.0 %)
Strong Intonation/Tone	272	62.4 %
Strong Accent	171	39.2 %
Speed	261	59.9 %
Slow	(183)	(42.0 %)
Fast	(78)	(17.9 %)
Gender Association	144	33.0 %
Aegyo	(75)	(17.2 %)
Feminine	(38)	(8.7 %)
Masculine	(31)	(7.1 %)

we further divided into *Aegyo*[6], *Feminine*, and *Masculine*. Using ArcGIS 10.0, we created composite maps for each of the categories listed in Table 6.3[7]. In the following sections, we focus on the analysis and discussion of the most salient feature, *Standardness*, both *Standard* and *Non-Standard*, and how this perception is stratified by the demographic feature of urbanicity.

4.2.1 Perceptions of *Standard* vs. *Non-Standard*

An overwhelming 96.3 % of respondents labeled at least one area that was categorized as either *Standard* or *Non-Standard*. Keywords for *Standard* included words and phrases such as *standard language, speech of popular media, mainstream, cultured, educated, academic, professional, official,* and *correct*. Keywords for *Non-Standard* included words and phrases such as *different from standard, substandard, illiterate, uncultured, uneducated,* and *incorrect*. Many respondents also wrote examples of non-standard forms or indicated areas on their maps that were different from the standard; these were also coded as *Non-Standard*.

A comparison of the composite maps for *Standard* and *Non-Standard* in Figure 6.8 reveals that (1) most respondents perceive the speech region labeled *Standard* to be located primarily in Seoul and the surrounding Gyeonggi province; (2) a smaller number of respondents also perceive their home region and North Korea as *Standard*[8]; (3) the speech region labeled *Non-Standard* is heavily associated with the Southeast dialect region, the darkest area on the right map, as well as with the dialects spoken in Gangwon, Jeolla, Jeju and North Korea[9];

---

6 *Aegyo* is a behavior and speech style typically used by younger females in Korea to appear cute, innocent, and child-like, sometimes for a manipulative purpose (Moon 2010). It is characterized by a higher pitched voice and utterances with a melodic sweep over the vocal folds. Often, child-like dress, mannerisms, and facial expressions also characterize *aegyo* speakers.

7 A discussion of each of the resulting maps is beyond the focus of this paper. See Jeon and Cukor-Avila (2015) for this analysis and discussion.

8 Preston (1996) suggests that non-linguists view dialects as sub-varieties of a super variety. Similar results in Long and Yim (2002) indicate that perhaps Koreans view the linguistic varieties spoken in their local area or in North Korea in a parallel sense to the standard variety – different but equal.

9 Another category related to *Non-Standard* that we have not included in the present analysis is *Unintelligibility*. Keywords for this label included *unintelligible* and words and phrases such as '*mwo?*' ('what?'), *foreign, can't/difficult to understand, doesn't sound like Korean, words you don't know or will hear for the first time,* and *like a different language/country*. The composite map of areas most frequently identified as *Unintelligible* suggests that the Korean spoken in Jeju province is very hard to understand, to the extent that many consider it to be another language. This is in direct contrast to the capital city of Seoul and the surrounding Gyeonggi province, which none of the 436 respondents perceived as *Unintelligible*.

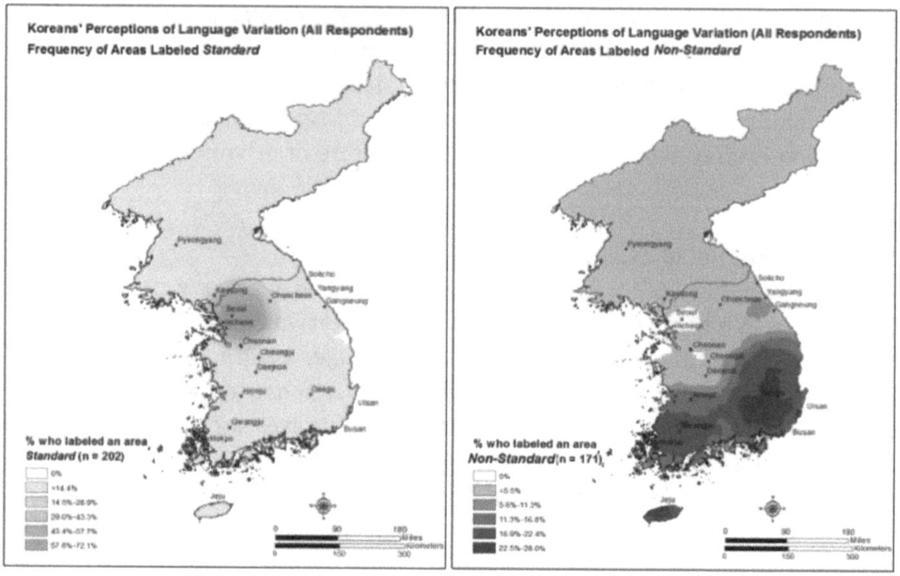

Fig. 6.8: Composite maps of *Standard* (left) vs. *Non-Standard* (right) adapted from Jeon & Cukor-Avila (2015).

and (4) *Non-Standard* is not associated with Seoul and parts of the greater Gyeonggi province areas, the white areas on the map.

4.2.2 Perceptions of *Standardness* stratified by *Urbanicity*

To analyze whether dialect perceptions would differ if urbanicity were taken into account, we also stratified the data by respondents' self-identification as urban, suburban, or rural. Figure 6.9 shows composite maps indicating the frequency of areas identified by urban (left), rural (middle), and suburban (right) respondents who labeled an area as *Standard*. As Figure 6.9 illustrates, there is a general consensus that the speech region labeled *Standard* is located in Seoul and the surrounding Gyeonggi province, indicated by the dark areas on the map. This suggests that the majority of respondents, regardless of urbanicity, buy into the notion of "one country, one language". However, 17.2 % of urban respondents and 16.7 % of rural respondents identified several areas other than Seoul and the Gyeonggi province as *Standard*, which also suggests that some perceive *Standard* to be located in other areas, including in their home regions.

Stratifying the data by urbanicity for perceived dialect areas labeled as *Non-Standard* yielded opposite results to the findings for perceived dialect areas labeled as *Standard*, with no respondents identifying Seoul or Gyeonggi

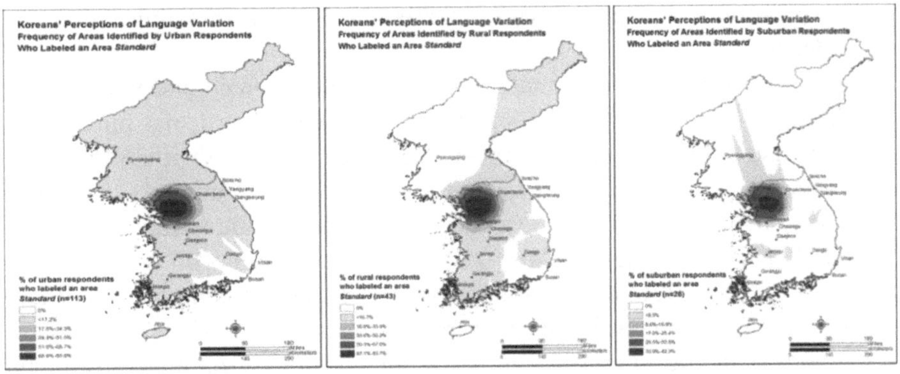

Fig. 6.9: Composite maps of *Standard* stratified by respondents who identified as *Urban* (top left), *Rural* (top right), and *Suburban* (bottom) adapted from Jeon & Cukor-Avila (2015).

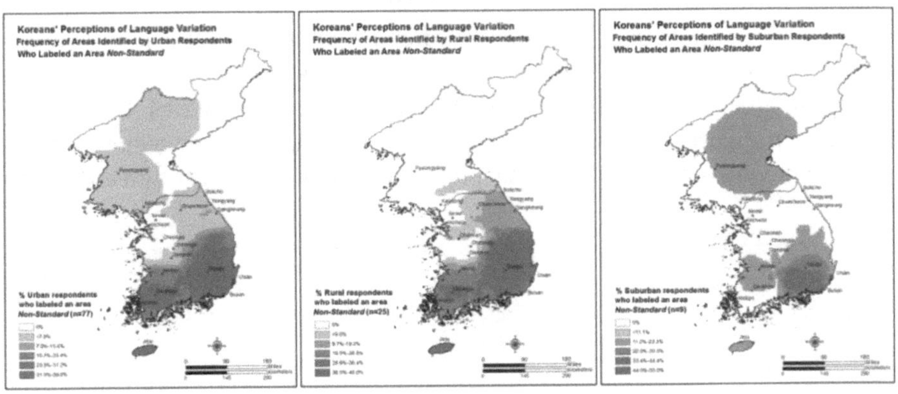

Fig. 6.10: Composite maps of *Non-Standard* stratified by respondents who identified as *Urban* (left), *Rural* (middle), and *Suburban* (right).

province as *Non-Standard.* Figure 6.10 shows composite maps indicating the frequency of areas identified by urban (left), rural (middle), and suburban (left) respondents who labeled an area *Non-Standard.* These maps suggest, however, that perceptions of the location of *Non-Standard* do vary somewhat by the urbanicity of the respondent. For instance, urban respondents were the most prescriptive about identifying *Non-Standard* speech in all areas outside of Seoul and Gyeonggi province, especially in the Southern dialect regions. Rural respondents had similar perceptions, though they omitted North Korea entirely from their perceptual dialect maps. Finally, suburban respondents identified fewer areas as *Non-Standard* overall and highlighted the Gyeongsang region as its locus.

4.2.3 Perceptions of Standardness stratified by Self-Identification with a Place

We also stratified the data to analyze whether dialect perceptions would differ according to whether respondents self-identified with a particular province or city, which is a notable cultural concept for South Koreans. Figure 6.11, Figure 6.12, and Figure 6.13 display the results of that analysis. Figure 6.11 shows two composite maps indicating the frequency of areas identified and labeled as Standard (left) and Non-Standard (right) by respondents who self-identified as Gyeonggi. As these maps suggest, there is a general consensus by these respondents that Standard is located in the central region of the peninsula – concentrated in their home region of Seoul and the surrounding Gyeonggi province. By contrast, they identified Non-Standard to be located everywhere except in Seoul and the Gyeonggi province and in North Korea. For these respondents, Non-Standard Korean is found throughout the southern regions of the peninsula and particularly in Jeju Island.

Stratifying the data by self-identification as *Jeju* and *Jeolla* (the two southernmost dialect areas that were often reported to be highly stigmatized by respondents) for perceived dialect areas labeled as *Standard* yielded the composite maps in Figure 6.12. These maps suggest that although the majority of respondents who self-identified with places outside of *Gyeonggi* agree that the

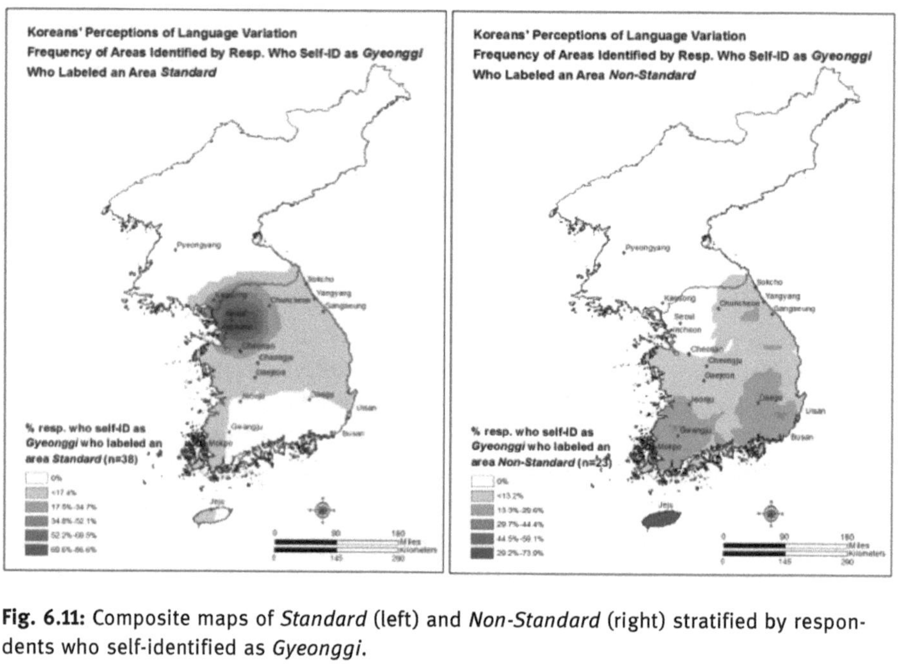

Fig. 6.11: Composite maps of *Standard* (left) and *Non-Standard* (right) stratified by respondents who self-identified as *Gyeonggi*.

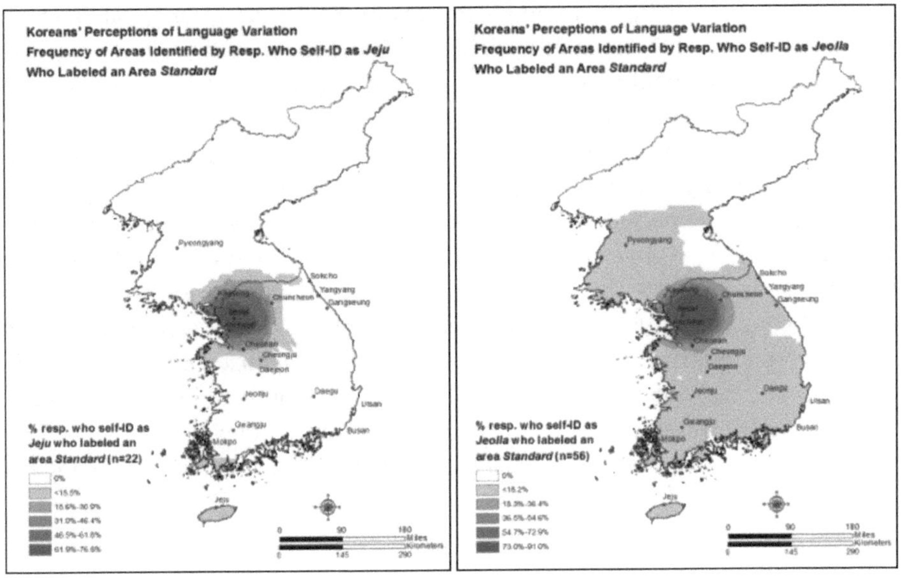

Fig. 6.12: Composite maps of *Standard* stratified by respondents who self-identified as *Jeju* (left) and *Jeolla* (right).

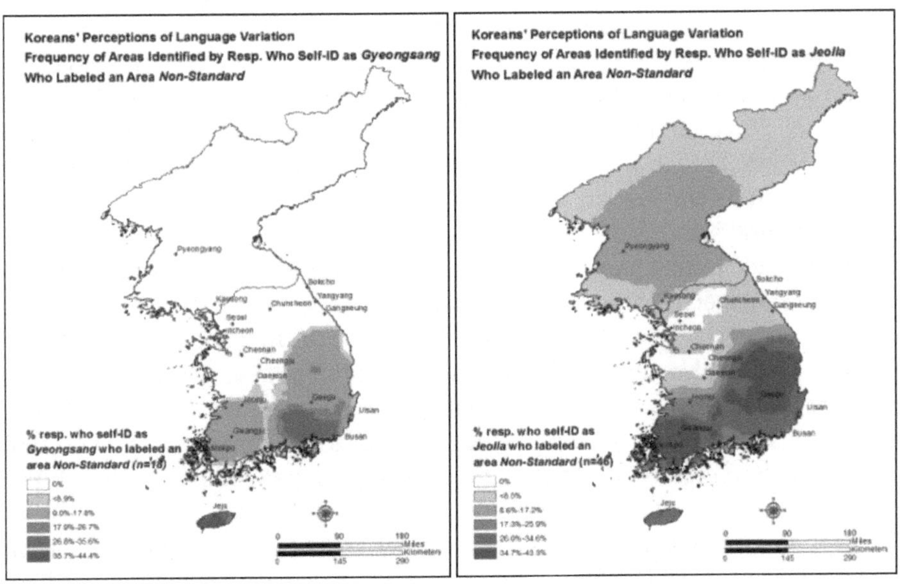

Fig. 6.13: Composite maps of *Non-Standard* stratified by respondents who self-identified as *Gyeongsang* (left) and *Jeolla* (right).

speech region labeled as *Standard* is located in the Seoul/Gyeonggi area, some respondents perceive the speech of the region they identify with as well as other areas on the peninsula (as in the case of respondents who self-identified as *Jeolla*) to be *Standard* as well.

Figure 6.13 shows a comparison of perceptions of where *Non-Standard* Korean is spoken by respondents who self-identified as *Gyeongsang* (left) and *Jeolla* (right). As expected, none of these respondents labeled Seoul or Gyeonggi province as *Non-Standard.* The respondents who self-identified as *Jeolla* were the most prescriptive – they perceived *Non-Standard* speech to be everywhere except Seoul and Gyeonggi province. Respondents who self-identified as *Gyeongsang* identified fewer areas overall as *Non-Standard,* and like the *Jeolla* respondents, they perceived the Southern dialect regions to be the most non-standard areas. Interestingly, neither group included their home region in a standard area, and both perceived Jeju Island as the locus of non-standard Korean.

5 Discussion

Results of this study suggest that Koreans do not perceive the peninsula as a homogenous speech community, despite the widespread notion that there is a national standard modeled after the Korean spoken in Seoul and throughout the Gyeonggi province. We suggest that the merged Central dialect region perceived by our respondents may simply be a reflection of the linguistic consequence of population movement from Seoul to more rural areas, incentivized by the South Korean government. These policies, implemented in the early 1970s, provide economic incentives for people to move out of Seoul and into the more rural areas, such as in Chungcheong province, which is only a one-hour train ride from the capital. The dialect differences between Seoul/Gyeonggi and Chungcheong may continue to diminish as more and more people leave Seoul to settle in this outlying region (cf. Kerswill 2001, 2003; Milroy 2002).

Our results also suggest that the North-South political border plays a significant role in the perception of a homogeneous North Korean dialect distinct from the dialects perceived in South Korea. The lack of familiarity and personal contact between residents of the two Koreas following their division after WWII may also contribute to this perception.

Not surprisingly, the more salient dialect areas for our respondents seem to be concentrated in major urban areas that are densely populated and are connected by high-speed transportation. In contrast, less salient dialect areas cluster in regions that are not urban centers and are not easily accessed by the

Korean rail system, e.g., Gangwon, the coastal areas, and the interior regions of the peninsula[10]. The exception to this is the highly-stigmatized variety of Korean spoken on Jeju Island that was identified by most of our respondents as a separate dialect region. Many of the people we talked to both on the mainland and on Jeju Island remarked that because Jeju is sparsely populated[11] and is located far from Seoul and closer to Japan, it is culturally and linguistically different from the rest of South Korea. On the other hand, the mainland respondents often commented that Jeju was a nice vacation destination but not a place where educated Koreans would live and work.

The content analysis of the respondents' comments provides additional insight into how attitudes and beliefs associated with language variation in and across speech communities correlate with perceptual dialect boundaries. Almost all respondents (96 %) wrote comments on their maps that were related to *Standardness*. This overwhelming tendency for respondents to view language as either *Standard* or *Non-Standard* supports Preston's suggestion that "one of the dominating folk concerns in language is pre- (and pro-) scription" (Preston, 1999b: xxxvii). *Non-Standard* was slightly more salient than *Standard* and mostly associated with the southern regions of Gyeongsang and Jeju. *Standard* was overwhelmingly associated with Seoul/Gyeonggi province, perhaps a reflection of the wholesale adoption by Koreans of "one country, one language". Perceptions of *Standard* and *Non-Standard* stratified by respondents' identification with a city or province provide additional confirmation that Koreans buy into the conceptual myth that there is one standard dialect based on the variety spoken in Seoul and the surrounding Gyeonggi province. Further evidence for this comes from respondents who identify with the most stigmatized Southern dialect areas, especially Jeolla, Gyeongsang, and Jeju, who perceive their local dialects to be inferior to the Seoul dialect. These data confirm the observations from Long and Yim (2002) that South Koreans who identify with Southern dialect areas tend to be linguistically insecure.

Our perceptual data also reflect current demographic trends as areas on the peninsula outside of Seoul begin to urbanize. As shown in the population density map in Figure 6.14, mountains cover 70 % of the land area in South Korea, thus the majority of the population has historically been concentrated in Seoul and other smaller urban centers in the lowland areas. In an effort to

---

10 Currently, high-speed rail lines that transverse the peninsula between the west and east coast are very limited.

11 According to census figures from 2011, the province of Jeju ranks last in terms of population with 583,284 inhabitants. Compare this with Gyeonggi Province, which ranks first and has a total population of approximately 12 million, 10.5 million of whom live in Seoul.

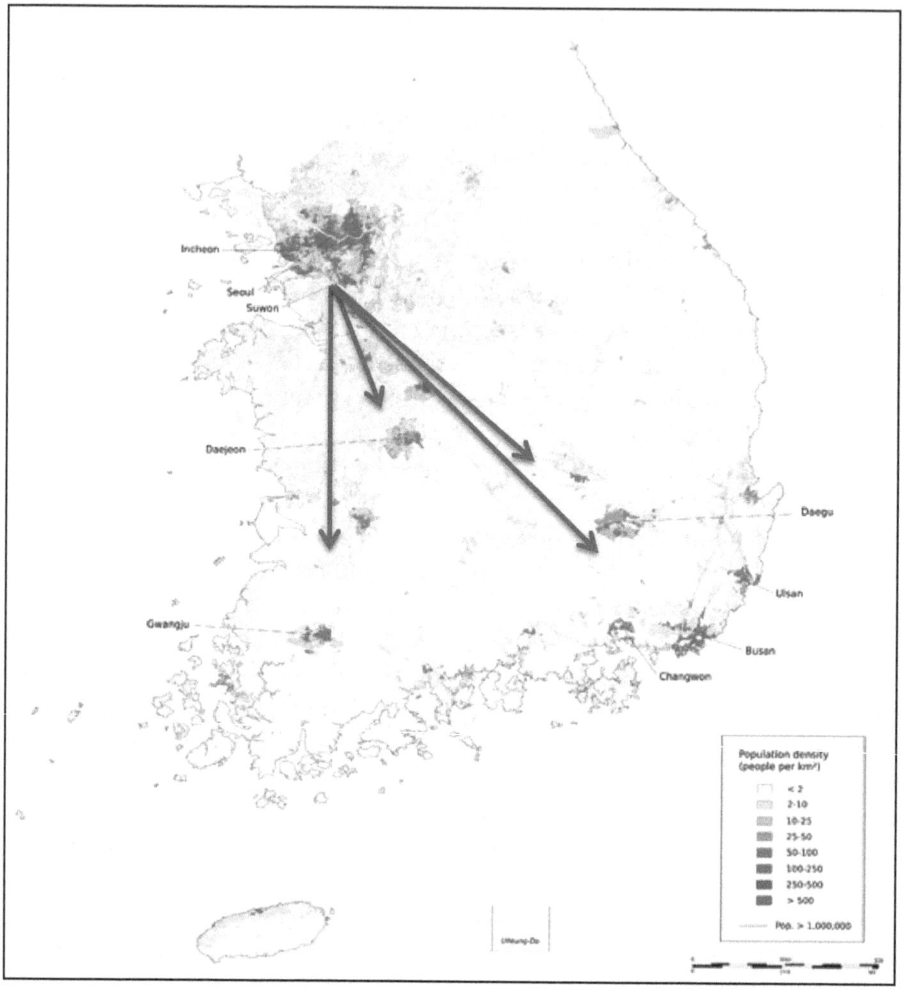

Fig. 6.14: Demographic changes in South Korea (Source: Asia Pop Project 2013).

curb the influx of people into Seoul, the government has been offering incentives for Seoul residents to relocate to rural areas outside of the city that are easily accessible by high-speed rail. In addition, there are plans to relocate several government agencies from the capital to the Chungcheong province south of Seoul (as shown by the arrows on Figure 6.14). Thus, the resulting demographic changes in Korea may well affect future perceptions of regional dialect boundaries, including where *Standard* and *Non-Standard* Korean is spoken.

Part II: **Processes of perception and language change**

Sylvia Moosmüller

Chapter 7 The strength of stereotypes in the production and perception of the Viennese dark lateral

1 Introduction

1.1 Background of the study[1]

In 2007, a project on dialect and sociolect synthesis was initiated by the Telecommunications Research Center Vienna (FTW). The aim of this project was to develop high quality speech synthesis for three varieties spoken in Vienna (the Viennese dialect, the colloquial Standard Viennese variety, and a dialect-based variety of adolescents, see Pucher et al. 2010). Among these, the Viennese dialect posed a major challenge, firstly, because a pronunciation dictionary had to be created, secondly, because the input orthography had to be processed by the synthesis system, and thirdly, because the unit selection technique (Hunt 1996) that was used for the synthesis necessitated a large amount of read material. Therefore, several female and male actors were invited for a casting and asked to translate sentences and text passages into the Viennese dialect. The speech material produced by these female and male actors was then compared to the speech of authentic Viennese dialect speakers. The actor whose speech came closest to the speech of authentic Viennese dialect speakers was then chosen for the synthetic voice.

1.2 Major language varieties in Vienna

The varieties spoken in Vienna belong to the Middle Bavarian dialects[2]. Labov has noted that in large metropolises, language use is "differentiated by social

1 This investigation was part of the project 'Viennese Sociolect and Dialect Synthesis' and was funded by the WWTF 'dialects. ftw.at'.

2 This does not hold for the immigrant varieties which emerged since the 1970s. The main immigrant groups which additionally contribute to the linguistic landscape of Vienna come from Germany, former Yugoslavia, and Turkey.

Fig. 7.1: Administrative divisions of Vienna (Foto: TUBS, source: Wikimedia [GNU free licence]).

class, ethnicity, gender, and race" (Labov 2001: 227) rather than by geographic limits, and the varieties of Vienna are no different. Nonetheless, some folk-linguistic beliefs circulate that the varieties spoken in Vienna are regionally distributed, i.e., that each district of Vienna shows specific linguistic traits. This belief might well be rooted in the fact that the diverse social classes of Vienna reside in specific districts. That is, the 10th, 11th, 12th, 16th, or 17th districts are among those traditionally populated by the lower social classes, so-called working class districts, whereas the 1st, 13th, 18th, and 19th districts are the traditional living environments of the upper social classes (see Figure 7.1).

The Middle Bavarian varieties of Vienna can be conceptualized as threaded on a line between two extreme varieties, the – negatively evaluated – Viennese dialect (VD) and the prestigious Standard Austrian German variety as spoken in Vienna (SAGV)[3]. The in-between varieties either gear towards SAGV (mostly)

3 Standard Austrian German is a Middle Bavarian variety spoken by the upper social classes of the main cultural centers, especially Salzburg and Vienna, to some extent also in Graz, although the varieties spoken in Graz belong to the transition zone between Middle Bavarian

or show more traits of the VD. Historically, in many respects, VD (and Austrian dialects in general) developed differently from Standard Austrian German (SAG) which in turn is geared towards Standard German German (SGG; see Moosmüller 2015 for a discussion). This differing historical development leads to opposite forms, i.e., forms which lack a phonological or phonetic connection between VD (or Austrian dialects in general) and SAG. These observations lead Dressler and Wodak (1982) to describe the interaction between the (Viennese) dialect and SAG by means of a two-competence-model. In this model, the (Viennese) dialect and SAG overlap to a great degree. However, similar to lexical differences (e.g., dialectal *gschroppn* vs. standard *Kinder* 'children'), certain phonological alternations stand in a distribution of mutual exclusion, e.g., dialectal [g̊uɐd̥] vs. SAG [g̊u:t][4] *gut* 'good'. Unlike phonological processes, such as, assimilation processes, these opposing forms have no intermediate forms; they stand in an "either/or" relationship. Such alternations have been termed input-switch-rules. They are highly salient both in production and in perception (Moosmüller 1991; Soukup 2009).

The phonology of SAG[5] is geared towards SGG, but shows many traits of its Middle Bavarian roots, e.g., neutralization of the high vowels /i, ɪ/ to [i], /y, ʏ/ to [y], and /u, ʊ/ to [u] (Moosmüller 2007; Wiesinger 2009; Brandstätter and Moosmüller 2015) or a quantifying temporal organization in stressed positions (Moosmüller and Brandstätter 2014). SAG can be described as a contact variety which has developed from contact with SGG, which is in many ways accepted as the prescriptive norm for Austria (Wiesinger 2009; Soukup and Moosmüller 2011).

A further trait of varieties spoken in metropolises concerns the specific interaction of a prestigious standard variety and a less prestigious dialect variety. In Vienna, this uneven assessment of the two varieties leads to the situation that a considerable amount of Viennese dialect speakers raise their children in a "dialect-based standard variety". Among the younger speakers of the lower social classes, this variety is prevalent and constitutes a newly emerged variety, clearly distinguishable from SAGV, especially by the maintenance of dialectal syntax and morphology, and by some specific phonological and phonetic features, e.g., the Viennese monophthongization, or longer durations of vowels.

---

and South Bavarian. Whereas SAG refers to Standard Austrian German in general, SAGV refers specifically to the standard variety as spoken in Vienna.

4 Both forms stem from Middle High German (MHG) *uo*.

5 Since the current study deals with the Viennese dialect, further varieties are only sketched here in order to give a survey of the sociolinguistic landscape of Vienna.

An "in-between" variety is also spoken by educated speakers who were raised in VD or any other dialect and had to acquire SAGV in school and at university. These speakers are aware of the morphological and syntactic differences, but show some insecurity with respect to phonology and phonetics, especially with respect to temporal organization.

1.3 The lateral in the Viennese dialect

The phoneme inventory of the Viennese dialect contains one lateral, an alveolar lateral approximant which is subjected to various allophonic processes (Wodak-Leodolter and Dressler 1978). First of all, vowels preceding a lateral are rounded, e.g., /ʃd̥ilɛr/ → [ʃd̥ylɑ] *stiller* 'quiet; COMP'. Word-finally and before consonants, the lateral is vocalized[6]. After back vowels, the lateral is vocalized to a palatal vowel, [e] in the case of VD, e.g., /ˈd̥ulb̥ɛn/ → [ˈd̥ue̯bm̩] *Tulpen* 'tulips'. After front rounded vowels, the vocalized lateral [e] is absorbed and transparent through compensatory lengthening, e.g., /fiːl/ → [fyːˑ] *viel* 'much,' /g̊ɛld̥/ → [g̊œːd̥] *Geld* 'money'.

In the remaining positions, i.e., in intervocalic position, in initial position, and after consonants, the lateral is realized in the Viennese dialect. In some of these positions, a dark lateral is produced.

Our analysis of the semi-structured interviews of Viennese dialect speakers showed that, contrary to other languages or language varieties which show a complementary distribution of the clear and the dark lateral, e.g., English (Sproat and Fujimura 1993; Narayanan et al. 1997; Carter 2003; Carter and Local 2007), Catalan (Recasens and Espinosa 2005), Czech (Šimáčková 2009), Italian (Marotta and Nocchi 2003), or the Rhenish dialect (Meuter 2008), in the VD, the dark lateral appears in initial position, e.g., /ˈlæːd̥ɛr/ → [ˈɫæːdɑ] *leider* 'unfortunately'. Since this position is usually prosodically strong, the realization of the dark lateral in this position is salient with respect to production and perception.

In addition, the dark lateral is produced after alveolar and postalveolar consonants, e.g., /ˈsæːd̥ɛl/ → [ˈsæːɫ̩] *Seidel* 'small glass of beer' or /ˈʃlɔːg̊ɛn/ → [ˈʃɫɔːŋ̍] *schlagen* 'to beat'.

After bilabial consonants, a clear or a retroflexed lateral is produced, e.g., /ˈb̥læːb̥ɛn/ → [ˈb̥læːm̩] or [ˈb̥ɭæːm̩] bleiben 'to stay,' and after velar consonants, the lateral is palatalized, e.g., /ˈg̊lɒːb̥ɛn/ → [ˈg̊lʲɒːm̩] *glauben* 'to believe'.

6 This process holds for Middle-Bavarian dialects in general.

In intervocalic position, rules are not fixed. However, the dark lateral preferably occurs between back vowels, e.g., /ˈholɛr/ → [ˈhoɫɑ] *Holler* 'elder'. Generally, it seems that in intervocalic position, the lateral rather adjusts to the following vowel, i.e., if the following vowel is a back vowel, a dark lateral is to be expected (Schmid and Moosmüller, forthcoming).

1.4 The sociolinguistics of the dark lateral in the Viennese dialect

The Viennese dialect is the most negatively evaluated dialect of Austria (Moosmüller 1991). Quite a few features either started or are restricted to the area of Vienna. Three main features are associated with the Viennese dialect by linguistically naïve speakers:

a) The Viennese monophthongization which was first observed around 1900 among speakers of the lower social classes (Gartner 1900; Luick 1904).
b) A long-drawn-out articulation which has its roots in additional monophthongization processes with concomitant compensatory lengthening from Middle High German diphthongs, e.g., /ɑ:/ (< MHG *ei) weiß* [vɑ: s] 'to know' or /ɑ:/ (< MHG *ou* preceding nasal consonants) *Baum* [b̥ɑ: m] 'tree'.
c) The dark lateral. The emergence of the dark lateral is less well documented in the literature. It might have emerged through contact with Czech immigrants living especially in the 10th District around 1900[7]. This would explain why the dark lateral is restricted to the area of Vienna and exclusively used by speakers of the Viennese dialect. However, these considerations do not explain the phonological distribution of the dark lateral, which occurs, contrary to languages or language varieties with a complementary distribution of clear and dark laterals, also in the prosodically strong word initial position.

While the processes of monophthongization in a) and b) are continuously spreading to other parts of Austria, the dark lateral is bound to Vienna. The combination of restrictions (the restriction to the area of Vienna, the restriction to a specific social group, and the perceptually salient word initial position) makes the usage of the dark lateral a social marker, which is highly stigmatized and subject to negative stereotyping. Consequently, the use of the dark lateral tends to be avoided by persons showing upward social mobility, and, among

7 Dialectologists report pervasive Czech influence on the Viennese dialect (see e.g., Steinhauser 1953; Kranzmayer 1953).

these, especially by women. The preference of women for linguistic forms associated with a standard variety has been proved in many studies and has been interpreted in various ways (see, e.g., Trudgill 1972; Gordon 1997; Yaeger-Dror 1998; Eckert 2000; Labov 2001; Sundgren 2001; Eckert and McConnell-Ginet 2003 for a discussion). The same trend has been documented in studies on language variation in Vienna (Wodak-Leodolter 1975; Moosmüller 1987, 1991). Since the dark lateral is associated with the Viennese dialect spoken by the lower social classes, especially female speakers of the Viennese dialect tend to avoid the realization of the dark lateral. This tendency can sometimes be observed in the speech behaviour of men as well, though to a much lesser degree, and predominantly in formal speech situations (Schmid and Moosmüller, forthcoming).

As will become evident in this contribution, the production and perception of the dark lateral is an excellent candidate to demonstrate the functions and fallacies of stereotyped production and perception.

2 Method

2.1 Production

For speaker selection for the dialect synthesis, nine female and male actors were invited for a casting. Of these, four candidates had to be excluded because they were not able to produce the Viennese dialect at all. Of the five remaining candidates, one female actor (AN) and two male actors (HP and CM) were born and raised in Vienna, one female actor was born and raised in Lower Austria, in an area close to Vienna (BD), and one male actor was born and raised in Graz (BJ), which is located about two hours' drive south from Vienna. Except for BD, none of the female and male actors had a lower social class background. None of the female and male actors was raised in the Viennese dialect. It has to be noted that dialectologically, the city dialect of Graz belongs to the transition zone from Middle Bavarian to South Bavarian dialects; as such the dialect spoken in Graz is quite different from the Middle Bavarian dialects spoken in Vienna or Lower Austria.

For comparison, three male speakers (sp194, sp195, sp198) and three female speakers (sp191, sp207, sp211)[8] of the Viennese dialect were considered

8 In our database on Austrian German varieties, speakers are numbered consecutively; the abbreviation 'sp' refers to 'speaker'.

for analysis. They all had a lower social class background. All speakers were older than 45 years, because, as outlined above, most of the younger speakers of the Viennese dialect use a dialect-based standard variety.

With all speakers, a semi-structured interview was conducted which, among other questions, included questions on biographical data and data on attitudes towards diverse varieties of Austria. In addition, the female and male actors were asked to transform 100 sentences and a text (*Der Geisterfahrer* 'The ghostdriver') into the Viennese dialect. The Viennese dialect speakers were also asked to transform the same text into the Viennese dialect; in addition, they performed a picture naming task. The text was presented in an orthography that approached the Viennese dialect. Speakers were allowed to read the text before transforming it into the dialect. They were also allowed to adjust the text and use different lexemes or expressions, if they wished.

For the current investigation, the production of the laterals in the task of transforming the text into the Viennese dialect was analyzed. Regarding the female and male actors, the transformation of the 100 sentences into the Viennese dialect was analyzed as well. The production data of the text to be transformed into the Viennese dialect was compared with the speech material gathered from the semi-structured interview.

All realizations of the lateral (224 in total) were segmented manually. F_1, F_2, and, where possible, F_3 were extracted by means of LPC. A 46 ms long gliding Hanning window was applied with an overlap of 95 %. Duration was measured as well. Depending on the duration of the lateral, the measurement procedure described rendered 20 to 150 measurements per lateral, i.e., the formant frequency contour of the whole lateral was analyzed. This method was chosen in order to obtain a view of formant frequency movement over time. In order to determine whether a lateral was produced with a velarized articulation, the value of F_2 was considered conclusive ($F_2 < 1200$ Hz for the dark lateral, (see, e.g., Andrade 1999; Carter and Local 2007; Recasens 2012).

2.2 Perception

For the perception test, 56 short utterances (comprising approximately an intonational phrase) were extracted from the read text transformed into the Viennese dialect by the five female and male actors and by the six speakers of the Viennese dialect. Except for six distractors, all utterances presented contained one lateral in word-initial position (*lauter* 'nothing but,' 2 x *Lastwagen* 'truck'), one after a bilabial plosive (*Platz* 'square'), and one after an alveolar plosive (*gemütlich* 'cosy').

An online perception experiment was designed. The question asked was as follows: "Is the sample you are going to hear produced in an authentic Viennese dialect?" Listeners were asked to rate each particular stimulus on a scale from 1 (= very authentic) to 5 (= not authentic) with respect to its Viennese dialect authenticity. For statistical analysis, these results were transformed into a trivalent scale (1 = very authentic, 3 = non authentic).

63 Viennese listeners and 81 listeners coming from other parts of Austria (= non-Viennese listeners) participated in the perception experiment[9].

For statistics, two sample t-tests were performed.

3 Results

3.1 Results of production study

In initial position (*lauter*, 2 x *Lastwagen*), all female and male actors plus all male and one female dialect speaker realized a dark lateral. Two female dialect speakers realized a clear lateral (see Figure 7.2).

In the Viennese dialect, the initial position demands the dark lateral. Therefore, from a phonological point of view, the process of velarization was correctly applied by the actors. However, it is interesting to note that only one of the female Viennese dialect speakers applied the process of velarization, the other two female speakers realized a clear lateral. Therefore, from a sociolinguistic point of view, the female actors used a stereotypical form.

After alveolar consonants, velarization of the lateral is also applied in the Viennese dialect. As a representative, the item *gemütlich* 'cosy' which is realized ['g̊miɐd̥ɫi] in the Viennese dialect, appears in the 'ghost driver' text (see Figure 7.3).

In the same way as in initial position, the dark lateral is realized by all female and male actors, plus by all male VD speakers and by one female VD speaker. However, now it was sp191 who realized the dark lateral.

After bilabial consonants, velarization is not applied; the lateral should either be realized as a clear lateral, or as a retroflexed lateral. All male and female VD speakers and all female and male actors realized the clear or retroflexed lateral in this position, in accordance with the pattern expected in the Viennese dialect.

9 Listeners were recruited via mailing lists and word-of-mouth advertising, in total, 144 subjects with an age range from 18–65 years, quite balanced for gender (75 men, 69 women), participated.

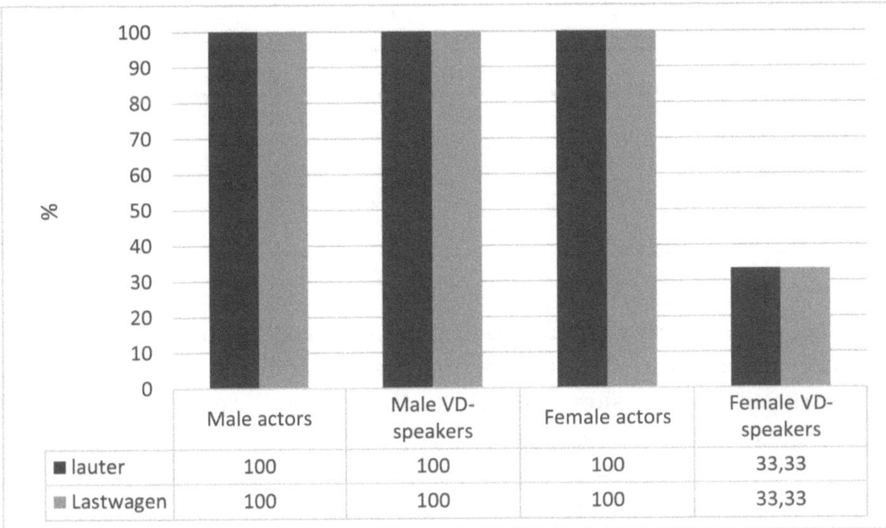

Fig. 7.2: Production of the dark lateral (in %) in word initial position (*lauter* 'nothing but' (dark grey) and *Lastwagen* 'truck' (light grey)).

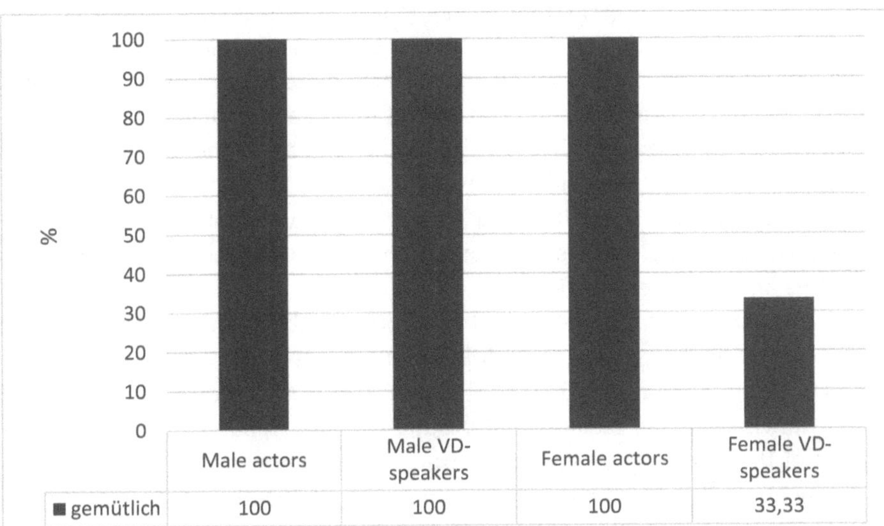

Fig. 7.3: Production of the dark lateral (in %) after an alveolar consonant (*gemütlich* 'cosy').

Tab. 7.1: Wrong applications of the dark and the clear lateral (in %).

Actors	[ɫ] instead of [l]	[l] instead of [ɫ]
AN (♀), Vienna	5	20
BJ (♂), Graz	21	11
CM (♂), Vienna	4	0
BD (♀), Lower Austria	9	0
HP (♂), Vienna	7	0

Since a tendency towards an overgeneralization of the application of velarization could be observed in transforming the "ghost driver story" into the Viennese dialect, the transformation of the 100 sentences performed by the female and male actors was analyzed as well.

Sociolinguistically, the realization of the dark lateral is stigmatized and negatively evaluated. Consequently, the dark lateral is hardly ever used by female Viennese dialect speakers. Male dialect speakers realize the dark lateral to a much greater extent, but tend to avoid it as well. Therefore, if a "misapplication" of the process of velarization can be observed in the speech of dialect speakers, it is rather the realization of the dark lateral that is suppressed. Conversely, velarization is never applied in contexts where a clear lateral is demanded.

The picture regarding the realization of dark and clear laterals looks different for the transformation of the 100 sentences by the female and male actors. As can be seen from Table 7.1, all female and male actors applied the dark lateral instead of the demanded clear lateral in some instances.

With 21 % of wrong applications of the dark lateral, the actor from Graz performed worst with respect to Viennese dialect authenticity. This is not surprising, since he was raised furthest from Vienna. All other female and male actors misapplied the process of velarization to some degree, which has to be judged as an overgeneralization.

In this table, women were rated according to the phonology outlined in 1.3. However, if we speak in sociolinguistic terms, the female actors (AN and BD) should predominantly realize clear laterals. AN produced at least 20 % of the dark laterals as clear laterals, but BD never applied the clear lateral instead of the dark lateral. Therefore, from a sociolinguistic point of view, the female actors outperformed the male actors with respect to overgeneralization.

3.2 Results of perception study

3.2.1 Viennese listeners vs. non-Viennese listeners

Figure 7.4 reveals that non-Viennese listeners gave higher scores to male and female Viennese dialect speakers than Viennese listeners ($p = 0.002$ and $p = 0.003$, respectively), i.e., non-Viennese listeners judged Viennese dialect speakers as less authentic than Viennese listeners.

However, no statistically significant differences between Viennese listeners and non-Viennese listeners emerged with respect to judging both Viennese and non-Viennese actors ($p = 0.27$ and $p = 0.43$, respectively).
Compared with the result on Viennese dialect speakers, this result also implies that non-Viennese listeners rated Viennese actors more authentic than male and female Viennese dialect speakers ($p < 0.01$) and male Viennese dialect speakers the same as non-Viennese actors ($p = 0.35$). Therefore, non-Viennese listeners rely on a stereotypical representation of the Viennese dialect to a higher degree than Viennese listeners.

The result on the assignment of actors also holds when broken down for the individual male and female actors. Figure 7.5 displays the ratings concerning these assignments.

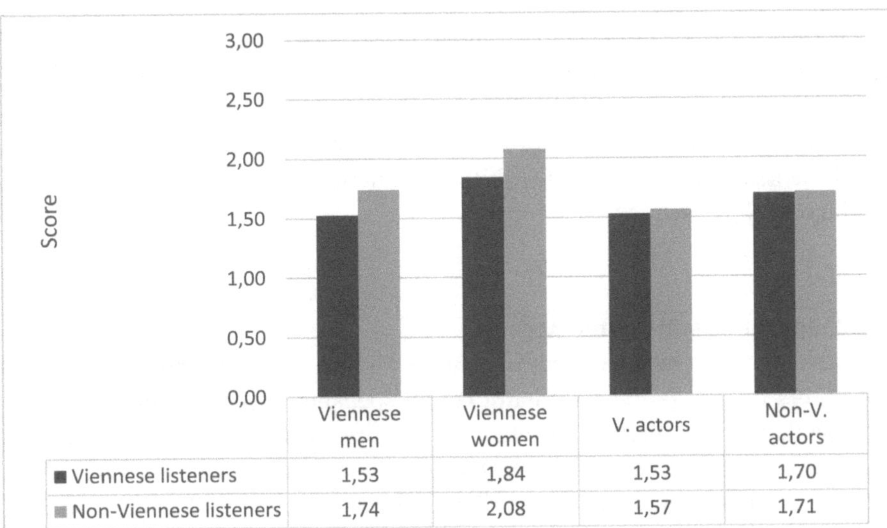

	Viennese men	Viennese women	V. actors	Non-V. actors
Viennese listeners	1,53	1,84	1,53	1,70
Non-Viennese listeners	1,74	2,08	1,57	1,71

Fig. 7.4: Overall results of the perception test. VD = Viennese dialect, V = Viennese[10].

10 Female and male actors were assigned the label 'Viennese' and 'non-Viennese', since they are not genuine Viennese dialect speakers.

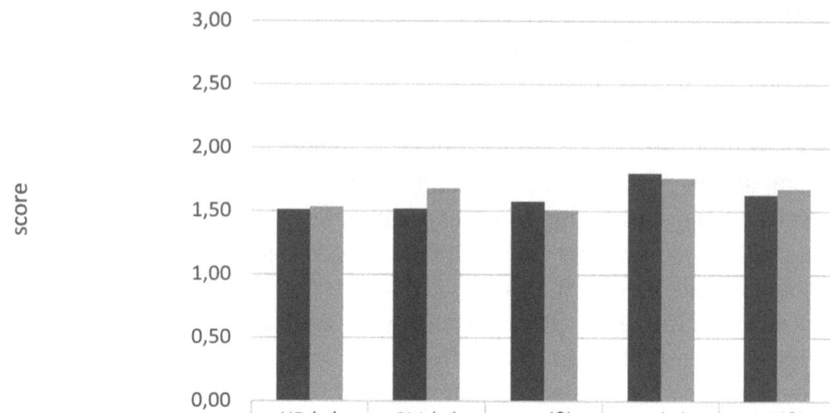

	HP (♂)	CM (♂)	AN (♀)	BJ (♂)	BD (♀)
■ Viennese listeners	1,51	1,51	1,57	1,80	1,63
■ non-Viennese listeners	1,53	1,68	1,50	1,76	1,68

Fig. 7.5: Ratings of the individual female and male actors. HP, CM, and AN = Viennese, BJ and BD = non-Viennese .

Figure 7.5 reveals that HP was judged the most authentic by all listeners, closely followed by AN, whereas BJ, the actor born and raised in the city of Graz, was judged worst. Except for AN, whose application of the dark lateral is not in accordance with the speech behaviour of VD women, as discussed in 3.1, these results correlate well with the production data. HP performed best in production, BJ worst.

3.2.2 Gender-specific differences

A further striking result concerns gender: Female Viennese dialect speakers were rated highest (= least authentic) of all speakers, by both groups of listeners (see Figure 7.3). Again, they were rated more authentic by Viennese listeners than by non-Viennese listeners; however, even Viennese listeners assigned a higher score to female Viennese dialect speakers than to all other groups of speakers. This means that female Viennese dialect speakers are also perceived as less authentic than non-Viennese female and male actors, and this last observation is statistically significant as well ($p = 0.04$).

This result needs more in-depth delineation. Figure 7.6 reveals that female Viennese dialect speakers were rated less authentic than the two female actors in our sample.

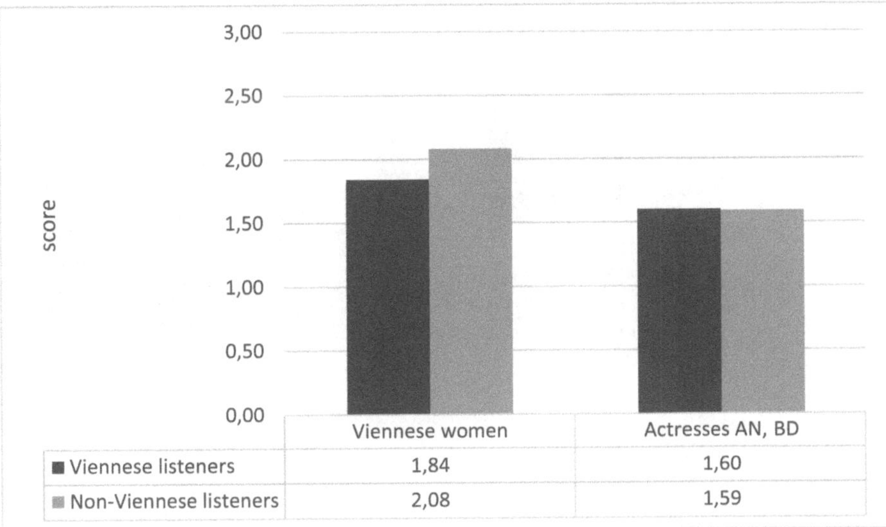

Fig. 7.6: Perception of female speakers. VD = Viennese dialect, AN = actor born and raised in Vienna, BD = actor born and raised in Lower Austria.

Again, it is noteworthy that with respect to the perception of the female actors, no differences can be observed between Viennese listeners and non-Viennese listeners ($p = 0.45$). But, the most striking outcome is definitely that female Viennese dialect speakers are judged as less authentic Viennese dialect speakers than female actors who were not raised in the Viennese dialect ($p < 0.01$).

The results on production revealed that female Viennese dialect speakers hardly ever used the dark lateral, whereas both AN and BD consistently made use of the dark lateral when reading the "ghost driver story". This means that the use of the dark lateral is ascribed to the Viennese dialect. Not realizing this salient feature entails that a speaker is not associated with the Viennese dialect.

3.2.3 Best raters

The overall results demonstrate that male Viennese dialect speakers are rated the same as Viennese female and male actors with respect to dialect authenticity (see Figure 7.3). This outcome is rather surprising, since it shows that even Viennese listeners are ultimately unable to distinguish between genuine dialect speakers, and female and male actors who imitate the Viennese dialect.

This result raises the question of whether there are any raters who are able to differentiate between Viennese dialect speakers and Viennese female and

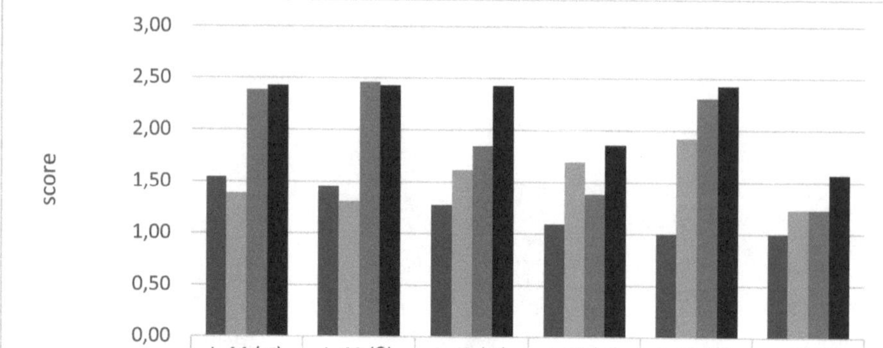

	L-44 (♂)	L-41 (♀)	L-45 (♂)	L-32 (♂)	L-21 (♂)	L-23 (♀)
■ VD men	1,55	1,45	1,27	1,09	1,00	1,00
■ VD women	1,38	1,31	1,62	1,69	1,92	1,23
■ V actors	2,38	2,46	1,85	1,38	2,31	1,23
■ Non-V actors	2,43	2,43	2,43	1,86	2,43	1,57

Fig. 7.7: Results of the best raters. VD = Viennese dialect, V = Viennese, non-V = non-Viennese, L = listener, numbers after L indicate the age of the listener.

male actors who imitate the Viennese dialect. This question can be answered positively: there are some listeners who are able to differentiate these two groups of speakers. In addition, they also spotted the female Viennese dialect speakers as dialect speakers!

However, none of the best raters came from the group of non-Viennese listeners; none of them were able to differentiate between Viennese dialect speakers, male or female, on the one hand, and female and male actors on the other. That is, when male or female Viennese dialect speakers were assigned a low score (= very authentic), the female and male actors were assigned a low score as well. Therefore, the best raters presented in Figure 7.7 are all Viennese listeners.

Figure 7.7 shows that especially L-44 and L-41 scored both groups of female and male actors as less authentic than both the male and the female Viennese dialect speakers; they clearly differentiate Viennese dialect speakers and female and male actors. L-45 and L-21 still evaluate Viennese dialect speakers as more authentic than female and male actors. However, the values increase progressively from "very authentic" to "not authentic," i.e., male Viennese dialect speakers are rated better than female Viennese dialect speakers who are again rated better than Viennese female and male actors. Non-Viennese female and male actors are rated worst. Finally, L-32 and L-23 have in common that at least male Viennese dialect speakers are rated best. However, they did not identify female Viennese dialect speakers as such. Therefore, we can summarize the following:

a) All best raters assigned the lowest scores to male Viennese dialect speakers and rated them as more authentic than female and male actors, and
b) four out of the six best raters judged female Viennese dialect speakers as more authentic than female and male actors.

4 Discussion

For the current study, Viennese and non-Viennese female and male actors as well as female and male Viennese dialect speakers were asked to transform a text into the Viennese dialect. As concerns the VD speakers, a comparison of the speech production data gathered from the task of transforming the text into the Viennese dialect with the speech production data of the semi-structured interview and the picture naming task confirmed that the female and male Viennese dialect speakers produced an authentic version of the text transformed into VD. Both in the semi-structured interview and in the picture naming task, male Viennese dialect speakers applied velarization in the contexts outlined in 1.3; female Viennese dialect speakers hardly ever produced the dark lateral.

All female and male actors transforming the text into the Viennese dialect, i.e., in imitating a variety in which they had not been raised, overgeneralized the production of the dark lateral to a greater or lesser extent. In imitating a variety, several strategies can be deployed, and differences in the ability to imitate a given variety are observed as well (see Evans 2010; Lindsey and Hirson 1999; Markham 1999). Overgeneralization of prominent features is one characteristic of imitated speech (Torestensson, Eriksson, and Sullivan 2004; Neuhauser 2008) and has been applied by the female and male actors of this study. Overgeneralizing specific features also implies that in imitating a variety, the imitator relies on (his/her) stereotypes of that imitated variety.

Stereotypes contain both positive and negative aspects. On the one hand, stereotyping derives from a cognitive need for categorization and systematization, which is a simple and highly efficient way to categorize our social world. From this functional aspect of stereotyping it follows that linguistic heterogeneity is needed in order to differentiate specific social or regional groups. On the other hand, by doing so, a gap occurs between the real and the stereotyped worlds, i.e., stereotypes are imprecise (for an in-depth discussion on the functions of stereotypes, see Kristiansen 2003). Honey, in his definition of stereotypes, precisely summarizes stereotypes as "popular and conscious but imprecise general characterizations of the speech forms of particular social groups". (1997: 99). Here, we are concerned with the imprecision-side of this categorization, that is, with the negative aspect of stereotypes which not only trap us and lead to incorrect assignments, but affect much deeper cognitive layers.

In the last years, many theoretical models on linguistic heterogeneity advocate the concept of style in order to "overcome the shortcomings of single variable studies" (Auer 2007: 11).

> The sociolinguistic analysis of style claims, then, that the social meaning of linguistic heterogeneity does not (usually) reside in individual linguistic features but rather in constellations of such features which are interpreted together. [...] we do not interpret single variables but a gestalt-like stylistic expression. (Auer 2007: 11)

In the same vein, Kristianssen (2003) conceives linguistic stereotypes "as formed by clusters of perceptual salient and contrastive features" (2003: 76f) and points out the gestalt perception. This theoretical approach is corroborated by the results of a study on the identification of African American Speech (Thomas, Lass, and Carpenter 2010). They conclude that listeners base their identification task not on a single cue, but on different cues in dependence, among others, on the sex of the speaker. The results of the current study do not fully support the findings of Thomas, Lass, and Carpenter (2010). Rather, the results presented in 3.2 point in the direction that specific, highly salient and therefore stereotyped features are able to override the gestalt-like interpretation of a given sequence of speech. This concerns especially the non-application of velarization. Female Viennese dialect speakers, as has been stated, hardly ever use the dark lateral. However, even if the remaining segmental and prosodic features were correctly employed, female Viennese dialect speakers were not judged as such. Therefore, solely the absence of the dark lateral caused a wrong assignment.

Conversely, the female actors AN and BD were judged as authentic Viennese dialect speakers, although they "misapplied" the dark lateral in sociolinguistic terms. Therefore, the dark lateral is strongly associated with the Viennese dialect.

This leads us to the question: What is a dialect? Is it the speech behaviour of men of lower social classes that is given the term "dialect"? Is one to accommodate to their speech in order to be judged an authentic Viennese dialect speaker? Is the dialect stereotyped as a variety without variation? Is it a mere accumulation of the most salient features as outlined in 1.2? Results strongly suggest these interpretations. Hence, these questions necessitate further studies on how sequences of speech which lack any of these highly salient features (dark lateral, monophthongization, long vowels) will be assessed.

On the other hand, there is extensive evidence from the literature that women's speech is geared towards a more close-to-standard pronunciation, also in Vienna. As concerns female Viennese dialect speakers, the exact reasons for avoiding salient dialect features still needs to be subjected to thorough analysis, by applying in-depth or narrative interviews and surveys by means of questionnaires. Since especially the Viennese dialect is negatively evaluated all over Austria (Moosmüller, 1991), it is obvious that women do not want to be judged as dialect speakers. From that standpoint, it is good news that female Viennese dialect speakers are not judged as such. Whether, on the other hand, female Viennese dialect speakers are indeed given a higher score with respect to education or social status cannot be answered from this study. However, a study on Japanese vowel devoicing and non-devoicing revealed that women who applied the standard process of devoicing were more likely judged as speakers from Tokyo, because women are stereotyped as being more likely users of a standard variety (Yonezawa-Morris 2010).

Can we conclude, therefore, that the female Viennese dialect speakers are correctly judged as non-authentic dialect speakers, since women are striving for a more close-to-standard pronunciation and are therefore possibly speaking another, maybe in-between variety? If we review the results of the best raters, the question has to be answered in the negative. Four out of the six best raters judged the female Viennese dialect speakers as more authentic than the female and male actors, classifying them in the same way as the male Viennese dialect speakers. Therefore, a few listeners have an accurate knowledge of the Viennese dialect, including knowledge not only about linguistic (in this study phonological) variation, but also knowledge about sociolinguistic variation. Conversely, this means that listeners, and not only non-Viennese listeners, but also listeners born and raised in Vienna, have a very restricted, i.e., stereotyped conception of the Viennese dialect, associating the dialect with the most salient features.

It is a well-known fact that stereotypes, as well as prejudices, are not easily discarded. They not only guide our perception, so that the non-application of a stereotype leads to a wrong assignment, they also guide our expectations. This seems to be the most relevant outcome of this study. The association of the dark lateral with the Viennese dialect is so strong that its use is also expected from women, who generally do not use this feature, otherwise they are not judged as authentic Viennese dialect speakers. Moosmüller (2010) reports on a female lay actor, raised in the Viennese dialect, who never made use of the dark lateral in the semi-structured interview, whereas in the task of reading the 'ghost driver story' and in the picture naming task she consistently applied velarization. Thus, when asked to produce the Viennese dialect, she switched to a completely exaggerated, stereotyped mode of presenting her own variety. The professional female actors discussed in this study performed better, however, in order to be perceived as a Viennese dialect speaker, a woman also has to apply this feature. The results of this study once more points out the double bind situation of women.

In the light of these results, what recommendation can be given for dialect speech synthesis? The safest way is to choose a male speaker, because a female speaker using the stereotypical features associated with the dialect might be accepted by one part of the listeners, and rejected by the other part of listeners[11].

11 It has to be emphasized that this recommendation holds only for urban dialects in metropolises. In rural areas, things are quite different, as has also been exemplified in many studies (see Labov 2001 for an overview).

Christoph Hare Svenstrup

Chapter 8
Access and attitudes: A study of adolescents' metalinguistic awareness

1 Introduction

As a part of a larger language attitudes study a number of German adolescents were interviewed about their metalinguistic awareness of the local use of different social styles and varieties in interaction. The aim is to reveal the adolescents' perspective on and attitudes towards ways of speaking in the area, e.g., the German standard variety *Hochdeutsch* and the local variety *Schwäbisch*. Here I will focus on two interviews and describe how the adolescents negotiate and construct linguistic labels and local identities associated with different ways of speaking, how they negotiate access to these local identities, and how they position themselves in the linguistic setting of the area.

1.1 Language attitudes in Southwest Germany

The majority of this chapter deals with a qualitative investigation of metalinguistic awareness, but I start out by briefly discussing the quantitative element of the study. This consists of two questionnaires regarding respondents' conscious attitudes (a label ranking task – LRT) as well as their subconsciously offered attitudes (a speaker evaluation experiment – SEE). The evaluation process behind language attitudes is illustrated in Figure 8.1.

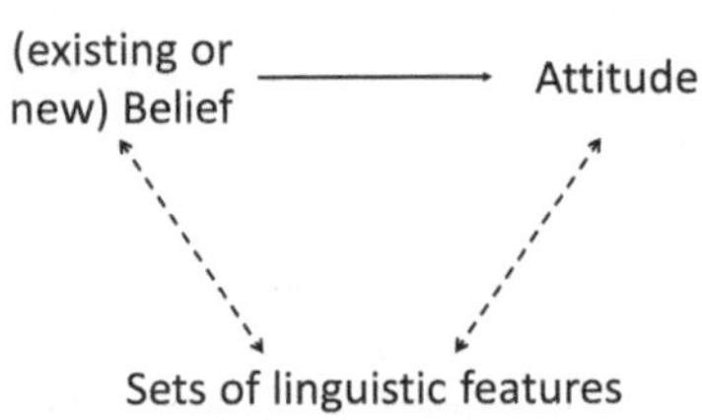

Fig. 8.1: Evaluational process behind language attitudes (adapted from an illustration done by Dennis Preston (September 22nd 2011) in his course "Language in America" at Oklahoma State University).

Following Krosnick, Judd and Wittenbrink (2005) I suggest dividing the process illustrated in Figure 8.1 into three phases: 1) the automatic activation/construction of a belief, 2) the deliberation phase, and 3) the response phase. The difference between conscious attitudes and subconsciously offered attitudes is contingent on the execution of the deliberation phase. If executed, the attitudes are consciously offered. Conscious and subconscious attitudes are assumed to represent two different kinds of attitudes: a) the overt attitudes shared in public discourse, and b) the covert attitudes kept from public discourse but disclosed through language use (Kristiansen and Coupland 2011), with the latter reflecting the ongoing change in language use. In the study here, in order to eliminate the deliberation phase, respondents were kept unaware of the object of investigation (geographically bound linguistic differences) and put under time pressure when answering the SEE.

The results of the SEE show that the respondents are less positive towards local, *Schwäbisch*-influenced speakers than towards outside speakers evaluated to be more standardized. The results of the LRT show that they are more positive towards *Schwäbisch* than towards outside labels, e.g., *Berlinerisch* or *Bayrisch*, and that they are most positive towards *Hochdeutsch* (Svenstrup 2014). A look at the respondents' self-reporting may explain part of the high ranking of *Hochdeutsch* – the respondents consider it to be an in-group label equal to *Schwäbisch*: 34 % report speaking a mix of *Schwäbisch* and *Hochdeutsch*, 28 % *Hochdeutsch*, and 26 % *Schwäbisch* (7 % report another variety, 5 % gave no answer). I consider these quantitative results to be a manifestation of the ongoing standardization in Germany (Spiekermann 2004; Auer and Spiekermann 2011).

Auer and Spiekermann (2011) consider the result of the standardization to be a (nationwide) spoken standard (with room for variation) at everybody's disposal and to be used in the private as well as the public spheres:

> [...] for many Germans, the standard is the language they grew up with (not the dialect). What they learn in school is literacy, i.e., the written variant of the spoken standard they are already familiar with. (Auer and Spiekermann 2011: 174)

Seen from an attitudinal perspective this ubiquitous (spoken) standard appears to enjoy high prestige in the German society and Auer and Spiekermann (2011) speculate that this applies to both overt and covert prestige. The quantitative element of this study confirms this. The layperson's label for the German standard is *Hochdeutsch*. *Hochdeutsch* is the stereotypical label ranked highest in the conscious attitudes, and the speakers deemed to sound most *Hochdeutsch* are the ones ranked the highest in the subconsciously offered attitudes (Svenstrup 2014). Thus, the setting for my study is a rather standardized group of adolescents from an area which is traditionally regarded as dialectal (Ruoff 1997; Eichinger et al. 2009).

1.2 Analyzing metalinguistic awareness

Metalinguistic awareness is about how speakers negotiate and construct labels and local identities based on ideologies behind linguistic differentiation:

> [...] the ideas with which participants and observers frame their understanding of linguistic varieties and map those onto people, events and activities that are significant to them. (Irvine and Gal 2000: 35)

A speaker can perform a whole range of different contextual identities in an interaction (Blommaert 2005), and if a speaker wants to be regarded as a member of a particular group this can be signaled through the use of linguistic resources associated with this group. This group membership or identity is negotiable as the other interlocutors have to validate and acknowledge it for the speaker to gain access to it. Performing an identity through linguistic resources and gaining the membership of a desired group also means that the speaker is part of defining the group. A speaker can also be assigned a group membership from others as a result of social evaluation based on the employment of linguistic resources (Coupland 2007). Consequently, a group membership, and the inherent identity, can be attained through a speaker's own performance and/or through the attribution by others with linguistic resources being both a tool for and source of such social evaluation and categorization.

On the normative level sets of linguistic resources are associated with social groups. This also means that a set of linguistic resources can be perceived to be linguistic resources that belong together. As member of a certain group a speaker can claim ownership of the set of linguistic resources associated with that particular group (Jørgensen 2010). Such a claim means that the speaker can function as a "gatekeeper" of that particular set of linguistic resources. But, it can also restrict access to other sets of resources. The membership of one group may prevent access to linguistic resources associated with another group, and the use of certain linguistic resources may prevent access to become a member of a particular group.

Jørgensen (2010) emphasizes that the difference between two linguistic units is only meaningful as a difference when agreed upon as such (Jørgensen 2010: 20). The same applies for the definition of the labels used to in the interviews. Labels, like *Schwäbisch*, are first of all determined by the participants agreeing on a definition. *Schwäbisch* is what the participants negotiate it to be in the given situation. This entails that the labels are open to negotiation and can vary from situation to situation. What is agreed upon as *Schwäbisch* in one situation may be contested as such in another.

> The general sociolinguistic term used to refer to ways of speaking that are indexically linked to social groups, times and places is dialects. Dialects are social styles. (Coupland 2007: 2)

I propose to use *social style* over *variety* because the former is capable of capturing ways of speaking that are not geographically bound but still are part of the participants' repertoires. The entire potential of linguistic resources available to a speaker is what Blommaert (2005) calls a *repertoire*.

> [...] no two human beings, even if they speak the same 'language', have the same complex of varieties. Their *repertoire* is different; they will each control a different complex of linguistic resources which will reflect their social being and which will determine what they can actually do with and in language. (Blommaert 2005: 13)

In addition to this, and following Auer (2012), I consider social style to be able to incorporate variation within a variety, e.g., age-grading, whereas the opposite is not the case. I do not fully abandon the term variety, though. I regard the labels ranked in the LRT as varieties because they are stereotypical labels representing geographically bounded ways of speaking German. The term is also used when the participants or the interviewer refer to such labels. When the talk is about ways of speaking employed by the participants themselves, I consider these to be social styles. I also use the term *stylization* for when a speaker reproduces linguistic resources associated with a particular group, of which the speaker is not or does not necessarily consider herself to be a member, but do so in an intentionally exaggerated (Rampton 2009) or a creatively portraying way (Bakhtin 1981).

I apply a sequential analysis based on conversation analysis in line with Auer's (1995) work on code-switching. A given utterance is embedded in a sequence of interaction. This means that both the preceding and the following utterances contribute to the meaning making of the given utterance (Auer 1995: 116). In general Auer focuses on the speakers' interpretative realities as central to the sequential analysis, as it is the speakers who assign value and make meaning in interaction (Auer 1995: 117).

2 The interviews

Interviews were undertaken in schools and the participants were chosen either by them volunteering or being hand-picked by their teacher. The interviews were semi-structured (Kvale 2005), informal group interviews/discussions (Kruse 2008). The aim of the interviewer in a semi-structured group interview is to guide the participants to talk about their own situation (Kvale 2005) – here metalinguistically speaking. This way they can apply categories and terms of their own. However, two components in my approach differ considerably from the traditional approach. First, Kvale argues for avoiding the participants' own reflection on their accounts (ibid.: 136), but I find such reflections to be valuable information about the motivation behind the attitudes they account for. Second, as interviewer I abandon the neutral role and play the part of the devil's advocate to incite the participants to actually talk about and discuss language use – something that is not a given. The fact that I am not German (but Danish) proved to be an advantage. As a foreigner I was allowed to ask "stupid" questions (such as those that a German would regard as common or presupposed knowledge – Fairclough (2001) calls such constructions "common sense assumption," an inference without empirical facts to support it but nevertheless taken to be true).

A total of 14 interviews with 59 participants (30 female and 29 male), aged 14 to 19, were carried out. The interviews vary in length from about 36 to 75 minutes. Here I will focus on two: ST-02-INT (recorded on May 5th 2010) and RE-06-INT (recorded on November 24th 2010). ST-02-INT was recorded in a *Gymnasium*[1] in Stuttgart with four participants from the 9th grade. Benjamin is originally from Berlin but moved to Stuttgart as a child. He reports growing up in both places and that he speaks "*Hochdeutsch/Märkisch-Brandenburgerisch*". Bastian is from the Stuttgart area and reports speaking "*Hochdeutsch mit vielen schwäbischen Merkmalen*" ("*Hochdeutsch* with a lot of *Schwäbisch* elements"). Beate is also from the area and reports speaking *Hochdeutsch*. Bruno is from the area too, and reports speaking *Hochdeutsch*. RE-06-INT was recorded in a *Hauptschule* in Reutlingen, near Stuttgart, with four participants from the 9th grade. Felicitas lives in a small village outside Reutlingen. She used to live in Tübingen, was born in the Dominican Republic, and reports speaking *Schwäbisch*. Franziska lives in Reutlingen, was born in Croatia, and reports speaking "Hochdeutsch / wie Ausländer halt so sprechen" ("*Hochdeutsch* / how foreigners generally speak"). Neither Felicitas nor Franziska reported when they moved to Germany but they both gave the impression of growing there during the interviews. Felix lives in a small village near Reutlingen, grew up in the area, and reports speaking *Schwäbisch*. Florian is from Reutlingen and reports speaking *Schwabendeutsch* ("*Schwäbisch*-German").

---

1 After elementary school (ends after the 4th grade) the students are allocated to three different school types according to academic ability. Those with the highest academic proficiency continue in the *Gymnasium*, those with the lowest academic proficiency continue in the *Hauptschule* and those in between continue in the *Realschule*.

3 What I speak?

All the interviews open with two questions concerning the questionnaire for the conscious attitudes: 1) the LRT, and 2) the self-reporting. Here the interviewer asks Bruno, Beate, and Bastian about their answers in the self-reporting task. Bruno's answer is here of particular interest, since he answers *Hochdeutsch* and *Schwäbisch* in the interview. In the questionnaire, however, he only wrote *Hochdeutsch.*

(1) Self-reporting ([ST-02-INT])[2]

a01 EX: und was habt ihr geschrieben unter was ihr spricht
a02 Bruno: was wir sprechen
a03 EX: ja
a04 Bruno: ähm auch # hochdeutsch und schwäbisch also hochdeutsch mit
a05 schwäbischem ähm # akzent so bisschen also ja # gerade wie ich
a06 vorher gesagt habe weil einfach mein vater auch schwabe ist
a07 und von daher ist es vielleicht so ein bisschen auf mich über
a08 gegangen und von meiner umwelt auch
a09 EX: mh=hm
a10 Beate: also ich habe nur hochdeutsch weil ich kann gar nicht
a11 schwäbisch glaube ich also bisschen aber # eigentlich nur
a12 hochdeutsch
a13 EX: und # was hast du geschrieben
a14 Bastian: ich habe auch hochdeutsch mit schwäbischen elmenten
a15 geschrieben |>|# weil einfach weil # sa/ es gibt enfach
a16 sachen so dieses # XXX s-t als [ʃ] betonen solche sachen das
a17 das mache ich halt einfach weil ich hier weil es alle hier

2 Transcription conventions can be found in Appendix A.

a18 machen deswegen
a19 Beate: |<| ja
a20 EX: wo machst du das zum beispiel
a21 Bastian: ähm # [geʃ] du nach hause oder sowas # solche sachen |>|# das
a22 sind einfach das habe ich mir einfach angewöhnt und deswegen
a23 und das kommt ja vom schwäbischen auch allermeiste deswegen #
a24 habe ich das geschrieben
a25 EX: |<|mh=hm
a26 EX: ja

(translation)
a01 EX: and what did you writer under what you speak
a02 Bruno: what we speak
a03 EX: yes
a04 Bruno: ehm also hochdeutsch and schwäbisch that is hochdeutsch with
a05 schwäbisch ehm accent a little like yes like i just said
a06 before i have because my father is schwäbisch too
a07 and therefore it has perhaps rubbed off on me
a08 and from my surroundings as well
a09 EX: mh=hm
a10 Beate: i just wrote hochdeutsch because i can't speak schwäbisch
a11 at all i think that is a little but in effect just
a12 hochdeutsch
a13 EX: and what did you write
a14 Bastian: i too wrote hochdeutsch with schwäbisch elements
a15 because simply because thi there are just things you
a16 like this s-t pronounce as [ʃ] things like that that
a17 i do simply because i because everybody does it
a18 that's why
a19 Beate: ja
a20 EX: where do you do that for instance
a21 Bastian: ehm [geʃ] (are you going) home like such things
a22 they are just i just got used to that and that's why and
a23 it also comes from schwäbisch primarily that's why

a24		i wrote that
a25	EX:	mh=hm
a26	EX:	yes

The interviewer asks about the participants answers to the self-reporting task (li. a01). Bruno repeats the last part of the question for confirmation (li. a02), and the interviewer responds with a confirming "*ja*" ("yes") (li. a02). When revealing what he reported Bruno immediately adjusts his answer from *Hochdeutsch* and *Schwäbisch* (li. a04), which indicates that he speaks both equally well, to *Hochdeutsch* with a *Schwäbisch* accent (li. a04–a05). He hesitates when defining *Schwäbisch* as an accent and uses the modifier "so *bisschen*" ("a little") after the definition. This suggests that maybe his access or claim to *Schwäbisch* is limited or restricted in some way. He legitimizes his claim to *Schwäbisch* through the fact that his father is Swabian and that he lives in a *Schwäbisch* speaking area (li. a06–a08). This legitimization is diminished by his use of the modifiers "*vielleicht so ein bisschen*" ("maybe a little") when he talks about the transfer of *Schwäbisch* to himself. It seems that Bruno does want to lay claim to *Schwäbisch* but he is not confident that he can do so, which makes him modify his utterance. In line 05–06 he refers to the very beginning of the interview and the question about what he ranked the highest in the LRT. Here he answered *Schwäbisch* and legitimized it with his father's heritage and the area he lives in. This answer is very likely influencing the answer he gives here, causing him to mention both *Hochdeutsch* and *Schwäbisch*. This may add to his uncertainty, as he is probably well aware that he only wrote *Hochdeutsch* in the questionnaire. Bruno demonstrates that *Schwäbisch*, to him, is a desirable identity but that it is not easily obtained, whereas *Hochdeutsch* seems to be "up for grabs".

Beate answer that she just wrote *Hochdeutsch* (li. a10) and explains that she does not speak any *Schwäbisch* (li. a10–a11). She modifies this with "*glaube ich also bisschen*" ('I think well a little') (li. a11), but then she states that in effect she only speaks *Hochdeutsch* (li. a11–a12). In Beate's answer there is no hesitation. Her modification could be an indicator of her awareness of living in a *Schwäbisch*-speaking area, which means that she probably has some competence (at least on the stereotypical level). Her confirmation of only speaking *Hochdeutsch* emphasizes that this is the identity she claims for herself and her interest in claiming a *Schwäbisch* identity is minuscule (but nevertheless present).

Bastian reports speaking *Hochdeutsch* with *Schwäbisch* elements (li. a14–a15). He explains that certain things he just pronounces in *Schwäbisch* and mentions that he pronounces s-t as [ʃ], and that is the way everybody does it (li. a15–a18). When asked for an example (li. a20) he suggest the phrase "*gehst nach hause*" ("are you going home") where "*gehst*" is pronounced [geʃ] (li. a21). This very specific example indicates that the difference in pronunciation between *Hochdeutsch* and *Schwäbisch* is known to him, and his argument that everybody uses it shows his awareness of just how widespread it is. The palatalization of [s] to [ʃ] is in fact a stereotypical feature of the area:

> The palatalization of /s/ is a typical feature of the Alemannic area as a whole, and is often associated with the state of Baden-Württemberg by outsiders. (Auer and Spiekermann 2011: 169)

He explains that he has gotten accustomed to using elements like that, stresses that it comes from *Schwäbisch*, and that is why he answered like he did in the questionnaire (li. a21–a24). This roots him in the area and substantiates his competence in the use of *Schwäbisch*. Bastian is very well aware of the fact that the palatalization of [s] indexes *Schwäbisch*, and when he says [geʃ] he expects his surrounding to perceive this as being *Schwäbisch*. By this he demonstrates his metalinguistic awareness on the level of single linguistic resources categorized by him as belonging to *Schwäbisch*, and his awareness of this particular resource having a (stereotypical) value indexing *Schwäbisch*, too.

4 Do you speak *Schwäbisch*?

I will now focus on Beate and Bruno and their interaction on access to *Schwäbisch* and claiming a *Schwäbisch* identity. As it has just been shown *Schwäbisch* is available to these adolescents as negotiable identity. Bruno already tries to negotiate his way into a *Schwäbisch* identity when he claims having written *Hochdeutsch* and *Schwäbisch* in the self-reporting task, although he actually only wrote the former. During the interview he repeatedly tries to lay claim to an identity as *Schwäbisch* speaker. At the same time Beate tries her best to negotiate her way out of one. She emphasizes her status as standard speaker when she explains why she only wrote *Hochdeutsch* in the task. Nevertheless, she positions herself as an expert in and gatekeeper of *Schwäbisch* throughout the interview. On several occasions during the interview Bruno's efforts to negotiate and constitute an identity as *Schwäbisch* speaker are hampered by Beate who restrict Bruno's access to it. Her use of stylized *Schwäbisch*, e.g., "*heimatländle*" ("home/native country") where the last part, "*ländle*," is stylized *Schwäbisch*, is an example of such positioning. Or situations like this:

(2) Kaum jemand ([ST-02-INT])

b01	EX:	aber kann man sich nicht dann in auf schwäbisch in der schule

b02 gewöhnen
b03 Beate: in der schule redet kaum jemand schwäbisch

(translation)
b01 EX: but can't you get used to then in the in schwäbisch in
b02 the school
b03 Beate: in school hardly anybody speaks schwäbisch

On the question, whether it is possible to get used to the use of *Schwäbisch* in school, Beate simply answers that in school "*redet kaum jemand schwäbisch*" ("hardly anybody speaks *Schwäbisch*"). This is uncontested by the other participants and her position as expert in the subject is consolidated through the fact that her utterance implicitly labels the interviewer's question as irrelevant.

In this excerpt Bruno is asked to describe in which situations he speaks *Schwäbisch*, but Beate interferes and contests his competence in *Schwäbisch*:

(3) Fetzen Schwäbisch ([ST-02-INT])
c01 EX: wo redest du schwäbisch
c02 Bruno: eigentlich ähm # schwierig also ich glaube ich rede auch vor
c03 allem auch mit meinen freunden eher schwäbisch und zu hause #
c04 mit meinem vater vor allem auch und # ich glaube sch/ nicht
c05 schwäbisch rede ich # eher # eher bei formellen anlässen so
c06 auf dem amt im rats/ im Rathaus # oder # in der sch|>|ule oder
c07 so
c08 Beate: |<|aber du
c09 redest ja auch nicht richtig schwäbisch eigentlich
c10 Bruno: |>|nein nur nur so kleine # |>|kleine sachen manchmal|>|fetzen
c11 so
c12 EX: |<|ab/
c13 Beate: |<|das hört man
c14 Beate: |<|hört
c15 man kaum raus

(translation)

c01	EX:	where do you speak schwäbisch
c02	Bruno:	actually ehm difficult well i think i speak that is
c03		especially with my friends more schwäbisch and at home
c04		with my father especially too and i think sch non
c05		schwäbisch i speak like more more on formal occasions like
c06		in the administrative office in the tonw in the town hall or
c07		in school or so
c08	Beate:	but you kind of don't speak proper schwäbisch
c09		actually
c10	Bruno:	no just just like small # small things sometimes snippets
c11		like
c12	EX:	bu/
c13	Beate:	you hardly
c14	Beate:	you hardly
c15		hear it

The interviewer asks Bruno where he speaks *Schwäbisch* (li. c01), indicating that Bruno employs *Schwäbisch* in some situations but not in others. Bruno hesitates a little in the beginning of his answer and points out that it is difficult (to say) (li. c02). He explains that he speaks *Schwäbisch* with his friends and at home with his father (li. c02–c04). This shows that Bruno employs *Schwäbisch* in informal settings, with friends and family, and in interaction with interlocutors who are *Schwäbisch*, such as his father. He continues to explain that in more formal situations, like in public administrative offices and even in school, he does not speak *Schwäbisch* (li. c04–c07). Here he constitutes a rather sharp divide in contexts relevant to his everyday life: the private sphere, where *Schwäbisch* is employed, and the public sphere, where it is not employed. Towards the end Beate interrupts him and points out that he does not speak authentic *Schwäbisch* (li. c08–c09). She does use the modifier "*eigentlich*" ("kind of") but she clearly positions herself as an expert in *Schwäbisch* and gatekeeper of it. According to her Bruno does not belong to the in-group of authentic *Schwäbisch* speakers. Bruno acknowledges this but does not agree completely. He negotiates Beate's refusal by modifying his own use of *Schwäbisch* to, sporadically employed fragments "*kleine sachen manchmal fetzen so*" ("small things sometimes snippets like") (li. c10–c11). He does not give up his claim to a *Schwäbisch* identity easily. Simultaneously with the beginning of Bruno's utterance the interviewer tries to initiate what could be an objection or a question (li. 12). Due to the simultaneity this is probably directed at Beate's utterance. However, the interviewer does not finish the first word, "*aber*" ("but"), before he cuts himself short. Beate first tries to interrupt Bruno and fails (li. c13) but she succeeds in her second attempt (li. c14). Her interruption is a reaction to Bruno's negotiation of her rebuttal of his access to authentic *Schwäbisch*. She states that the fragments Bruno mentions hardly are recognizable as *Schwäbisch*. Again she constitutes herself as an expert and gatekeeper and restricts Bruno's access to *Schwäbisch*.

This short excerpt is another example of the ongoing negotiations and power relations between Beate and Bruno. The interviewer asks about Bruno's "*favoritdeutsch*" ("favorite German") and suggests *Schwäbisch* as a possibility. Bruno answers that he favors the way he speaks himself, but before he gets a chance to respond to the suggestion of *Schwäbisch*, Beate interferes and states that he does not speak *Schwäbisch*. For the rest of the excerpt, Bruno tries to answer both the interviewer and Beate, whereas Beate makes fun of his competence in *Schwäbisch*.

(4) Kackschwäbisch ([ST-02-INT])

d01	EX:	was ist dein favoritdeutsch
d02	Bruno:	das deutsch was ich spreche
d03	EX:	schwäbisch
d04	Beate:	der redet kein |>|schwäbisch
d05	Bruno:	|<|ja ma ich red|>|e kein schwäbisch direkt
d06		|>|aber ich glaube mein mein hochdeutsch mit schwäbisch #
d07		bisschen
d08	astian:	|<|(lacht)
d09	Beate:	|<|er redet kacke kackschwäbisch

(translation)

d01	EX:	what is your favourite german
d02	Bruno:	the german that i speak
d03	EX:	schwäbisch
d04	Beate:	he speaks no schwäbisch
d05	Bruno:	yeah well i don't speak schwäbisch as such
d06		but i think my my hochdeutsch with schwäbisch
d07	Bruno:	a little

d08	Bastian:	(laughs)
d09	Beate:	he speaks shitty shit-schwäbisch

The interviewer's question about Bruno's favorite way of speaking German (li. d01) presupposes that Bruno is aware of more than one German variety and that he prefers one of these to the rest. Bruno answers, quite logically, that it is his own way of speaking German (li. d02). This answer does apparently not satisfy the interviewer who then suggests *Schwäbisch* (li. d03). Here the interviewer draws on the fact that Bruno earlier in the interview has made an effort to establish an identity as a speaker of *Schwäbisch*. The interviewer's suggestion causes Beate to interfere and point out that Bruno does not speak any *Schwäbisch* (li. d04), and thereby constitute herself as a gatekeeper of the access to it. Bruno does accept this interference to a certain degree and admits that he does not speak *Schwäbisch* as such (li. d05). However, he does try to reclaim a limited access to *Schwäbisch* by suggesting that he uses a little *Schwäbisch* mixed with his *Hochdeutsch* (li. d06–d07). Bruno acknowledges Beate as a greater authority on the matter of *Schwäbisch* than himself, but does not agree completely with her. Instead he makes an effort to establish an identity as a speaker of mixed *Hochdeutsch* and *Schwäbisch*, with the former being the predominant of the two. In doing so he constitutes *Hochdeutsch* as his own way of speaking with *Schwäbisch* being available as small fragments mixed into it. During Bruno's utterance Bastian laughs as a reaction to Beate's utterance (li. d08). His support seems to encourage Beate and she starts making fun of Bruno's competence in *Schwäbisch* by labelling what he speaks as "*kack-schwäbisch*" ("shit-*schwäbisch*"). This utterance is rather insulting, even when uttered as jokey comment, which indicates that Beate's status as gatekeeper of *Schwäbisch* (and probably her social status in the group as such) has a rather extensive potential in terms of what she is allowed to do, in relation to Bruno at least.

5 What we speak

In the following the topic is *Schwäbisch* in Stuttgart. The participants report that authentic *Schwäbisch* is being spoken in Stuttgart, but that they themselves speak something else, which they make an effort to label.

(5) Neuschwäbisch ([05. 05. 2010-ST-02-INT])

e01	Bastian:	ja es ist so es gibt ein stuttgarter schwäbisch also
e02		|>|wirkliches schwäbisch was wirklich nach schwäbisch klingt

e03		aber es gibt auch das was wir reden eben |>|diesen
e04	Beate:	|<|ja eben ja |>|eben dieses
e05	Benjamin:	|<|ja
e06	Bruno:	|<|dieses|>| neu
e07		neuschwäbisch ich glaube dafür gibt es noch keine
e08		definition oder
e09	EX1:	|<|was ist
e10		das
e11	EX:	|>|neuschwäbisch cool das ist ein cooler ausdruck
e12	Beate:	|<|das ist auch so bisschen so ist auch so |>|eine
e13		jugendsprache |>|schon
e14	Bastian:	|<|ja es ist halt
e15		XXX
e16	Benjamin:	|<|jugendschwäbisch

(translation)

e01	Bastian:	yes there is kind of a stuttgart schwäbisch that is
e02		proper schwäbisch that really sounds schwäbisch
e03		but there is also that which we speak right this
e04	Beate:	yes right yes right this
e05	Benjamin:	yes
e06	Bruno:	this new/
e07		neuschwäbisch i think there is no
e08		definition yet or
e09	EX1:	what is
e10		that
e11	EX:	neuschwäbisch cool that is a cool term
e12	Beate:	that is also kind of is also kind of
e13		a youth language already
e14	Bastian:	yes it is just
e15		XXX
e16	Benjamin:	youth-schwäbisch

Bastian explains that there is a kind of Stuttgart *Schwäbisch* that really sounds *Schwäbisch* (li. e01–e02) and then there is what they speak themselves. He sets out to label the latter but does not get to finish because he is interrupted (li. e03). Bastian confirms that there are speakers in Stuttgart who uses what he categorizes as authentic *Schwäbisch* "*wirkliches schwäbisch also was wirklich nach schwäbisch klingt*" ("real *Schwäbisch* that really sounds *Schwäbisch*") (li. e02), and then distances himself and the other participants from these authentic *Schwäbisch* speakers. In line e04 Beate supports his utterance about the existence of a "*stuttgarter schwäbisch*" (li. e01) as she unsuccessfully tries to get the attention of the others. Benjamin also shows his support (li. e05), although it is unclear who and what he supports. Bruno interrupts Bastian and completes the labelling with "*neuschwäbisch*" ("new-*schwäbisch*") (li. e06–e07). The label itself, "*neuschwäbisch,*" shows Bruno's awareness of his own and his peers use of a "newer" or "younger" version of what Bastian defined as "*wirkliches schwäbisch*" (li. e02), which also indicates that those who employ authentic *Schwäbisch* have access to an older and perhaps more original version of it. Bruno goes on to point out that this label has not yet been defined (li. e07–e08). He does this with some uncertainty, though, shown by his use of "*ich glaube*" ('i think') (li. e07) and by his alternation of the utterance to a question by adding "*oder*" ('or') at the end of it (li. e08). This indicates that the labelling positions Bruno in a role (as expert) with which he is not entirely comfortable, and that the suggestion, "*neuschwäbisch,*" is entirely his creation and not generally recognized.

During Bruno's utterance the interviewer makes what seems to be a simultaneous start where he asks "*was ist das*" ("what is that") (li. e09–e10). This seems to be directed at Bastian's utterance. After Bruno finishes the interviewer shows his support by commenting on how thrilled he is by Bruno's label (li. e11). Simultaneously, Beate elaborates on the subject and brings the label "*jugendsprache*" ("youth language) in play (li. e12–e13). She does say "*eine jugendsprache*" ("a youth language"), which indicates that she considers it one of several varieties of youth language. It is not clear whether the utterance is in support of Bruno or meant as a competing label (and a finish to her initial efforts in li. e04). Either way, she uses modifiers and repeats herself, "*das ist auch so bisschen so ist auch so*" ("that is also kind of is also kind of ") (li. e12), before getting to the point, which indicates that she is a little uncertain. Bastian makes an overlap at the very end of her utterance with what seems to be a comment in support of either Bruno or Beate. The ending of the utterance is unintelligible (li. e14–e15) so there is no way of telling who. Benjamin overlaps both of them and suggests "*jugendschwäbisch*" ("youth-*schwäbisch*") (li. e16) which could be an effort to elaborate on or emulate and modify both the previous labels. Both Beate and Benjamin's suggested labels, "*jugendsprache*" and "*jugendschwäbisch,*" signal that the style being labelled is employed by young speakers exclusively. This categorizes the divide presented by Bastian at the beginning as an age divide between authentic *Schwäbisch* speakers (who are not young) and young people employing their own version of *Schwäbisch*, or as Bruno puts it "*neuschwäbisch*".

6 *Straßensprache* is not for everybody

Now I will turn to the second interview, which was recorded in Reutlingen. The excerpt below is the interaction following an argument between Franziska and Felicitas. The topic of their argument is whether a certain location is used by what the girls describe as one of the major gangs in Reutlingen (called *HKS*) or not. During the argument Franziska and Felicitas completely ignore the two other participants, Felix and Florian, as well as the interviewer, and they switch to a more aggressive (quarreling) style of language use, raising their voices and using a more staccato intonational pattern. For the sake of the analysis the excerpt is divided in two parts.

(6) *Straßensprache* part I: *Arschkarte* ([RE-06-INT])

f01	EX:	was war das
f02	Franziska:	wir haben gerade diskutiert
f03	EX:	nein aber was habt ihr |>|ge/ gesprochen
f04	Felicitas:	|<|strassensprache (laughs)
f05	Florian:	diskussionsschwä|>|bisch
f06	EX:	|<|was hab|>| was habt ihr ihrer meinung nach
f07		gesprochen
f08	???:	|<|XXX
f09	Florian:	(laughs) wir haben auch diskutiert (laughs)
f10	???:	|>|(laughs)
f11	Felix:	|<|so strassensprache so |>|redet man halt
f12	Florian:	|<|ja
f13	EX:	aber was ist strassenspra/ wo wie ist
f14		strassensprache ähm unter|>|schiedlich von kanaksprache
f15	Franziska:	|<|das ist keine strassen-sprache
f16	Florian:	ha doch |>|# strassensprache |>|ist das was man in der region
f17		so spricht
f18	Felix:	|<|XXX # ich habe nicht mal richtig zugehört
f19	Felicitas:	|<|äh_ das ganze hat alles damit
f20		zu tun wenn du in reutlingen jemand |>|sein willst oder hier
f21		leute hinter dir haben willst dann musst du dir

f22 respekt verschaffen wenn du diesen respekt nicht hast
f23 dann hast du einfach die arschkarte gezogen
f24 Florian: |<|XXX was hast du gesagt
f25 Felix: |>|ja

(translation)
f01 EX: what was that
f02 Franziska: we just had a discussion
f03 EX: no but what did you spe speak
f04 Felicitas: strassensprache (laughs)
f05 Florian: discussion-schwäbisch
f06 EX: what di what did you in your own opinion
f07 speak
f08 ???: XXX
f09 Florian: (laughs) we discussed as well (laughs)
f10 ???: (laughs)
f11 Felix: like strassensprache you just talk like that
f12 Florian: yes
f13 EX: but what is strassenspra where how is
f14 strassensprache ehm different from kanaksprache
f15 Franziska: that is no strassensprache
f16 Florian: of course it is strassensprache it is just what you speak in
f17 the area
f18 Felix: XXX i did not really pay attention
f19 Felicitas: eh_ the whole thing has everything
f20 to do if you in reutlingen want to be somebody or you here
f21 want to have people backing you then you have to earn
f22 yourself respect if you do not have this respect
f23 then you are just fucked
f24 Florian: XXX what did you say
f25 Felix: yes

After Franziska and Felicitas finish their argument the interviewer asks what they were speaking, and he makes an effort to get Felix and Florian's perspective on it as well. The four participants discuss whether the style should be labelled *Straßensprache* or not and Felicitas concludes this by giving a definition of what employing this style entails in Reutlingen. The interviewer then asks Felix and Florian directly if they speak *Straßensprache*. Florian initiates a hesitant answer but is interrupted by Felicitas who explains that they, Felix and Florian, are not the right people to ask that question, and she continues to explain why she and Franziska are better qualified to answer it.

The interviewer asks "*was war das*" ("what was that") (li. f01) referring to the preceding argument between Franziska and Felicitas. Felicitas answers that they just had an argument (li. f02), which compels the interviewer to reformulate and specify his question to explicate that he meant their way of speaking (li. f03). During this the interviewer is interrupted by Felicitas, who answers "*straßensprache*" and laughs (li. f04). Felicitas demonstrates that she understands the interviewer's intention of inquiring into the language use by labelling this *Straßensprache*. However, considering her subsequent laughter she could just be joking around with the label. Florian suggests "*discussionsschwäbisch*" ("argument-*schwäbisch*) (li. f05) to add to the amusement. The suggestion shows his awareness of the interview situation and the topics being discussed. He uses the situation to make a parody of the focus on metalinguistic awareness and the interviewer's efforts to elicit this awareness from the participants. The interviewer interrupts Florian to reiterate and explicate for the second time (li. f06–f07). Florian adds that he and Felix were also arguing and laughs (li. f09). This is directed at Franziska and Felicitas' argument. He makes fun of their obvious violation of the interview setting, stating that he and Felix were quarreling, which they were not. During this passage there is a lot of laughter from the participants which indicates that they find the situation amusing.

Felix "comes to the rescue" of the interviewer by answering his question seriously, stating that it was *Straßensprache*, and that is "just how you speak" (li. f11), and Florian confirms (li. f12). The interviewer then asks about the difference between *Straßensprache* and *Kanaksprache* (wog/foreigner language (derogatory)), but is interrupted by Franziska who reacts to Felix' utterance (li. f015) and claims that it was not *Straßensprache*. The fact that Franziska here claims that she and Felicitas were not speaking *Straßensprache* could indicate that she does not acknowledge Felix (and Florian) as qualified to make that call. Earlier she did not dispute it when Felicitas suggested the label *Straßensprache*, which either indicates her awareness of Felicitas making fun, or that she acknowledges Felicitas as qualified to label their language use in the argument. Florian contradicts Franziska and supports Felix' statement (li. f16–f17). Felix accepts Franziska's rebuttal of his statement and tries to save face under the pretense that he did not really pay attention to what was going on (li. f18). During Florian and Felix' utterances Felicitas starts her own utterance with a prolonged voiced hesitation, "*äh_*," (li. f19). She continues to explain (li. f19–f23) that it is all about ("*das ganze hat alles damit zu tun*" li. f19–f20) attaining status in Reutlingen and gaining respect, and that if you do not gain this respect you are "fucked" ("*dann hast du einfach die archkarte gezogen*" li. 23). Felicitas' account is an affirmation of her own status as an expert in the subject of *Straßensprache* and on how to gain access to the style. This indicates that her earlier suggestion of the label *Straßensprache* (li. f11) was not just a joke. In line f24 and f25 Florian and Felix are talking in the background during and after Felicitas' utterance.

(7) *Straßensprache* part II: *Nicht böse gemeint* ([RE-06-INT])

f26 EX: |<|das heisst ihr könnt auch strassensprache sprechen |>|#
f27 nicht nur schwäbisch
f28 Florian: |<|nja
f29 |>|# XXX
f30 Felicitas: |<|hh hh nicht schlecht verstehen aber ich/ von/ # nicht böse
f31 nicht böse gemeint |>|# aber von den beiden ist nicht eher
f32 so gemeint |>|weil
f33 Florian: |<|ich weiss
f34 EX: |<|warum warum nicht
f35 Felicitas: wir beide kennen kennen die zwei großen gangs hier #
f36 wir sin/ |>|haben freunde da drin #
f37 wir ha/ teilsweise haben wir uns beweisen müssen #
f38 also in der |>|hinsicht wie halt wie die als menschen
f39 drauf sind
f40 EX: |<|XXX
f41 Florian: |<|ja das XXX
f42 Felix: |<|hello kitty # hello kitty for life
f43 Florian: (laughs)

(translation)
f26 EX: that means you can speak strassensprache too
f27 not just schwäbisch
f28 Florian: well
f29 XXX
f30 Felicitas: hh hh do not understand this the wrong way but i of not to be

f31		rude not to be rude but you could not talk about
f32		these two because
f33	Florian:	i know
f34	EX:	why why not
f35	Felicitas:	the two of us we know the two major gangs here
f36		we ar have friends among them
f37		we ha to a certain degree we had to prove ourselves
f38		that is in the sense that just how they as persons
f39		are
f40	EX:	XXX
f41	Florian:	yes that XXX
f42	Felix:	hello kitty hello kitty for life
f43	Florian:	(laughs)

The interviewer begins the second part of the excerpt by asking Felix and Florian directly if they have access *Straßensprache* (li. f26–f27), knowing that this implicates that they have earned the "status and respect" referred to by Felicitas. Florian answers with a vague "*nja*" ('well') followed by something incomprehensible (li. f28–f29). This indicates that he may be able to employ linguistic resources categorized as *Straßensprache* but that he is somewhat reluctant to lay claim to the style. His utterance is almost immediately interrupted by Felicitas, who makes two voiced breaths before uttering the first word (li. f30). These two voiced breaths could be an effort to get the attention of her interlocutors, but it could also indicate that she is about to say something which has to be very carefully formulated. After this Felicitas emphasizes that what she is about to say should not be taken in the wrong way. She makes a couple of false starts with self interruptions and finishes with emphasizing that she does not mean to be rude to Felix and Florian (li. f30–f31). Florian makes a confirming overlap "*ich weiss*" ("i know") (li. f33) which affirms Felicitas' status as gatekeeper and expert on the subject. Felicitas explains that when it comes to *Straßensprache* Felix and Florian are not the right people to speak to (li. f31–f32). Here, Felicitas positions herself as gatekeeper and categorizes Felix and Florian as out-group with respect to *Straßensprache*. Her cautious approach indicates that her utterance is a potential face-threat to Felix and Florian. The interviewer interrupts to ask why Felix and Florian are not the right people to speak to (li. f34). Felicitas responds and explains that she and Franziska know the two major gangs in Reutlingen, that they have friends among the members, and that they have had to prove themselves to the gang members (li. f35–f39). Felicitas categorizes herself and Franziska as in-group with respect to *Straßensprache*, and she legitimizes the claim with their connections to two gangs in Reutlingen. This legitimation indicates that the gangs are users and gatekeepers of *Straßensprache* and that access can only be gained by association or affiliation with them. Florian deals with the out-group categorization (and the inherent face-loss) by making fun of Franziska and Felicitas' in-group status and their claimed gang association with the utterance "*hello kitty hello kitty for life*" (li. f42), to which Felix responds with laughter (li. f43). The comical aspect of Florian's utterances lies in the reference to the Hello Kitty character[3]. This is generally perceived as something cute and girlish, which is the exact opposite of the tough front Felicitas is trying to constitute.

7 Conclusion

In the analyzed excerpt the adolescents clearly reveal a keen metalinguistic awareness. They demonstrate that they are fully capable of differentiating between sets of linguistic resources associated with different ways of speaking. For instance, Bastian exhibits a rather detailed linguistic knowledge in his account of the use of palatalized [s] as an example of a linguistic resource traditionally associated with *Schwäbisch*. Later in the same interview the participants show their awareness of the linguistic situation in Stuttgart as Bastian gives an example of age grading in relation to speaking *Schwäbisch* in Stuttgart. They make it clear that they themselves are not part of the group of authentic *Schwäbisch* speakers there. Instead they make an effort to enregister (Agha 2007) a label for what they consider their own way of speaking (with "*neuschwäbisch*," "*jugendsprache*," and "*jugendschwäbisch*" suggested). This show that they are well aware that certain stereotypic variety labels are associated with particular groups, and they are rather creative in their attempt at enregistering an in-group label of their own.

The power relationships between Beate and Bruno and Franziska-Felicitas and Felix-Florian demonstrate the importance of local identity work, social categorizations, and group membership in interaction. The analysis shows how the adolescents' social relations and their language use are connected, and that this connection is a complex matter of access, authenticity, and power relations. Having lived in a traditionally *Schwäbisch* speaking area with a Swabian father does not necessarily entitle a claim to be an authentic speaker of *Schwäbisch*, as it is the case with Bruno. Whereas Beate achieves the status as gatekeeper of *Schwäbisch*, and the power to restrict Bruno's access to it, even though she distances herself from the label. This status and power is granted to her by the other partici-

3 Hello Kitty is a fictional character, with roots in the Kawaii segment of the Japanese popular culture that has become a world-wide brand.

pants in the interview without her having to negotiate or justify these positions. She simply starts wielding the power procured through the expert status.

Franziska and Felicitas show how group membership and association with particular groups can warrant a status as gatekeeper of the set of linguistic resources associated with the group. Their status as experts and gatekeeper is justified by their affiliation with the proper groups, or gangs, and the set of linguistic resources in question is labelled *Straßensprache*. The two girls make it abundantly clear that the other two participants, Felix and Florian, have no access or claim to it – simply because they do not mix with the right people.

The adolescents demonstrate how they have several different social styles at their disposal and how they negotiate access to these through group memberships that go beyond mere geography. The access to the sets of linguistic resources associated with *Schwäbisch* and *Straßensprache* in these excerpts may be negotiated in a local setting, in a part of Germany where *Schwäbisch* is considered to be spoken, but the social categories being negotiated here are on a social level that eludes the mere geographical demarcation of the variety *Schwäbisch*. The power struggles between Beate and Bruno and between Franziska-Felicitas and Felix-Florian and the discussion of "real *Schwäbisch*" vs. "*neuschwäbisch*" does not involve actual references to or examples of particular linguistic resources associated with "*Straßensprache*" or "*Schwäbisch*", as is the case with Bastian's reference to the palatalization of [s]. These instances negotiate variation that is determined by social rather than geographical factors. This supports the use of the term "social style" as a more comprehensive alternative to "variety" (Auer 2012) in the analysis.

Appendix A

Transcription key

|>|	=	the speaker is interrupted
|<|	=	the interruption
#	=	a pause
/	=	a self interruption
s-t	=	the spelling out or mentioning of singular letters
XXX	=	incomprehensible
_	=	prolonging
()	=	commentary
[IPA]	=	transcribed with IPA

Médéric Gasquet-Cyrus

Chapter 9
The accents of Marseille: Perceptions and linguistic change

1 Introduction

Arguing for the combination of urban sociolinguistics, perceptual dialectology and ethnographic approaches, this text focuses on how the perceptions of city dwellers are shaping the sociolinguistic landscape and reflecting the ongoing linguistic change and its social meaning in Marseille, a major town in the South of France, on the Mediterranean coast.

1.1 Representations and place: A theoretical framework in French sociolinguistics

Folk linguistics in general, and Perceptual Dialectology (PD) in particular are not prevalent fields in French linguistics, not to say they are almost absent. The recent evolution of (a part of) French (Romance) traditional dialectology towards sociolinguistics does not include PD as a branch used to describe regional varieties. With its own history and particularities, French sociolinguistics, mainly based on discourse analysis and qualitative or ethnographic approaches, is hardly variationist, in spite of increasing variationist works in recent years, concerning among others the undergoing leveling in Northern France (Armstrong and Pooley 2010). Nevertheless, the point of view of non-linguists is seriously taken into account since the pioneering works in social psychology in the 1970s (and the 1990s in the francophone field) and the development of a solid body of studies, both theoretical and empirical, concerning language attitudes (see Preston 1999c and Montgomery, this volume), commonly labeled in the francophone tradition as *représentations* (Boyer 1990; Castellotti and Moore 2002; Petitjean 2008). These are articulated with the *pratiques* ('usages') of the speakers through their discourse about language and languages. Different from the metalinguistic descriptions of linguists or grammarians, this *epilinguistic* activity (i.e., comments about languages from non-experts; see Canut 2000a) is often studied through processes of categorization, linked to issues of identity (Blanchet 2004). For instance, language labeling has produced several major works in French sociolinguistics (Tabouret-Keller 1997; Canut 2000b and 2001), concerning different topics and "varieties" such as: "youth language" (see below) or regional languages.

This now common combination of *pratiques* and *représentations* (see Calvet 1999 for instance) leads to new analyses of notions such as norm(s), linguistic loyalty or linguistic insecurity, and allows going above the concept of variation too often based on a supposed homogeneous and stable language (Gadet 2012). This kind of sociolinguistics goes beyond the *langue/parole* dichotomy and integrates the speakers as actors with specific socio-cultural and psychological features in the description of language practices. In this sense, Gadet (2008) advocates a "speaker linguistics" while for Branca-Rosoff (1996: 104), the time has come to "take seriously not only the scientific theories of the 'professionals of language'"[1], but also "the 'epilinguistic' representations of ordinary speakers," in order to investigate what she calls "the social efficiency of representations".

This framework is used to investigate variations and multilingualism in France. Although it is still known for its strong monolingual ideology (Spolsky 2004), France has begun to explore systematically its own linguistic diversity. As pointed out by historical sociolinguistics (Lodge 1996), the French language has been constructed to become the homogeneous language of a powerful kingdom (from Middle Ages and Renaissance to King Louis XIV), then the language of the French people, then of the French republic (from the Revolution of 1789 to the 21st century). The ideology of the Norm, embedded in a highly standardized language, is still deep-rooted in the minds, and the so-called "standard French" still associated with "Parisian French," that is to say with the political locus of power. Nevertheless, sociolinguistic studies have been led since the 1980s and regional or immigrant languages, accents and dialects (Gadet 2007; Durand, Bernard and Lyche 2009), phonetic leveling (Armstrong and Pooley 2010; Durand et al. 2013) constitute contemporary objects of research. The variationist paradigm has been used to describe social and regional variations in French, as in the "Contemporary French Phonology Project"[2]; however, according to one leader of this project (Durand 2008: 164), there is a need of "large sociolinguistic inquiries allowing to draw a precise cartography of the systems in competition in the South of France, particularly in the great cities". In fact, recent works concerning variations of French in France have been conducted, especially in cities like Rouen (Bulot 2002), Grenoble, (Trimaille 2005),

---

1 All translations from French to English are personal.

2 www.projet-pfc.net

Toulouse (LTVI Project[3]), Orléans (ESLO Project)[4], Paris (Multicultural Paris French Project[5]) and Marseille (Binisti et Gasquet-Cyrus 2003). A part of these works is focused on youth language in multiethnic neighborhoods and suburbs, and has tackled with the problem of labelling these practices: *langue des banlieues, langage des cités, parlers jeunes* (Auzanneau and Juillard 2012; Fagyal 2005). This need to investigate the specific contacts and variations occurring in urban centers or suburban areas has increased the theoretical reflections on space and language.

1.2 Urban sociolinguistics

Indeed, the question of place is also taken into account in French sociolinguistics. While space has been for a long time reduced to no more than a simple backdrop, it is now a real object of sociolinguistic investigation, with recent focuses on interface between social mobility and identity. Space is now fully recognized as something more than just a physical place, but the combination of material features and socio-psychological elements, including social and "lived spaces". As Eckert (2010: 163) says, "space is central to the study of social indexicality, because space is imbued with social meaning, and the distribution of linguistic forms across space is key to the construction of meaning in variation". Thus, the semiotic meaning of space is more or less integrated in the descriptions of contextualised practices, especially in urban areas. But once again, in France, PD is not mobilized to investigate the relationships between space, language and identity. The researches on these themes are conducted in the field called *urban sociolinguistics* (Calvet 1994; Bulot 2002, 2009; Gasquet-Cyrus 2002), which brings together sociolinguists, (social) geographers and sociologists to describe discourse about language, speakers and linguistic landscapes.

The main aim of urban sociolinguistics is to analyze the impact of urbanization on language(s), its effects on discourse and the way it shapes divisions among individuals and groups. Among several tasks (study of variation, emerging norms, discrimination), urban sociolinguistics analyzes space as a semiotised space through the notion of *territory*. For Bulot (2009: 68), "the sociolinguistic territorialization is thus the way how, in discourse, the speakers of a town appropriate and establish a hierarchy among places depending of

3 www.projet-pac.net/index.php/frfr
4 http://eslo.huma-num.fr
5 www.mle-mpf.bbk.ac.uk/Home.html

the ways of speaking (real or stereotyped) associated with themselves or others to make sense of their own identity".

In accordance with this theoretical background, this paper investigates how the perception of accents, through epilinguistic discourses, shape significant sociolinguistic territories embedded in the more general frame of an urban identity made of both identification (to a large community: Marseille and the "Marseillais") and differentiation (among sub-groups in the city). The study of this sociolinguistic landscape should lead to a better understanding of the ongoing linguistic change and its social meaning as it is impacted by recent changes in the urban environment.

Usually, PD focuses on dialect areas (see Jeon and Cukor-Avila about South Korea, this volume) or on an accent associated with a particular town (see the case of Nottingham studied by Barber, this volume); an exception in this volume is Lonergan's study of variations in Dublin. In this paper, I try to investigate not only the accent of Marseille in the national landscape, but also its subtle variations as they are perceived by the city dwellers. The way to describe properly these "accents" is a big issue combining theoretical and methodological aspects: a description of linguistics features (provided by dialectological or phonetic approaches) should be linked with the study of how these accents are perceived and with an analysis of the social meaning of these variations and perceptions (through the broader paradigm of the sociology of language). The present study will focus on perceptions only, but as stated by Cramer (this volume), "studies that combine the methods of perceptual dialectology with traditional dialectology can provide important insights into the realities of linguistic variation". Such a fundamental research program was already claimed by Preston (1999b) whose ambition was, among others, to establish connections between PD, sociolinguistics and the study of language attitudes. Following these insights, the object of this research is to show the impact of urban changes on sociolinguistic perceptions and to provide explanations about what these perceptions tell us about society.

1.3 Methodology

From what has been said above, it follows that the present study is at the crossroads between perceptual dialectology and urban sociolinguistics even if its methods come mainly from the latter. That's why from a methodological point of view, as it is the case in the French tradition, draw-a-map-tasks are rarely used, one exception being the pioneering work by Spini (2013). Instead of mapping, the way the speakers associate places with specific accents or ways of speaking is investigated through their epilinguistic discourses and categorizations. Blommaert (2005) sustains that discourse is "language-in-action" where issues of power are debated, and as Darquennes (2013: 110) writes: "in democratic societies conflicts in general, and language conflicts in particular, are 'battled out' on a discursive level".

The data has been mainly collected through a non-stop multifaceted research from 1999 to the present days in various social settings of the town, and through a general ethnographic approach, including fieldwork and participant observation, direct observations, free conversations, utterances caught on the fly, recordings of ecological productions, more or less guided sociolinguistic interviews, written questionnaires. The corpus also includes literature (novels, short stories), media, songs, and in general, all traces of discourses with epilinguistic comments.

2 The (singular) accent of Marseille

2.1 In the national landscape

France's second largest town with about one million of inhabitants, Marseille is well known for what is considered the most famous accent in the country, the so-called *accent marseillais*, given that the category "accent" corresponds actually – for the speakers – to various "ways of speaking" (Hymes 1974) including elocution, prosody, phonetic and phonological features, words, phrases, etc. The prototype of the French southern accent ("accent méridional' or "du Midi"; see Durand 2008; Gasquet-Cyrus 2012a), this accent is often described stereotypically in the media and in popular discourse. Qualified as "sunny," "singing" or "funny" by both city dwellers and strangers, it seems to remind "holidays" or the pleasant way of life in the touristic region of Provence. In this sense, it generates positive attitudes, as in Kuiper's study (2005) who shows that the variety of French spoken in Provence is ranked by all the judges as "the most pleasing for the ear". De Robillard (2003) places Southern French at the top of a hierarchy of "enchanted Frenchs" contrasting to the "tasteless" perception of the French norm. This accent is invested positively by many city dwellers and is brandished as an emblem by numerous artists (singers, writers, movie makers). The specific local speech (labeled "parler marseillais"), recognized and known far beyond the boundaries of the city, generates dozens of books, in particular novels written with local features but also a huge amount of glossaries and dictionaries, both professional or amateur, which exhibit the differences between the local language and the other varieties of (spoken) French, especially Parisian (Gasquet-Cyrus 2012c).

But this representation is quite ambiguous: a "funny" accent can quickly turn into a "non serious accent". In spite of the positive attitudes associated with this accent, many people confess that it is sometimes lived as a handicap. Gasquet-Cyrus (2012b) has collected evidences of this feeling through the interviews of more than a hundred of speakers from South of France who often present themselves as victims of a certain kind of discrimination, especially in jobs where language plays a crucial role (journalists, actors, singers, teachers, etc.). In this sense, the title of the essay by Weck (2008) illustrates the general feeling of linguistic insecurity more or less openly shared among many Southerners: *Fucking Accent! How Southerners live their language.*

This local accent is considered as very different from the "Parisian accent," often mocked by local humorists but above all in everyday conversations (Gasquet-Cyrus 2004a). The rivalry between Marseille and Paris is the heritage of a long history, mainly based on a common province-capital rivalry, but it has increased recently since the cultural development of Marseille in the 1990s, and the emergence of a rivalry between two major football clubs of France, l'Olympique de Marseille and Paris Saint-Germain.

As seen above, at a general (national) level, the accent of Marseille can be associated with the Southern accent. But at a regional scale, it is more often considered as a very peculiar one: as pointed out by Montgomery (this volume), proximity conditions the way in which speakers perceive dialect variations.

2.2 In the regional landscape

A first perception draws an opposition between the accent of Marseille vs a "Provençal" accent. The latter is often described as "singing," "sunny" and "soft" while the former can be qualified as "more pronounced," "harsh" or "vulgar".

(1) M: me I have the Provençal accent not the accent of Marseille
E[6]: what is the difference between the accent of Marseille and the Provençal accent?
M: the accent of Marseille it's an emphasis of the Provençal accent (Spini 2013)
{M= Female, born in Lyon, living in Marseille since 37 years}

All the samples are originally in French, only the translation is given here.

6 In Spini's corpus, 'E' is the interviewer (Spini herself).

As Martel (1999: 25) points out about Marseille's speech seen by the Provençal people: "its general perception is quite negative". This distinction lies on the opposition between rural and urban worlds (see Cramer, this volume, for a study of the rural vs. urban perception through the case of Louisville). According to the ideology spread by a certain regionalist folklore, the values of the "authentic" Provence would be degraded in the mixed urban center of Marseille, not to talk about its bad reputation. Trying to compare the accents of three majors towns of the department, Le Douaron (1983) concluded that each of them was associated with clichés related to the sociology of the town: Aix-en-Provence, the "bourgeoise" town, would have a normative pronunciation; the speech of Arles would be "manly" and the accent of Marseille more "populaire" ('working-class').

A particular opposition between Aix and Marseille is to notice. Viewed from the working-class town of Marseille, the rich city of Aix is perceived as a town whose speakers try to conform themselves to a standard or a Parisian pronunciation, erasing their (supposed) Provençal accent or switching it for "un accent pointu" (a 'sharp' accent) –which means locally posh or stilted. That explains why they are considered are traitors regarding the regional identity. This opposition appears in discourses collected among people from Marseille.

(2) the bourgeois of Marseille always tend to talk rather um "Paris 16th district" or aixois this accent thus trying to eliminate the Marseille accent but there are bourgeois living in Marseille who are proud of their accent
{Male, 39, born in Marseille, formative of social agents for the municipality; La Plaine}

There is also this riddle destined to mock the people from Aix: "What's an Aixois? A Marseillais who sees himself as a Parisian".

The accent of Marseille emerges as specific in the Provençal background, but the differences are above all sketched in discourse, when the people can hardly list effective linguistic differences (an aspect underlined by Lonergan, this volume). Thus, while it is common to distinguish the accents of Marseille and Toulon (a city distant from less than 70 km with a strong local identity), it's almost impossible to find a single feature showing effective or significant linguistic differences.

3 The (plural) accents of Marseille

As seen above, at a national or even regional scale, people outside the city recognize or evoke one singular "accent of Marseille" (as provincials talk of one "Parisian accent" while the linguistic diversity of the capital is evident). But the social complexity of Marseille, its strong identity, a long history resulting from the presence of immigrants, and the configuration of its legendary 111 districts has had an impact on the sociolinguistic landscape (Gasquet-Cyrus 2004b). Indeed, there is not one single "accent of Marseille" but a continuum of varieties, with many differences as to how they are perceived among the population. As in every speech community – a complex system of individual and collective variations – people from Marseille can rely on a set of resources belonging to a general repertoire "from Marseille".

Previous studies (Binisti and Gasquet-Cyrus 2003, 2007) have outlined the existence of at least three accents distinguished by the population interviewed or in everyday talk: (i) the (so-called) "real" or "traditional" accent of Marseille; (ii) the accent "of the upper middle class" or the accent of "la *bourgeoisie* marseillaise"; and (iii) the accent "of northern neighborhoods" (*quartiers Nord*). This tripartition is well summarized in the words of a photographer who was asked to identify the different accents of Marseille:

(3) it depends o::f the places it depends of the districts you've / the accent of the fishwife it's not the same accent than the accent of the suburbs or the accent of the 8th (district) where the guy wants um absolutely struggle against his accent
{Male, 45, photographer, born in Marseille; Saint-Barnabé}

The same division in three major accents appears, in a more recent collect, in this interview made by Spini (2013) with a woman coming from Lyon, but living in Marseille since 37 years:

(4) E: what do you recognize as accents in Marseille?
M: um me I recognize them difficult- / oh yes yes yes I recognize / the accent of the Old Port yes / the accent of the Northern neighborhoods yes / the accent of boulevard Périer yes (Spini 2013)
{M= Female, born in Lyon, living in Marseille since 37 years}

The data show that each accent is more or less accurately associated with specific areas or districts of the town, which are not objective places, but social areas functioning as symbolic *territories*, in the sense of urban sociolinguistics. Of course, one could "hear" such accents everywhere in Marseille, but the epilinguistic discourses draw of a sociolinguistic landscape in which the territories have a specific social significance, analyzed in the sections below.

3.1 The "real" accent of Marseille

The very first accent mentioned by people when they are asked to say whether there are several accents in Marseille is what is often called "the real" or "the pure accent of Marseille". Of course, the adjectives "real" or "pure" are not scientific descriptions, but the words coined by most of the speakers to qualify what is considered as the traditional accent of the town, the "authentic" way of speaking of the legitimate people from Marseille, which benefits from the most positive perceptions: "singing," "funny," "picturesque," "tasty," etc. In the hierarchy of the local accents, it can be considered as the endogenous norm (Manessy 1997), i.e., the conscious representation of the current use admitted by all (local) speakers as neutral, "normal" in the local context. This accent – corresponding to the prototypical accent of Marseille at a more general level – is somehow the reference accent, from which the other accents are evaluated.

Older speakers are considered as having this accent when, as seen below, young people are supposed to have a new accent, in contrast nay in opposition with this "authentic" one. Associated with a golden age of the traditional Marseille, it is evocative of the 1930s and especially the cinematographic imagery of Marcel Pagnol, a famous writer and filmmaker from Marseille. People often label it as "Pagnol's accent," that is to say the accent one can hear in those films, especially "the Marseille trilogy": *Marius* (1929), *Fanny* (1931) and *César* (1936). The famous actors of the movies, still considered as emblematic figures of Marseille, are often convoked as linguistic models when speakers talk about the "Raimu's" (1883–1946) or the "Fernandel's" (1903–1971) accent". That's why, in the following comment, the accent is weirdly associated with a norm situated "fifty years ago":

(5) on the Old Port all that well there are still people who still live downtown and who who really have the accent of of fifty years ago
{Male, 40, "Pied noir"; Saint-Giniez}

This accent is also linked to masculine local stereotypes, especially the fisherman. Corresponding to this imagery are the places of the town which evoke this universe, especially the seaside. That's why this accent is mainly associated with places like the Old Port (*le Vieux Port*, at the very center of the town) or *Le Panier* (on the north bank of the port).

(6) F: it's true that for instance in the inner city / at the Panier on the Port people have more accent (Spini 2013)
{Female, about 40, born in Marseille, living in the 9th district, municipal agent}

Other places are the peripheral neighborhoods almost seen as specific "villages" like *L'Estaque* (North) or *Les Goudes* (South), districts owning their own small port and having kept a certain old-fashioned way of life. With a particularly strong identity, L'Estaque is the neighborhood most frequently associated with this accent.

(7) I think about L'Estaque because well me I have my aunts who live at L'Estaque it's true that they have the deep accent of Marseille so um (laughs) [...] but it's really THE ACCENT OF MARSEILLE oh gàrri![7] can you see?
{Male, 45, photographer, born in Marseille; Saint-Barnabé}

(8) the accent of Marseille of the old Marseillais of L'Estaque / of the old fisherman
{Male, about 45, living in Vitrolles, a small town close to Marseille; he know Marseille since about twenty years}

As belonging to the golden age of Marseille, this first accent is said being less and less used, nay, for some people, close to the extinction. Following that idea, the "new" accents (in fact, the "different" accents, or just the variations) can be perceived as distortions with respect to the original. As being the reference positively evaluated by most of the Marseillais, this accent concentrates all the good perceptions and implies that the other accents could be negatively evaluated.

3.2 The "bourgeoisie" accent

For most of the Marseillais, the upper middle-class accent is not considered as the prestige norm. While some "bourgeois" defend a less "vulgar" or at least "lighter" local accent that the marked "working class" accent, this accent "bourgeois" is mainly perceived as "posh," "stilted," "artificial" and the speakers are easy targets of mockeries (see Gasquet-Cyrus 2013 for a historical explanation); a young Marseillais interviewed qualified it as

(9) the accent of "ceux qui s(e) la pètent" ('them who swank').

In the following quote extracted from a novel by a local writer, this accent is qualified with pejorative adjectives:

7 Local interjection, which can be translated like "Hey guy!".

(10) Alexandre has increasingly the small accent of the bourgeoisie of Marseille, an awkward, dummy, mannered accent (André Remacle, *La calanque de Maldormé*, Les Editeurs Réunis, 1971, p. 103)

Another mark of this accent can be found in the text of a play in which a local hairdresser imitates her customers from different social backgrounds. The singularity of this accent appears through an eye-dialect process showing the attenuation of the nasal vowels (ɔ̃ for ɑ̃) of a rich woman of the Southern districts, a pronunciation contrasted with the local one (as this effect is hard to realize in an English translation, the French text is presented with boldface for involved vowels).

(11) Josy, vous pouvez me pr**on**dre assez rapidem**on**t ? Juste pour un coup de peigne ? [...] Ma belle-mère était d'une vieille famille prov**on**çale et ses par**on**ts avaient cette vieille baraque [...] Oh, vous savez, c'est imm**on**se, mais il n'y a aucun confort[8].

Very often, this "light" accent is negatively perceived as a mark of treason from people who would want to erase their (supposed) "natural" accent, denying at the same time their cultural roots. This attitude is associated with a (supposed) will to imitate a Parisian norm; a 23 years old student nurse evokes, through a mocker imaginary reported speech,

(12) a sharp speech like "I'm Marseillais but I try to do the Parisian"

The bourgeois of Marseille are historically suspected to be ashamed of their belonging to a working-class city; for this photographer already quoted (see (3)),

(13) it is the accent of the Marseillais who does not want to show that he has some accent

Let's notice that it could be impossible to mask totally the local accent which could be back, sooner or later, to reveal the (supposed) "real" identity of the renegade. The two trends concerning the behavior of the local bourgeoisie are

8 "Josy, could you take me quickly? Just for a combing?... [...] My mother-in-law was from an old Provençal family and her parents had this old house... Oh, you know, it's vast, but there is no comfort at all". (Anne-Marie Ponsot, *Josy Coiffure*, Marseille: Les Editions du Fioupélan, 2010, p. 56)

summarized in the words of this formative (see (2)) talking about the accent of the mayor, Jean-Claude Gaudin:

(14) yes when I speak of Gaudin for instance it's really the accent of Mazargues / an accent of Marseille / which remains an accent / nevertheless working-class but rather um pop- well working-class but bourgeois / bourgeois and member of the city of Marseille / the bourgeois of Marseille always tend to talk rather um "Paris 16th district" or aixois this accent thus trying to eliminate the Marseille accent but there are bourgeois living in Marseille who are proud of their accent but it's still an accent um that I call the "Mazargues accent"

This stiff accent is sometimes labeled (negatively) "accent jambon" ('ham accent'),[9] but the etymology is lost. It is also associated with the accent of Aix-en-Provence which is, as seen above, negatively perceived as an accent close to the Parisian pronunciation. In the sociolinguistic interviews, this accent is located in the (supposed) rich districts of the town, mainly constituted by residential areas, with lower and upper middle classes: the southern neighborhoods in general, and some districts in particular (*Saint-Giniez, Mazargues, Périer*), some bourgeois avenues (*le Prado, le boulevard Michelet* or *la Corniche* on the seaside) or some emblematic streets like the "rue Paradis," considered as the most bourgeois, with luxury brands and chic stores; one can talk of "*l'accent de la rue Paradis*" ('the accent of Paradis street'). In the following extract (see sample (1)), the speaker labels this accent with the name of a rich district (*boulevard Périer*) and to show its differences with the "authentic" accent of Marseille compares it to a "Breton accent" and imitates it with the same process seen with the example of the hairdresser (see above).

(15) M: the accent of the boulevard Périer yes
E: what is the accent of the boulevard Périer?
M: it's an a- it's an accent of *Bretons* an accent of *Bretons* [...]
M: {laughter} yes it's the it's the accent if one can say of the good society of Marseille
E: and how do they speak / can you imitate them?
M: they talk "like that" {phonetic imitation}
E: "like that"?

9 A radio report about the 'accent jambon' has been broadcasted; http://arteradio.com/son/616508/l_accent_jambon.

M: yes / no they don't talk like that they talk / how could I say? / they they they don't emphasize as the Marseillais do / and they try to speak without without too much of of of accent / without too many inequalities um {gesture showing a plane line of prosody, without emphasis} (Spini 2013)

Thus, the accent of the middle-class and upper-middle class is considered as local, i.e., recognized among the local accents, but the bourgeois associated with this accent are designated as traitors or, at least, as people who are ashamed of their local identity. The prestige of the districts, avenues and houses owned by the middle-class and upper-middle class residents is not transferred on their way of speaking. They constitute a group at the margins: they belong to the community, but their behavior does not reflect a sufficient investment in the local identity.

3.3 The Northern districts accent

In Marseille, the opposition between south and north is clear-cut: when, as seen above, the rich residential areas and avenues are mainly found in the South, the North concentrates most of the "cites," "grands ensembles" or "quartiers difficiles," more than 2/3 of the public housing of the town, a very high rate of unemployed persons, especially among young people, with most of the immigrants or of people with "foreign origins".

Several generations of immigrants have integrated Marseille before World War II: Italians, Armenians, Spaniards, Greeks, etc. But many arrived during large-scale migrations after World War II and the process of decolonization in the 1960s. Many Marseillais have their direct or indirect roots in Maghreb with a lot of Algerians or people with Algerian origin (including Berbers), and a waste community of "Pieds-Noirs" (former French colonial settlers of Algeria repatriated in 1962), but there are also Moroccans, Tunisians, Comorians (about 80.000 for a bit more than 800.000 inhabitants), Africans of various Sub-Saharan countries, Turks, Gypsies, people from Eastern Europe, Asians (from Vietnam, Cambodia, China, etc.), and a majority of them is concentrated in the northern neighborhoods of Marseille which have become (and remain), as pointed out by sociologists and historians, a segregated territory. It's on this soil that has emerged in the late 1980s the "quartiers Nord" accent (now QN).

(16) well the Northern districts they have an an accent that you can recognize easily
{Male, about 40, newspaper vendor; La Plaine}

Assigned to the young of the northern districts, especially of Maghreb origins, this accent is commonly stigmatized when some deny its legitimacy as being an authentic local accent, as seen in the distinction made by this man living in a small town close to Marseille:

(17) the accent the accent of / of the strain Marseillais the [...] and the accent of / the Marseille accent of the Marseillais from the North / the Northern neighborhoods
{see sample (8)}

The French phrase "de souche" ('strain') suggests that the "real" accent should be the prerogative of people born in Marseille; implicitly, the residents of the northern districts are not supposed to be "from Marseille," in spite of the fact that many of them are born in Marseille and can objectively be considered as "Marseillais". In the previous epilinguistic discourses, the QN accent is presented as "not really" local, or less authentic that the "real" accent. In the example below, this accent, presented as a "mix," is considered as "not quite" "marseillais".

(18) yeah it's rather the Northern neighborhoods accent / it's it's a mix it's a mix say it's a mix of accents not quite / the accent of Marseille huh
{see (3)}

It's interesting to notice that during the inquiries, when people are asked how many accents they perceive in Marseille, the QN accent appears often first, but quickly, the speakers try to reduce its legitimacy or to stigmatize it. Anyway, the simple fact that it is mentioned shows clearly that it is undoubtedly a part of the sociolinguistic landscape.

(19) me I talk about the accent of Marseille [...] that of Pagnol one huh not the one of the Northern neighborhoods
{Male, about 50, Parisian origins, florist; Saint-Giniez}

In this discourse, the man makes a clear distinction between two accents: "that of Pagnol" and the QN; but even if his preference goes to the first one, the two are considered as accents of Marseille.

For many speakers, the QN accent "sounds Arabic". A doctor of a town at about 50 kilometers from Marseille labels it "marseillais beur[10]," when in a

10 In French, the adjective *beur* (from the reverse *arabe* > *be-ra* > *beur*) qualifies any French person of Magheb origins.

national journal a local writer evokes a contemporary speech he calls "pseudobeur" (*Le Point*, 23/11/06, p. 4). A journalist (without any Arab origin) told me:

(20) "if I listen to my son without seeing him it's as if I was speaking to an Arab".
{Male, about 50, born in Marseille, journalist; phone conversation}

Indeed, some phonetic and lexical features of the QN accent may have some Arabic influences (Binisti and Gasquet-Cyrus 2001), but it also shares a lot of features with the common accent of Marseille (regional variations in French). Yet, the "foreign" elements are often underlined, as if it was important to maximize the differences instead of maximizing the similarities.

(21) it's not the real accent of Marseille [...] it has Arabic tones
{Male, 28, student in Literature, Aix-en-Provence University}

While some evocations are quite neutral, some are more pejorative.

(22) oh the Northern neighborhoods oh fuck! so / it's different / it takes after to to the Algerian accent / this this this this has deviated like that
{see sample (1)}

The word chosen indicates that the accent is not perceived as "just a variation," but as a "deviation," i.e., something not correct, not pure.

(23) E: and what this accent looks like in the Northern neighborhoods?
L12: well the accent of the accent of the Maghreb people Black and so on
L11: yes but they have the accent of Marseille too
L12: oh not very much (Spini 2013)
{Male, about 80, Pied-noir }

The phrase "and so on" ("et compagnie" in French) indicates a certain contempt for these populations. Very often, this accent is qualified as "vulgar".

(24) LA: do you hear in the bus and in the subway / these girls / who yell who who who:: and who have an accent: so vulgar
E: how is this accent.
LA: an accent so vulgar / a vulgar accent an accent half half Arabic half French [...] at the beginning maybe it's for um fun then it re-

mains huh / it remains / because me in our block there's a girl who is decent the parents are decent and all you hear her speak you say "but wait where did she get this accent?" / it has an effect / but yeah she has bad acquaintances (Spini 2013)
{Female, about 80, Pied-noir, lives in the neighborhood of Dromel}

We know how language attitudes indicate are more than judgments about language, but judgments about the people associated with some linguistic features. For some people, having a QN accent indicates some kind of handicap.

(25) E: in the 9th (district) how do they speak?
LA: ah no / it's better yet / better than in the 14th (district) huh / in the 14th / no there is too much of of I think there is too much of carelessness and I think that the parents are at fault (Spini 2013)
{see sample (24)}

The adjective "better" used by this speaker is indicative of a certain level associated with the accents: while the accent of the 9th district, the one of the 14th, in the Northern neighborhoods, is associated with a non-correct way of speaking. From a judgment concerning language, the epilinguistic discourse slides to a moral level.

(26) I know that the "cité accent" upsets me / the fact that they cannot anymore / that this- that they are not able to say a "d" / it's a "tch" / it's not "ce que je te dis" it's "ce que je te" [tʃi] it sounds to me like an infirmity / that is to say um / you are someone locked in a language you see? (Spini 2013)
{Male, 55, sociologist, Pied noir origins, living in Marseille since he is 18; lived in La Plaine, Noailles then Longchamp}

In media discourses, Marseille is often presented as a model of integration, a multicultural city where all the cultures would live in peace; a comparison is frequently established with the urban riots occurring from times to times in the suburbs of Paris or Lyon, while on this point, Marseille is quite calm (Heyer 2011). But one must not forget that many other aspects of violence exist here, especially in the Northern districts well known for drug trafficking which impacts the lives of hundreds of young men and girls. More, Marseille establishes records of votes for the extreme right party in national and local elections. During the municipal elections of 2014, the candidate of the extreme right party has obtained 23.16 % of the votes, quite far ahead the socialist candidate

(20.77 %). Thus, a latent (or sometimes overt) xenophobia (nay racism) can be expressed in the epilinguistic discourses. In the following quotes about the QN accent, the precautions of the speakers cannot mask such judgments.

(27) it's a Marseille accent but:: tainted of:: "exoticism" for some of them (laughs)
{see sample (3)}

The "exoticism" evoked by this speaker is a way for him to avoid talking about "Arabs" or "Black" people, but there is no doubt at all about this allusion.

Thus, the QN accent generates many negative judgments and many of them because it is associated with Arabic and, beyond that, people of Arabic origin and Muslims. Knowing the issue of immigration in France – especially for Maghreb people – it's easy to understand why many speakers focus on or exaggerate the differences of this accent compared to a reference accent: if this accent is not "authentic," than the persons who have it are not legitimate, and by this way symbolically excluded of an idealized or "sanitized" community.

However, this accent has spread above the northern area (and even outside Marseille) and among oldest speakers of different cultural and social settings (including middle-class) (Gasquet-Cyrus 2009; Binisti and Gasquet-Cyrus 2007). Comparative works show that some similar linguistic features can be found in varieties spoken by young speakers and young adults in the suburbs of other French cities (in particular the affrication of the coronal consonants /t/ and /d/ ahead of front close vowels like /i/ and /y/; the voiceless and constrictive articulation of /r/; a stress on the penultimate syllable). This convergence in divergence (Jamin, Trimaille and Gasquet-Cyrus 2006) is one aspect among others of the ongoing linguistic change in contemporary French.

In Marseille, this diffusion suggests a certain covert prestige and indicates that this accent could have already become an endogenous norm for the local and regional young people, whatever their origins. Still being depreciated by a large part of the population, the QN accent plays an increasing role in the sociolinguistic landscape and can be considered as the emblematic speech of a generation, a counter-norm in a socially tensed context. This accent should become the future reference accent of Marseille, and this linguistic change has to be accurately observed. But the fact that many discourses still locate it in the northern segregated area indicates why this accent should be stigmatized too: the ones who are supposed to have it are not judged "authentic" enough to be members of the community, and they are associated with stigmatized territories of the city. Through epilinguistic comments emerge serious social nay ethnic tensions.

4 Gentrification of space and the potential emergence of a new accent

4.1 Gentrification: A spatial, social, and political phenomenon

A global and massive phenomenon affecting numerous cities in the world especially in Europe (Van Criekingen 2008), gentrification can be defined as "the processes through which inner city, formerly working-class areas are profoundly transformed by the arrival of new residents belonging to middle and upper classes" (Fijalkow and Préteceille 2006: 5). This spatial process involves an increasing mobility, but also the displacement of populations: the increase of the rents due to the presence of these richer newcomers entails the fact that some working-class families have to find cheaper housing, when they are not forced to leave or outright expelled by developers. This has happened with the "Euroméditerranée" program: the old buildings of the seaside have been converted into modern luxury flats for rich tourists and new inhabitants, and the way some previous residents have been more or less forced to leave is discussed in numerous papers (Jourdan 2008; Collective 2009; Ruffin 2010). As a reaction to these policies several groups campaign against the negative effects of gentrification and what is considered to be neo-liberal and inhumane processes, denying the working-class character of Marseille and discriminating against its poorer residents.

Indeed, since the 1990s increasing changes aim at creating a new urban identity for Marseille. The policy of the municipality, supported by private funds, is to raise the city to international status within the Euro-Mediterranean space, which meant giving it a new, "cleaner" identity in line with the standards of rich European cities. However, the transformations of urban space linked to this political project also change the social background of the city and the relationships that existed between different categories of residents. These urban policies took the form of a "reconquest" of inner-city areas, especially through the process of gentrification. As it was planned, these urban policies have attracted a new category of immigrants whose socio-cultural and socio-economic profiles unbalance the "traditional" sociolinguistic patterns: the "neo-Marseillais" (NM).

4.2 The "neo-Marseillais"

According to the data of AGAM (2009), some 100.000 people settled in Marseille between 2001 and 2006, of which 44 % came from France (outside Provence), 32 % from the Ile-de-France and 18 % from major towns such as Lyon, Lille, Toulouse or Bordeaux. This mobility has been facilitated by the development in 2001 of a TGV railway line linking Paris and Marseille in three hours. 23.7 % of these new residents were white-collar workers compared to a local average of only 14.7 %. 1 NM in 5 is a homeowner, and students represent 16 % of the NM. They settled predominantly in three inner-city districts (the Southern, more affluent ones, rather than in the northern more "working-class" neighborhoods – for example in some southern districts there are twice as many NM managers as Marseille-born managers). Such statistics indicate that the typical profile of a NM is that of a young person (less than 40 years), a student or a professional of a relatively high socio-economic status.

The sociolinguistic aspects of gentrification are studied since a few years only and a theoretical frame and deeper analyses concerning Marseille can be found elsewhere (Trimaille and Gasquet-Cyrus 2013; Gasquet-Cyrus 2015). I will focus here on the perception of a potential emerging accent caused by gentrification.

4.3 The perception of the NM

Since 2000, the NM have gained such visibility that they are now identified as a specific group; thus the label "neo-Marseillais" can be found in various media (websites, guides, humorous texts). Because of their cultural capital and their qualifications, the NM play an increasingly prominent part in the city's life and they are concretely involved in Marseille's social networks and attendees at many fashionable events. In some domains (culture, sports), they can be considered as new leaders, and some are nationally known figures, such as writers and sportsmen. For Peraldi and Samson (2006: 116), "Marseille attracts that minority that make or break fashionable places, the avant-garde that builds the reputation of Parisian districts" and that shapes the likes of the "trendy gentry".

But these newcomers who claim to become "marseillais" are not always welcome, and their arrival has instilled a fear of a "sanitized" identity. In this sense, the comments of some locals about their language are all ways to sketch boundaries between "us" and "them". Here is how in an interview a local author describes the new town:

(28) On the one hand, we have the steel and glass façades of Euromed with a population of tie-sporting sushi eaters with the most refined elocution, and on the other, a downgraded city centre with the Marseille of the unemployed and the immigrants (quoted in Collective 2009: 49).

In a town where the local French is brandished as an emblem and seen as a counter-norm, a "refined elocution" is more the butt of jokes rather than a norm to imitate (see above). Indeed, the (supposed) speech of the NM is frequently the target of ironic comments or mockery as it is seen as posh, stilted or associated with the Parisian accent. An increasing corpus of such comments can be found on different media.

(29) It is true that you can hear people speaking more and more with a Northern accent in Marseille [...] with the arrival of the parigots[11], the Marseille accent is going down the pan [...] soon Marseille will just be a city like any other[12]

The description of the NM as "invaders" or "foreigners" is documented via some of the comments submitted following an article entitled "Has Marseille lost its accent?" in a regional online daily:

(30) These *parigots* who are invading the city
You hear more *parigot* in the streets than Marseillais
Everywhere you look in the city centre you find newcomers who see themselves as Marseillais, when they've only been here a few years, and who come and lecture us, Marseille isn't Marseille anymore
The néo Marseillais should have to take accent lessons[13]

The arrival of the NM has also an impact on linguistic productions, for instance accommodation or the adoption of linguistic phonetic features or some emblematic words (Trimaille and Gasquet-Cyrus 2013). But some locals criticize this kind of appropriation, commenting that the NM use the words artificially, without the "appropriate" accent:

(31) This new way of speaking is for the newly-settled *bobos*[14] who think they're "being" Marseillais and for tourists[15]

---

11 French pejorative word for "Parisian".

12 www.laprovence.com/article/a-la-une/marseille-a-t-elle-perdu-son-accent (accessed June 2012)

13 www.laprovence.com/article/a-la-une/marseille-a-t-elle-perdu-son-accent (accessed June 2012)

14 *Bobo* is an abbreviation of "bourgeois bohème", an expression referring to middle-class people living (what they consider to be) a bohemian lifestyle.

15 www.laprovence.com/article/a-la-une/le-nouveau-dico-du-parler-marseillais> (accessed June 2012)

The traditional opposition between "us" (the Marseillais) and "them" (the "outsiders," and more precisely "the Parisians") was quite simple, because the "Parisians" and more generally the tourists were really from outside the city. The new thing is that the NM are residents and involved in the city's life; they want to share the local identity, and even to preserve it, so as to become "real marseillais" and to index themselves as (new) locals. This of course questions the very nature of what it means to be a "marseillais" and the language practices and norms that correspond to this affiliation.

The linguistic behavior of the gentrifiers may play a crucial in the sociolinguistic landscape. They should adopt local phonetic or lexical features, but they are above all responsible for the spread of standard or leveled supra-local forms within the speech of the original residents (Trimaille and Gasquet-Cyrus 2013). There is a need for ethnographic and interactional approaches to analyze precisely the styling practices of newcomers and local in face-to-face encounters. Studying the negotiation of different social styles among different speakers (see a similar perspective in Svenstrup, this volume), Spini (2013) has begun to investigate these micro-level behaviors and their social meaning. The developments or urban studies should articulate macro-analyses including quantitative data (demographic movements, social mobility, territories, boundaries) and micro-analyses with more qualitative data (interactions, speech productions, language attitudes, language behavior).

4.4 Perception and conflict

The description of the sociolinguistic landscape of Marseille does not consist only in the description of perceptual dialect areas and of linguistic features allocated to different groups. It shows clearly that the perceptions and the way they are presented in the epilinguistic discourses illustrate some severe social or ethnic tensions between groups in contact. That's why the frame of conflict linguistics is interesting at this point (Gasquet-Cyrus 2015). Indeed, "language contact [...] exists only between speakers and language communities, not between languages" (Nelde 1997: 292; see also Nelde 2006). The "linguistic" conflict conceal (or convey) symbolic but also more concrete social conflicts, dealing with issues of power, territory and access to resources between dominant and dominated groups. Then, language can become an important issue when it is brandished as an emblem by a group: "language very often develops into a manifest significant symbol of social conflict in minority settings, even when it may not be the direct cause of the conflict" (Darquennes 2010: 344).

In the case of Marseille, the conflict does not concern "languages," but varieties or styles within a single language, French. Around these varieties and styles, the groups are mobilizing different practices and representations which can become antagonistic: "the study of language conflict certainly in the EU is more to be seen as the study of differences of opinion or incompatibilities between two or more opinions over language in society" (Darquennes 2013: 110). The tension between the "traditional" working-class character of Marseille and the new and "cleaner" identity planned by the municipality and supported by the gentrifiers corresponds to an opposition between the (supposed) local accent and a "new" (standardized or leveled) way of speaking. The sociolinguistics of mobility proposed by Blommaert offers a theoretical frame to investigate how the access to languages/varieties/styles/registers can support the domination of one group on another.

> Movement of people across space is therefore never a move across empty spaces. The spaces are always someone's space, and they are filled with norms, expectations, conceptions of what counts as proper and normal (indexical) language use and what does not count as such. [...] Big and small differences in language use locate the speaker in particular indexical and ascriptive categories (related to identity and role). (Blommaert 2010: 6)

As Darquennes (2013: 112) explains, "[t]he surface structure of the language conflict (i.e., the debate about language or language use) functions as a sort of "lightning rod" in that it distracts the attention from the real social motivations behind the conflict (i.e., the "deep structure" of the conflict)". That's why, within the frame of critical sociolinguistics (Bochmann 2003; Blommaert 2005; Boutet and Heller 2007), it is important to take seriously the questions of productions but also of perceptions of these ways of speaking in these particular urban settings, for what they have to teach about the political conflicts in these moving societies.

> [...] plurilingual conflict in Europe and more precisely in urban societies show clearly that the attempts of the dominant group to block the social mobility are the ones at the origin of linguistic conflicts. [...] it's precisely in these modern and urban societies that conflicts are the result of normative impositions from the most powerful and generally majority group. This one requires a linguistic adaptation and thus programs conflicts with speakers who refuse to adapt themselves. (Nelde 2006 : 2419)

The urban changes in Marseille address in a sensitive manner the question of adaptation, and above all the following one: "who is supposed to adapt his/her language use to whom?" (Darquennes 2013: 115).

5 Conclusion

The study of language representations through epilinguistic discourses collected by different means indicates the (perceived) existence of at least three accents in the sociolinguistic landscape of Marseille. One of them is put up as the endogenous norm nay the reference, while the others attract in general pejorative attitudes: the accent of the bourgeoisie is considered as "posh" and "stilted" and the QN accent "vulgar". Beyond the (supposed) linguistic differences, one can feel the social or ethnic tensions and the complexity of a city with its stratification and its dynamics. The maximisation of the differences between the accents is a way to sketch boundaries between groups – and to deny to some people the legitimacy of being members of the urban community. The methods of PD – especially draw-a-map tasks – would be welcome to complete this approach and to draw with more precision the boundaries of the dialect areas identified by the speakers.

The ongoing linguistic change concerns not only the spread of the QN accent which could become a new endogenous norm for younger speakers, but also the emergence of a potential "new" accent – or at least practices leading to levelling and the decrease of local features. The urban changes since the 1990s, and the massive arrival of a new category of people since 2000, the NM, have another social impact with its sociolinguistic correlates. The gentrification process put on the front of the stage a new group with new practices and the (linguistic and social) contact between these groups will sketch the lines of the new society.[16]

At this point, one can understand how important is the study not only of productions, but also of linguistic perceptions. As Preston (this volume) points out, PD is "a major way of solving the evaluation and social embedding problems of language variation and change". In this sense, it meets the main ambition of (general) sociolinguistics, which is not only the understanding of sociolinguistic behavior of individuals and societies, but also the understanding of social, political and anthropological processes through the study of language productions and perceptions. As Blommaert (2010: xiv) says: "We possess (if we use it well) a descriptive apparatus of unparalleled precision, capable of reading infinitely big features of society from infinitely small details of communication behaviour". These small details of communication can be collected by different means and within different theoretical backgrounds, among them discourse analysis, urban sociolinguistics and PD. Following Hymes (1974), I consider that "what's important for linguistics goes far beyond linguistics itself". The present study is just a modest attempt to go through and beyond linguistic productions and perceptions in order to understand these changing accents, these changing cities, and this changing world.

---

16 For further developments, see M. Gasquet-Cyrus & J.-M. Géa (eds.), Marseille: gentrification et ségrégation, *Langage et société*, in press (to be published in 2016).

Part III: **The relationship between perception and “reality”**

Chris Montgomery

Chapter 10
Perceptual prominence of city-based dialect areas in Great Britain

1 Perceptual Dialectology in Great Britain

Perceptual dialectology (PD) research in Great Britain does not have as long a heritage as that undertaken elsewhere in the world. Although non-linguists' perceptions were viewed as integral to understanding findings from dialect surveys undertaken in The Netherlands (Rensink [1955] 1999) and Japan (Sibata [1959] 1999), the Survey of English Dialects (SED), for example, (Orton and Dieth 1962) did not include any focus on perceptual aspects of the linguistic landscape. Such a focus solely on production data is not surprising, especially given the extensive historical interest in production (Bonaparte 1876; Ellis 1889; Wright 1905) in English dialects.

Such historical interest in production notwithstanding, non-linguists' perceptions have not been neglected. In fact, they have been extensively studied as part of language attitudes research in Great Britain (Giles 1970; Giles and Bourhis 1975; Giles and Powesland 1975; Edwards 1982; Ryan and Giles 1982; Bishop, Coupland, and Garrett 2005; Coupland and Bishop 2007). Perceptual dialectological approaches, which can be seen as bridging the "gap" between language attitudes research and social dialectological studies, have only relatively recently been applied to the linguistic situation in Great Britain. The wealth of data from production studies over the last 140 years, along with interest in perceptual dialectology since the early 1990s, means that stability and change in both production and perception can be examined in this chapter.

The first perceptual dialectology study to take place in Great Britain was undertaken by Inoue in the years 1989–1990 (Inoue 1999b) and further studies have taken place in Wales (Williams, Garrett, and Coupland 1996), Northern England (Montgomery 2007), the North-East of England (Pearce 2009), the Scottish-English border (Montgomery 2012), and the East Midlands (Braber 2014; this volume). In addition to this, the draw-a-map component (see Chapter 2 in this volume) of the PD method has been introduced to some studies using the SuRE[1] (Llamas 1999) methodology (e.g., Burbano-Elizondo 2006).

---

1 The 'Survey of Regional English' (SuRE) methodology was new package of fieldwork materials designed for use in a new nationwide survey of English in Great Britain. Designed with a

The studies performed by Pearce (2009) and Burbano-Elizondo (2006) are examples of small-scale studies which were designed to answer specific research questions about the nature of perception in a relatively small area (the North-East region in Pearce (2009) or the area immediately north and south of Sunderland in Burbano-Elizondo (2006)). The studies undertaken by Inoue (1999b), Williams, Garret and Coupland (1996), and Montgomery (2007; 2012) were national studies, and concern themselves with broader areas. Thus, these studies focus on perception in Wales (Williams, Garrett, and Coupland 1996), England and Wales (Inoue 1999b; Montgomery 2007), and England, Wales, and Scotland (Montgomery 2012). This chapter uses data and findings from Inoue (1999b) and Montgomery (2007; Montgomery 2012) to examine perception in relation to areal classifications of English variation in England and Wales.

2 Areal classifications of English dialects

Despite Bonaparte (1876: 577) claiming that "no real exact delimitation of English Dialects is [...] possible" due to the "fragmented" state of English dialects, it did not stop him trying to do just this. He therefore offered the "first classification of English dialects" (Viereck 1992: 20), which although overlooked by many subsequent dialectologists (see the discussion in Viereck 1992: 18–20), provides a systematic insight into the state of dialect variation in England. Bonaparte's classification was constructed on the basis of his own observations, as well as the translation of Ellis's "comparative specimen" (Ihalainen 1994: 233; Maguire 2012: 87) by a number of "native authors", and examples of dialect writing (Bonaparte 1876: 579). Based on 14 phonetic and grammatical features (Bonaparte 1876: 575–6), this classification resulted in 13 dialect areas.

Bonaparte's contribution to Ellis' work is paid homage to by Ellis (1889: 5), not least in the supply of some of comparative specimens used to construct his classification. However, despite Bonaparte's efforts, Ellis' work was "the first survey of English dialects based on rich, systematically collected evidence" (Ihalainen 1994: 232). His methods comprised three distinct elements (summarised in Maguire 2012: 86–7): a dialect test (a short paragraph containing 94 keywords along with a small number other variables), a comparative specimen (a longer paragraph containing a greater number of words and morphosyntac-

---

core set of tools with the possibility of adding additional components, the method has not been used as extensively as envisaged, although studies by Llamas (2000), Asprey (2006), and Burbano-Elizondo (2006) demonstrate data gathered using the method.

tic variables), and a classified word list (containing 971 words). On the basis of five phonological characteristics present in his data, Ellis was able to divide the dialects of England into five major divisions: Southern, Western, Eastern, Midland and Northern (with an additional division including the "lowland" area of Scotland) (Ihalainen 1994: 234), further subdivided into 23 smaller dialect districts.

After the early interest in dialect classification, it was over 50 years before the Survey of English dialects (SED) provided the opportunity to classify English dialects based on more numerous and systematically collected data than that relied upon by either Bonaparte or Ellis. The SED, which was undertaken in the 1950s and 1960s, surveyed 313 locations using a questionnaire of 1300 items. Data from the SED has been published in the Basic Materials (e.g., Orton and Barry 1969), in dictionary form (Upton, Parry, and Widdowson 1994), and as dialect atlases (Orton, Sanderson, and Widdowson 1978; Kolb et al. 1979; Upton, Sanderson, and Widdowson 1987; Viereck and Ramisch 1991; Viereck and Ramisch 1997). Although many of the maps in earlier atlases were criticized due to lack of interpretation (McDavid 1983), the maps I discuss below provide attempts to show classifications of dialect areas using various methods.

Viereck's three divisions of the country were based on data from the SED and used lexical (Viereck 1986a), phonetic and morphological (Viereck 1986b) data. In contrast to isoglossic maps, Viereck's dialect divisions of the country are based on heteroglosses. The heteroglosses drawn by Viereck used heavier lines to indicate more similarity, with lighter lines indicating less similarity. No hierarchy of features was assumed, which means that "the relative importance of the boundaries rests solely on the number of heterophones [or other types of heteroglosses] in any given bundle" (Viereck 1986b: 241). Viereck found that for the three different linguistic levels that he examined, there was "substantial agreement" with regard to the dialect boundaries found.

Trudgill (1999) offers two English dialect classifications, of traditional and modern dialects, both based on data from the SED. In common with the dialect divisions of Ellis, both of Trudgill's classifications are based on phonetic features, and have primary and secondary divisions. Trudgill's traditional dialects are described as those "spoken by a probably shrinking minority ... [found] in the more remote and rural areas of the country" (Trudgill 1999: 5). Based on eight phonetic features (each exhibiting an "older" and "newer" form), Trudgill's map of traditional dialect areas has three primary divisions (North, Central, and South) and 19 other divisions.

Trudgill claims that his modern dialects are now most widely spoken in England and that they "are particularly associated with those areas of the country from which Standard English originally came – the southeast of England; [and] places that have become English-speaking only relatively recently, such as the Scottish Highlands, much of Wales, and western Cornwall" (Trudgill 1990: 5). Created on the basis of seven phonetic features (each with two claimed variant realisations) Trudgill's modern dialect divisions has two primary divisions (North and South) and 17 secondary areas.

The areal classifications I have discussed above are all based on an isoglossic or heteroglossic approach to the areal delimitation of dialect areas. Maps generated in this fashion provide the impression (but only the impression) of dialect boundaries. Such "boundaries" most closely mirror the types of map perceptual dialectologists ask their respondents to draw (see Section 4 below). As well as these types of classification, there have also been great strides made in the quantification of dialect difference via the methods of diatectometry (see Nerbonne and Kretzschmar 2013). It is often the case that maps produced as a result of these methods are quite different to those that perceptual dialectologists wish to access (i.e those with boundary lines). Despite these differing methods some dialectometrically produced maps do make use of lines to separate "dialect areas".

Goebl's (1997) dendrographic classification of the SED data collected in the Computer Developed Linguistic Atlases of English (CLAE 1 and 2) is based on the MICRO-data from CLAE (Goebl 1997: 23), and as such contains a composite analysis of 321 maps from the atlases. The resultant classification displays 10 generally contiguous areas. Goebl and Schiltz (1997) offer another reading of the SED data, arrived at by different dialectometry calculations on the data from the two volumes of CLAE. After the creation of maps that seem "somewhat disappointing" (Goebl and Schiltz 1997: 15) due to a lack of clear boundaries between areas, Goebl and Schiltz overlay these maps and attempt to identify such boundaries, with some success. They found seven dialect areas using this approach, with a strong East-West dimension.

Maps based primarily on the concept of dialect boundaries, such as those discussed above, lend themselves to being examined together using Geographical Information Systems (GIS). Described further in Chapter 2 of this volume, a GIS is a system that integrates hardware, software, and data for "capturing, managing, analysing, and displaying all forms of geographically referenced information" (ESRI 2011). All lines from the classifications discussed above were traced and overlaid using ArcGIS 10.1. A line density analysis (Montgomery and Stoeckle 2013: 71) was subsequently applied to the lines in order to ascertain the extent of agreement amongst them (cf. Goebl and Schiltz 1997). The resulting line density map can be seen in Figure 10.1, with darker shading indicating greater agreement over line placement.

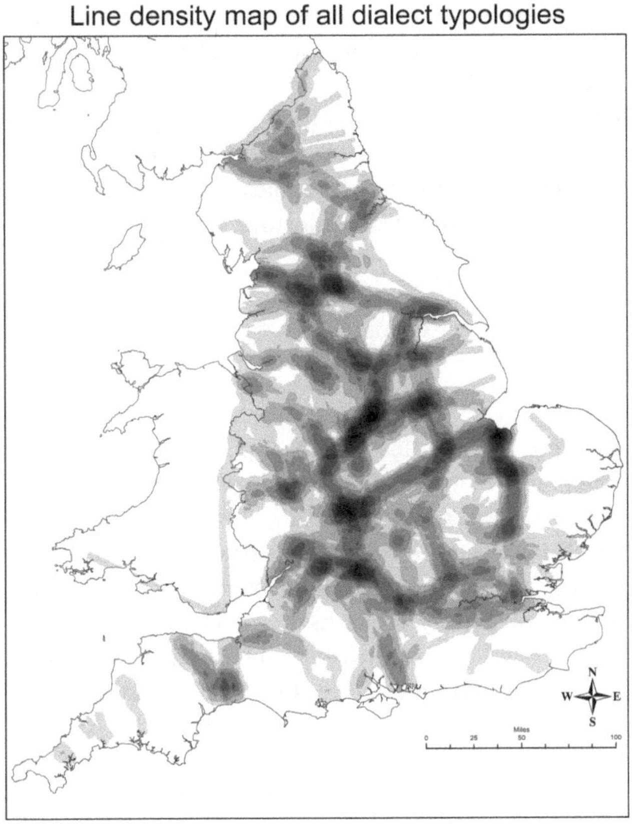

Fig. 10.1: All dialect area boundaries, after line density analysis[2].

As can be seen in the map, there is a good deal of disagreement about the boundaries of dialect areas in England. This is not too surprising, and again illustrates Trudgill's point that "there are no really sharp dialect boundaries in England" (Trudgill 1990: 6). However, there are clearly some areas in the map where lines converge. Therefore, there seems to be a good amount of agreement about a North East area, distinct from North and East Yorkshire to the

2 Figure 10.1–Figure 10.6 are based on data provided through EDINA UKBORDERS with the support of the ESRC and JISC and uses boundary material which is derived from National Imagery and Mapping Agency VMAP geospatial data along with boundary material which is copyright of the Crown. Further data is provided with the support of the ESRC and JISC and uses boundary material which is copyright of the Crown and the ED-Line Consortium. Location information is © Crown Copyright/database right 2011. An Ordnance Survey/EDINA supplied service.

South. The often mentioned Humber-Ribble line appears on the map, as does a darker line from the Humber to the approximate location of the River Severn. A further point of agreement is seen in around the Wash, with the line in this case termination towards the direction of Bristol. There is clear agreement that East Anglia should be separate from the area of the country to the West. Moving further South and West, there are a number of lines of agreement which indicate the separation of the South Western peninsular from other parts of England.

3 Inoue's study

Inoue (1999b) used two methods to investigate perceptions of English dialect variation in the country. The first, which asked respondents to engage in listening tasks to assess "dialect image" (Inoue 1999b: 161), is not relevant to this chapter. Inoue's second task involved the use of the draw-a-map task (see Chapter 2, this volume) to gain access to respondents' mental maps of dialect variation in the UK, and this will be my focus here. Inoue's fieldwork took place in 1989, whilst he was based at the University of Essex. Respondents were University students, and although Inoue states that "the collection of student data was carried out by staff members of several universities in Great Britain" (1999b: 163), it is clear that most respondents were mainly from Essex University, where Inoue was based for the duration of the fieldwork. Inoue's base map was not blank, nor minimally detailed, as seen in other studies. Instead, it had county boundaries marked. It appears from the map that Inoue produces as a demonstration of the way in which his respondents completed his draw-a-map task (Inoue 1999b: 164), that these county lines provided a guide to map completion but were not used to "trace" area boundaries.

Inoue did however use county boundaries in order to process his data and compile a composite map. Data were processed according to whether a county had been included within a particular area (Inoue 1999b: 164). Of course, this was a departure from methods used in studies both contemporaneous to Inoue's (e.g., Preston 1989) and subsequent to it (e.g., Long and Yim 2002) in which the actual positions of lines are used to assemble composite maps and composite maps are created at the pixel level. Unsurprisingly, Inoue's composite map (Figure 10.2) of the perception of dialect variation in the UK follows county boundary lines.

Although Inoue claims his map "does not show hierarchical relationships" (1999b: 168), the use of heavy and light lines (reproduced in Figure 10.2 from Inoue's original) does seem to indicate primary and secondary divisions. In-

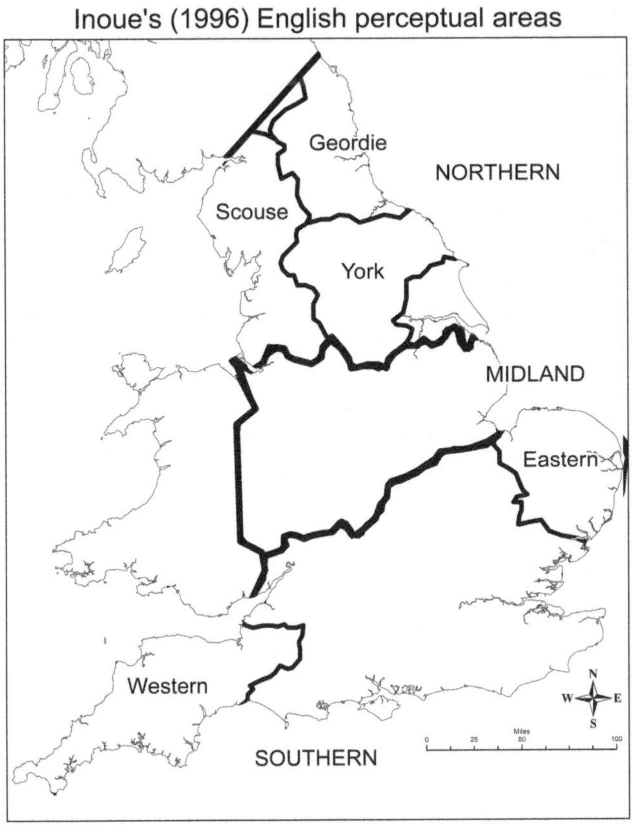

Fig. 10.2: Inoue's perceptual dialect areas in England (redrawn from Inoue (1999b: 167)).

oue's "primary" divisions (seemingly selected according to the area that they cover) are *Northern*, *Midland* and *Southern*. Secondary delimitations of these areas can then be found based on cities[3] in the Northern area and geographical extremes in the Southern division.

Inoue compared his composite map with other maps produced as a result of production studies (Brook 1963; Viereck 1986a; Trudgill 1990), and found that none of these corresponded to his composite map (Inoue 1999b: 167). This was a finding that Inoue found discouraging, although reappraisal of Inoue's comparisons with production-based map appears to indicate that there was more agreement than Inoue had initially thought between his map and others

3 *Scouse* is the colloquial name for the Liverpool dialect; *Geordie* is the colloquial name for the dialect of Newcastle upon Tyne.

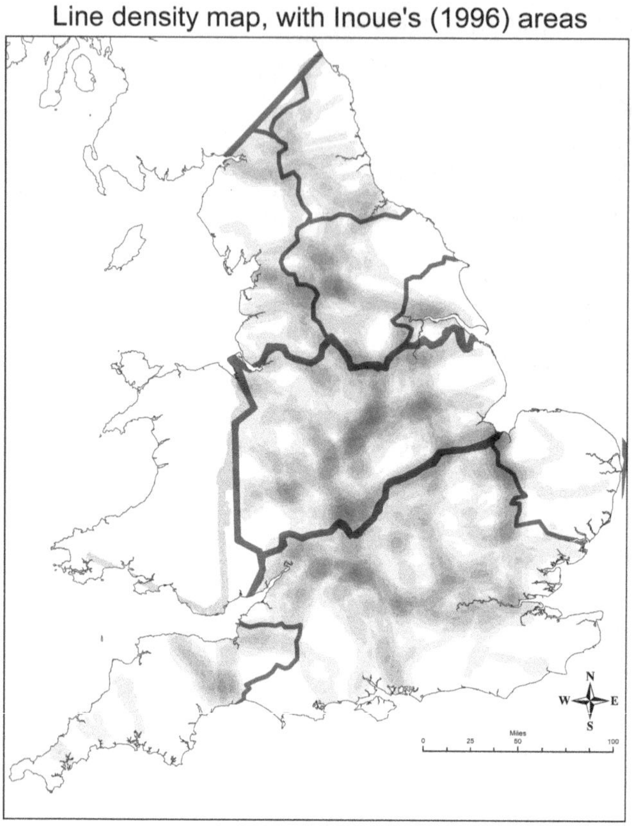

Fig. 10.3: Inoue's perceptual dialect areas, superimposed on the line density map based on historical typologies.

based on production data. Figure 10.3 shows the line density analysis from Figure 10.1 with Inoue's composite lines superimposed.

Although it must be stated that Inoue's perceptual map in no way captures the extent of variation suggested by the historical typologies of variation in England, in some ways (especially at the sites of the "primary" divisions), Inoue's map is similar to the line density map. In the far north, there is a coincidence of lines separating east from west, and although the line density map clearly indicates a distinction between the North and South of Inoue's *Scouse* area, the eastern boundary of the region is around the same as the line density analysis. Inoue's border between *Northern* and *Midland* is similar to the line density analysis, as is the boundary between the *Midland* and *Southern* area. In the *Southern* area, the *Eastern* and *Western* areas both correspond to areas suggested by the line density analysis. Overall, although much of the variation indicated by the line density analysis is not present in Inoue's map, it is closer to the production-based maps that Inoue had previously thought.

Although this means that Inoue should perhaps not have been as disheartened as he was after his perceptual study, it also has implications for the ability of his respondents to accurately perceive dialect variation at the time of the research. If it is the case that Inoue's respondents were perceiving dialect variation with some degree of accuracy in 1989, it allows their data to be compared with that gathered more recently in order to examine changing perception of the linguistic landscape of England.

4 Montgomery's studies and findings

The author of this chapter undertook two studies to investigate perceptions of national dialect variation from locations in the North of England (Study 1), and the Scottish-English border (Study 2). These studies, fuller findings from which are reported in Montgomery (2007, 2011a, 2012, 2014), were carried out in the eight locations indicated in Figure 10.4. Full details of the studies can be found in Table 10.1.

Respondents in each survey location were given a minimally detailed map[4] which contained information relating to country borders, along with city location dots[5] and were asked to add data to the map with a pen or pencil by responding to the following questions:

1. Label the nine well-known cites marked with a dot on the map.
2. Do you think that there is a north-south language divide in the country? If so, draw a line where you think this is.
3. Draw lines on the map where you think there are regional speech (dialect) areas.
4. Label the different areas that you have drawn on the map.
5. What do you think of the areas you've just drawn? How might you recognise people from these areas? Write some of these thoughts on the map if you have time.

4 It should be noted that the map given to respondents in Study 1 only displayed England and Wales, whereas the map used in Study 2 displayed the countries that constitute Great Britain (England, Scotland and Wales).

5 The decision to include these city location dots was made to ensure that respondents' geographical knowledge was consistent and the spatial data they provided could be treated as accurate (cf. Preston 1993: 335). Further details relating to this methodological decision can be found in Montgomery (2007).

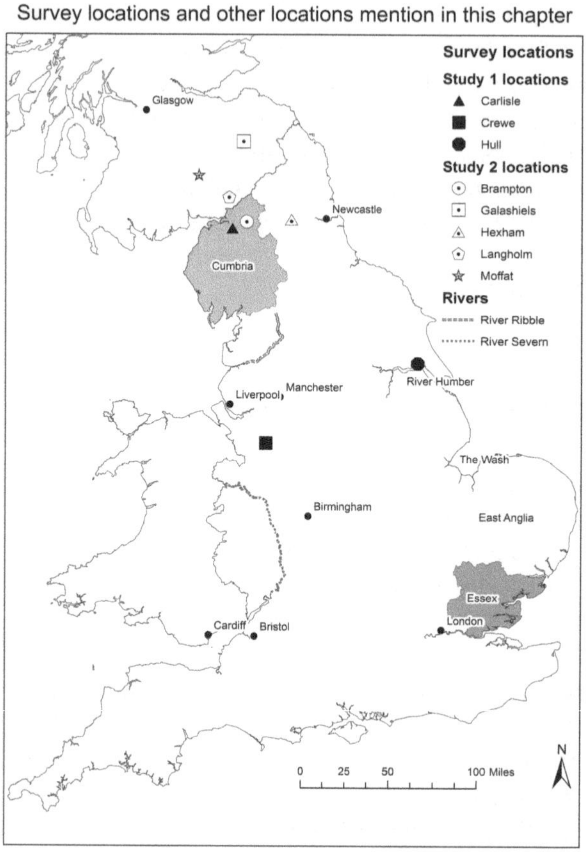

Fig. 10.4: Survey locations for Studies 1 and 2.

For both studies, the draw-a-map task lasted for 10 minutes. In order to assist respondents a location map which contained a number of cities and towns in England, Scotland (in Study 2) and Wales was projected during the first five minutes of the task.

Upon completion of the task, data collected as part of study 1 were processed using Onishi and Long's (1997) PDQ software (see Chapter 2 of this volume). These processed data were later added to a GIS in order to facilitate comparisons with the data from Study 2 which were originally processed using GIS following the methods outlined in Montgomery and Stoeckle (2013). Table 10.2 and Table 10.3 show the numerical data from Studies 1 and 2, and Figure 10.5 and Figure 10.6 show the geographical data.

As I deal with at more length elsewhere (Montgomery 2012) than I am afforded here, proximity appears to be the most important factor in conditioning

Tab. 10.1: Details of surveys and respondents.

Study	Locations	Respondents		Total	Mean age
		Male	Female		
Study 1 (data collected in 2004)	Carlisle	39	54	93	23.1
	Crewe	21	64	85	16.5
	Hull	34	62	96	19
			Total	274	19
Study 2 (data collected in 2008)	Galashiels	6	12	18	16.4
	Moffat	19	19	38	16.2
	Langholm	6	14	20	16.5
	Hexham	9	11	20	16.5
	Brampton	17	38	55	17
			Total	151	16.5
Grand total				425	

the way in which respondents add dialect areas to maps. That proximity is important in perception is quite expected, as Tobler (1970: 236) has previously stated that "everything is related to everything else, but near things are more related than distant things". As an example of this, the results for Study 1 show that for the Crewe-based respondents, the *Potteries*[6] areas ranks as the 7th most recognised area, but is recognised by nobody else in either of the studies discussed here.

In this case, it appears that there is a straightforward relationship between increased distances from locations leading to a decrease in the information available about them. Research in both perceptual geography (Gould and White 1986) and PD (Preston 1999b: xxxiv) has highlighted this effect: when respondents are asked to draw maps of regions or dialect areas , areas local to them have increased recognition (although see Braber, this volume, who finds that her participants from the East Midlands of England very rarely recognise their own regions).

Simple theories of the impact of distance do not however account for the many and varied ways in which respondents might draw dialect areas. Britain (2010: 70–1) notes that factors other than physical distance might influence the ways in which people use and perceive space. Imagining a simple relationship

6 *The Potteries* is a colloquial label for an area centred on the city of Stoke-on-Trent, which was for many years characterised by its principal industry, the production of pottery.

Tab. 10.2: Dialect areas and recognition rates (RR) for Study 1.

Rank	Study 1[7]								
	Carlisle (*n* = 93)			Crewe (*n* = 85)			Hull (*n* = 96)		
	Area	*n*	RR	Area	*n*	RR	Area	*n*	RR
1	*Geordie*	60	65.2	*Scouse*	67	78.8	*Geordie*	47	49.0
2	*Scouse*	55	59.8	*Brummie*	65	76.5	*Scouse*	46	47.9
3	*Brummie*	43	46.7	*Geordie*	56	65.9	*Brummie*	45	46.9
4	*Cockney*	39	42.4	*Manc*	48	56.5	*Yorkshire*	34	35.4
5	*Manc*	38	41.3	*Cockney*	43	50.6	*Cockney*	28	29.2
6	*Cumbria*	31	33.7	*Cornwall*	14	16.5	*Manc*	26	27.1
7	*Yorkshire*	10	10.9	*Potteries*	13	15.3	*South West*	10	10.4
8	*West Country*	10	10.9	*Yorkshire*	12	14.1	*London*	9	9.4
9	*Cornwall*	9	9.8	*Posh*	11	12.9	*Posh*	8	8.3
10	*Lancashire*	8	8.7	*West Country*	7	8.2	*East Anglia*	6	6.3
11	*London*	5	5.4	*London*	6	7.1	*Hull*	6	6.3
12	*North East*	4	4.3	*Welsh*	6	7.1	*Cornwall*	5	5.2
13	*Devon*	3	3.3	*Common*	5	5.9	*West Country*	5	5.2
14	*Posh*	2	2.2	*Chavs*	5	5.9	*Lancashire*	5	5.2
15	*East Anglia*		2.2	*Bristol*	4	4.7	*Welsh*	5	5.2
16	*Norfolk*	2	2.2	*Norfolk*	3	3.5	*South East*	5	5.2
17	*Gangster*	2	2.2	*Cheshire*	3	3.5	*Bristol*	4	4.2
18	*Leeds*	2	2.2	*Essex*	3	3.5	*Estuary English*	4	4.2
19	*Mackem*	2	2.2	*East Anglia*	2	2.4	*Common*	3	3.1
20	*Welsh*	1	1.1	*Somerset*	2	2.4	*Humberside*	3	3.1

between distance area recognition assumes a "homogenous cultural information space [the space in which we all live]" (Gould and White 1986: 153) in which there are no disruptions to the amount of information that might "flow" from one location to another. Britain (2002a; 2010) has argued that this is rarely the case, and that the heterogeneity of information space is one of the key reasons for variation in respondents' abilities to recognise home or near-to areas. This could help to explain why, for Preston's (1999b: xxxiv) respondents,

7 The dialect areas in Table 10.2 and Table 10.3 are referred to by the name most often given by those who completed the draw-a-map task. This means that colloquial names for dialect areas are given in some cases, these refer to locations as follows: *Geordie*: Newcastle upon Tyne; *Weeji*: Glasgow; *Scouse*: Liverpool; *Brummie*: Birmingham; *Cockney*: London; *Manc*: Manchester. Other labels given refer to counties, countries, or cities in Great Britain.

Tab. 10.3: Dialect areas and recognition rates (RR) for Study 2.

Rank	Study 2														
	Galashiels (n = 18)			**Moffat (n = 38)**			**Langholm (n = 20)**			**Hexham (n = 20)**			**Brampton (n = 55)**		
	Area	***n***	**RR**	**Area**	***n***	**RR**	**Area**	***n***	**RR**	**Area**	***n***	**RR**	**Area**	***n***	**RR**
1	*Geordie*	18	100	*Weeji*	27	71.1	*Welsh*	14	70	*Geordie*	13	65	*Scouse*	43	78.2
2	*Weeji*	16	88.9	*Scouse*	24	63.2	*Geordie*	13	65	*Welsh*	11	55	*Geordie*	42	76.4
3	*Scouse*	15	83.3	*Geordie*	22	57.9	*Scouse*	13	65	*Scouse*	10	50	*Brummie*	32	58.2
4	*Welsh*	12	66.7	*Welsh*	20	52.6	*Brummie*	12	60	*Cockney*	10	50	*Cockney*	30	54.5
5	*Manc*	10	55.6	*Brummie*	17	44.7	*Weeji*	10	50	*Manc*	10	50	*Manc*	30	54.5
6	*Brummie*	9	50.0	*Cockney*	14	36.8	*Manc*	7	35	*Brummie*	9	45	*Cumbria*	26	47.3
7	*Cockney*	9	50.0	*Manc*	10	26.3	*Cumbria*	6	30	*Yorkshire*	6	30	*Welsh*	25	45.5
8	*Farmers*	8	44.4	*Aberdeen*	9	23.7	*Scottish*	6	30	*Cumbria*	5	25	*Yorkshire*	17	30.9
9	*Posh*	7	38.9	*Broad Scottish*	5	13.2	*Posh*	6	30	*West Country*	5	25	*Scottish*	13	23.6
10	*Borders*	7	38.9	*Bristol*	5	13.2	*Yorkshire*	5	25	*Weeji*	4	20	*Weeji*	11	20.0
11	*Highlands*	6	33.3	*West Country*	4	10.5	*Broad Scottish*	5	25	*Scottish*	4	20	*Broad Scottish*	9	16.4
12	*London*	5	27.8	*Farmers*	4	10.5	*Cockney*	4	20	*Farmers*	3	15	*London*	9	16.4
13	*West Country*	5	27.8	*Gaelic*	4	10.5	*Borders*	3	15	*Midlands*	3	15	*Aberdeen*	8	14.5
14	*Gaelic*	4	22.2	*Cumbria*	3	7.9	*Gaelic*	3	15	*Aberdeen*	2	10	*Bristol*	6	10.9
15	*Aberdeen*	3	16.7	*London*	3	7.9	*Cardiff*	3	15	*Broad Scottish*	2	10	*Southern*	6	10.9
16	*Yorkshire*	1	5.6	*Posh*	3	7.9	*London*	2	10	*London*	2	10	*West Cumbria*	6	10.9
17	*Broad Scottish*	1	5.6	*Highlands*	3	7.9	*West Country*	2	10	*Bristol*	2	10	*Highlands*	5	9.1
18	*Cornwall*	1	5.6	*Skins*	3	7.9	*Farmers*	2	10	*Cornwall*	2	10	*Cornwall*	5	9.1
19	*Lancashire*	1	5.6	*Borders*	2	5.3	*Highlands*	2	10	*Posh*	1	5	*West Country*	4	7.3
20	*Cardiff*	1	5.6	*Cardiff*	2	5.3	*Edinburgh*	2	10	*Highlands*	1	5	*Northern*	4	7.3

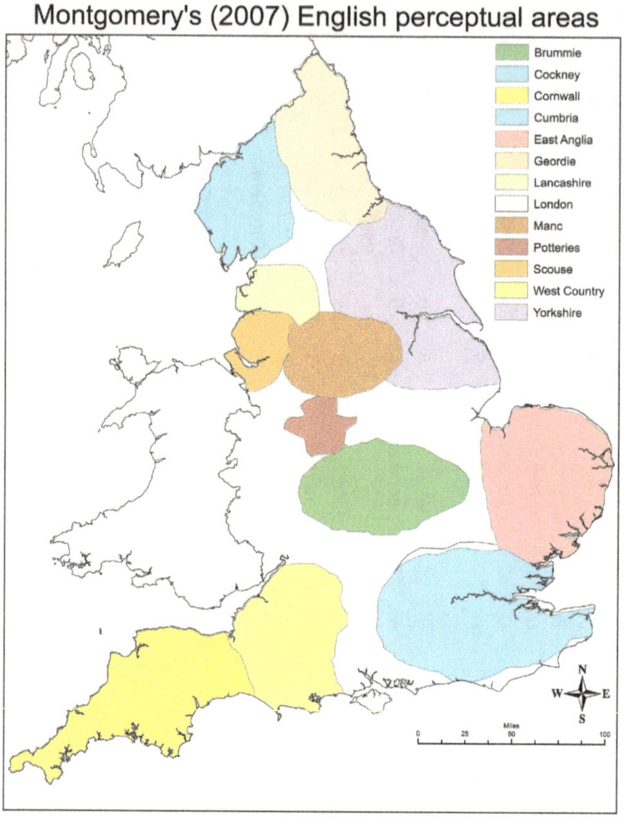

Fig. 10.5: Geographical results from Study 1.

proximity was not the most important factor (as respondents would first draw stigmatised, and then local areas).

Factors that interfere with the homogeneity of the information space could include barrier effects, such as obvious physical barriers like rivers and mountain ranges, along with national boundaries (such as the Scottish-English border) and other political/governmental boundaries. In the case of the research in Study 2, the impact of the Scottish-English border is not equal for both respondents who live on either side of it. Table 10.3 shows the area recognition rates for respondents north of the border (Galasiels, Langholm, and Moffat) as well as those living south of the border (Hexham and Brampton). The Figures show that whilst respondents living in Scotland have a rich and detailed knowledge of English dialect variation (at least as rich and detailed as their English counterparts) as well as Scottish variation, respondents living south of

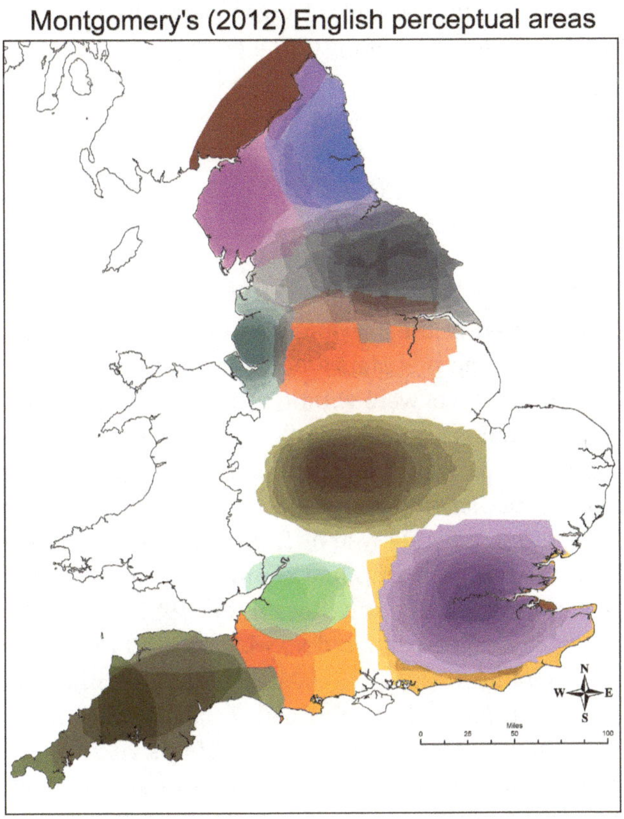

Fig. 10.6: Geographical results from Study 2.

the border have very little knowledge about variation in Scotland (cf. Inoue 1999b). This border effect leads to a unidirectional proximity effect, where Scottish respondents have a "normal" perception of the dialect landscape in Great Britain and areas further away from respondents' locations have lower levels of recognition, notwithstanding the impact of cultural prominence (see below). By contrast, English respondents have comparatively little knowledge of variation in Scotland, due to the effect of the border.

Cultural prominence (previously termed *cultural salience* (Montgomery and Beal 2011: 138)) is related to increased metalinguistic knowledge and refers to the boosting effect that the media or other mechanisms might have on the way in which a location is perceived. The phenomenon of cultural prominence is intertwined with proximity, and functions by virtually increasing it. Thus, "far-away" areas are brought "closer" to respondents through increased exposure in various forms of media and public discourse. I argue below that *cultural prominence* is at the heart of findings in relation to the emergence of a *Manc* (Manchester) dialect area.

5 City-based dialect areas

Perhaps the most striking difference between the dialect areas drawn by respondents in both studies discussed in the previous section and those added to maps by Inoue's respondents is the prevalence of city-based dialect areas. Although, of course, Inoue's respondents were willing to add *Scouse* and *Geordie* areas to their map in large numbers, all of the other areas referred to geographically bounded regions. In this way Inoue's respondents were echoing regions identified in earlier production studies.

My methods were different to those employed by Inoue, and these differences must not be overlooked. Inoue's base map had county lines present, and his data were processed using these same county lines (Inoue 1999b: 164–166). My maps were blank, save for some city location dots which served to provide some common geographical ground for respondents (see footnote iii). Data were then processed according to methods established in Preston and Howe (1987); further details are which are available in Chapter 2 of this volume.

It could be claimed the use of city location dots skews my data in favour of city-based dialect areas, although extensive piloting of the draw-a-map method in England during the research instrument design phase of fieldwork (reported in Montgomery 2007: 346–8) demonstrated no such effect. In a comparison on the ten most frequently recognised dialect areas in pilot studies with various amounts of geographical "help" of the type provided by city location dots, I demonstrate that city location dots did not change the recognition rank of dialect areas (i.e., the most frequently recognised dialect areas were still most frequently recognised whether dots were used or not). The city location dots did however increase the number of lines added to maps, as respondents who had the geographical "safety net" provided by the location dots were more confident about their geographical knowledge and therefore more likely to draw dialect areas.

Although it is possible that different levels of detail on maps do have the capacity to change respondents' reactions to tasks, the differences between Inoue's results and my results are particularly stark. The city-based dialect areas of *Brummie*, *Bristol*, *Cockney*, *London*, and *Manc* do not appear at all in Inoue's (1999b: 167) map. This suggests a significant shift in the perception of dialect variation in England since Inoue's data were collected. Although there is only a 15 year gap between the collection of Inoue's data and the collection of the data for Study 1 (and an 18 year gap for study 2), I argue that just as Inoue's perceptual data reflected historical dialect variation, the data from Studies 1 and 2 reflects contemporary variation in English dialects.

The maps in Figure 10.5 and Figure 10.6 are striking in their similarity to many aspects of Trudgill's (1999: 83) speculative map of the "future" of dialect variation in England. This map, which was created by Trudgill to predict some of the effects of regional dialect leveling, shows the emergence of large city-based dialect areas across England. Whatever the reality of linguistic contact on the ground (cf. Britain 2002b), there is no doubt that regional dialect leveling is occurring in many areas of England (e.g., Watt 2002), and it appears that respondents in my studies are as aware of this as linguists working in the field.

That non-linguists would notice a changing dialect landscape is of course not unexpected. However, the rapidity of the change in perceptions is something that needs accounting for. In the next section I use the concept of cultural prominence to suggest sociocultural reasons for the rapidly changing perception of the *Manc* dialect area, and argue that such reasons could contribute to the process of regional dialect leveling.

6 The emergence of *Manc*

Manchester is a city in the North West of England which played a particularly important role in the development of trade and industry in England. It was the centre of the cotton industry (earning it the nickname "cottonopolis" in the 1800s), and throughout the 19th century experience very rapid population growth. In the 20th century, the city retained its importance, and although experiencing industrial decline (along with many other Northern towns and cities), its population remained stable. At the end of the 20th century, the city experienced extensive urban renewal, kick-started after an IRA bomb damaged a large portion of the city centre.

In conjunction with this urban renewal, the city was also at the centre of a dramatic increase in popular culture in the late 1980s and early 1990s which earned it the nickname "Madchester". This cultural prominence was epitomised by the music emerging from the city by such bands as the *Happy Mondays* and the *Stone Roses*, both of which were major chart successes after album releases in 1989. Essentially a short-lived phenomenon which reached its peak in 1990, the "Madchester effect" none-the-less provided the platform for many more musicians and groups, including *Oasis* who remained particularly popular throughout the 1990s. The effect on the city is difficult to estimate; although there is no doubt that it became a focal point for the youth of the country and has remained a significant centre. The city is also host to two of the largest football clubs in England (Manchester United and Manchester City), with Manchester United now the most successful club in the country.

Despite its importance in the development of the contemporary British industrial and societal landscape, it is notable that the city is absent from many of the historical accounts of dialect variation in England. The city, although marked and delimited on Ellis' (1889) map, does not appear to have its "own" dialect area in any of the other typologies of variation. The composite map in Figure 10.6 sees the city as existing in a transitional area between other more established dialect areas. Partially, this could be due to historical and rural focus of much of the early dialect studies undertaken in England, although other city focussed areas do appear to exist in Figure 10.3 (such as London, Birmingham, and Newcastle-upon-Tyne).

Inoue's study also finds no *Manc* area. Instead, respondents in this research included the city in the *Scouse* dialect area. In contrast, it received a consistently high recognition amongst respondents in both Study 1 and Study 2. In Study 1, *Manc* had a mean 40 % recognition rate (i.e., 40 % of respondents drew and named the area on their maps), and in Study 2 the recognition rate was 44.4 %. This means that the recognition level increase from seemingly nothing in 1989 (when Inoue's data were collected), to a recognition level of nearly 50 % in the studies undertaken 15 and 18 years later. Such an increase is in step with Trudgill's (1999: 83) future map of dialect variation in England, where a *Manchester* dialect area suddenly appears (when compared with Trudgill's (1999) Traditional and Modern maps of variation). The increased cultural prominence of the area, and by extension the variety of English spoken there, can be used to explain why the area might be appear in non-linguists' mental maps of dialects as it similar appears in the classifications of linguists.

As I have mentioned above, a key factor in the perception of dialect variation is that of proximity. Although it is not surprising that local areas would be perceived more readily than others, it does not happen in all cases. Considering Table 10.2 again, it is clear to see that the Crewe-based respondents' recognition of a *Potteries* (Stoke-on-Trent) dialect area is primarily a function of their proximity to this city. A *Potteries* area does not appear for respondents in any of the other survey locations in either Study 1 or 2. In a similar fashion, examination of Table 10.2 and Table 10.3 reveals the recognition rates for the *Cumbria* dialect area are higher amongst respondents in locations closer to (or in) this area, again demonstrating the effect of proximity.

Proximity cannot account for the high recognition rates in Table 10.2 and Table 10.3 across all locations for the most frequently recognised dialect areas of *Scouse*, *Geordie*, *Brummie*, *Cockney*, and *Manc*. Here, one must look at other factors for explanation, and cultural prominence is a factor that can go some way towards providing such an explanation. As Johnstone observes, it is not only languages that "are created in discourse, so are dialects" (Johnstone 2011: 1). Referring to the "creation" of Pittsburghese by linguistic laypeople (Johnstone 2011: 6), she cites the important disconnect between the presence of a geographical area that can be identified by linguists as a "Pittsburgh dialect area" and the belief amongst the population of the city that they use a distinctive variety. This variety has particular enregistered features, which prominently feature on a range of dialect commodities that help to perpetuate the "idea that there is a Pittsburgh dialect in the first place" (Johnstone 2009: 159). Such "creation" of a dialect area by residents of a city or area is not a phenomenon limited to Pittsburgh, as Beal (2009) has noted. Beal's article, which discusses the enregisterment of "Geordie", notes that features that were more widespread in the north of England were "claimed" by the emergent dialect area, and became emblematic of it. Once enregistered and in existence as an identifiable entity (as evidenced by dialect commodities, for example), a "new dialect" can be boosted via various means. Such boosting effects mean that an area's dialect can be made more culturally prominent, and will thus appear more readily in the mental maps of non-linguists further afield.

Cultural prominence can be driven by many factors. Johnstone (2011) has examined local newspapers, a website, an email discussion forum, and a collaborative wiki page in order to further understand the creation of Pittsburghese. Other media can of course be important, and Stuart-Smith has claimed that "it is inevitable that the broadcast media will have an impact on metalinguistic awareness of linguistic varieties and variation, standard and non-standard, and the ideologies surrounding them" (2011: 3). Here, I examine the role of print media and the way in which it can boost an area's cultural prominence.

Print media have been chosen as they offer the ability to quantify the amount of exposure a place might receive, which can prove instructive when trying to measure the relative prominence of locations. LexisNexis library (2011) was used to search mentions of the cities or regions in which dialect areas mentioned by respondents in Studies 1 and 2 were located every year from 1988 (the year before Inoue's data were collected) to 2008, (when the data for Study 2 were collected). Due to the nature of the newspaper corpus provided by LexisNexis, only *The Times* and *The Sunday Times* offered the time-depth needed, so only mentions in these newspapers were searched for. This necessitated a further compromise as searching these newspapers meant that I was

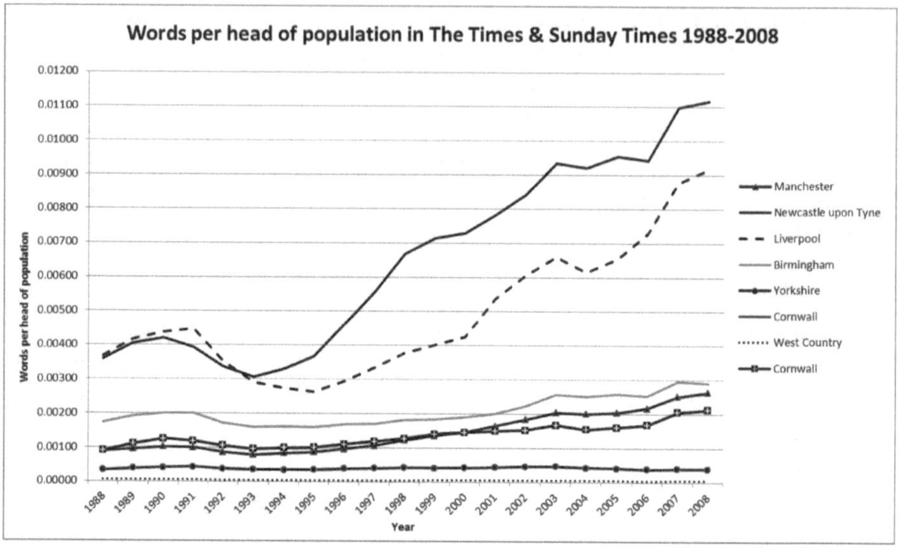

Fig. 10.7: Mentions per head of population for frequently recognised dialects areas in Studies 1 and 2.

not able to search for mentions of London (as the full name of *The Times* is *The Times of London*). Raw counts for the number of mentions in print media are interesting as far as they go, but they do not offer the researcher the opportunity to investigate whether a particular location is gaining greater prominence than its size would otherwise suggest it might. For this reason, a "mentions per head of population" calculation was made. This calculation simply divides the number of mentions a particular location received in each year between 1988 and 2008 by the population of the location for each year. The results of this calculation can be seen in Figure 10.7.

Figure 10.7 shows that Newcastle upon Tyne and Liverpool both consistently demonstrate more prominence than the other locations searched in the newspaper corpus. Most of the areas demonstrate an increase over time, although Figure 10.8 reveals, this increase is not uniform for all areas.

Figure 10.8 reveals the general upward trend in the mentions per head of population, with the exception of the West Country for 1989–2005. A striking feature of the chart is the very large increase in the mentions per head of population for Manchester, which is second only to the increase in mentions for Newcastle-upon-Tyne. This increase in mentions per head of population for Manchester is as a result of many of the cultural and economic changes that occurred in Manchester between Inoue's fieldwork and the fieldwork undertaken in Studies 1 and 2. These changes led to an increase in cultural prominence,

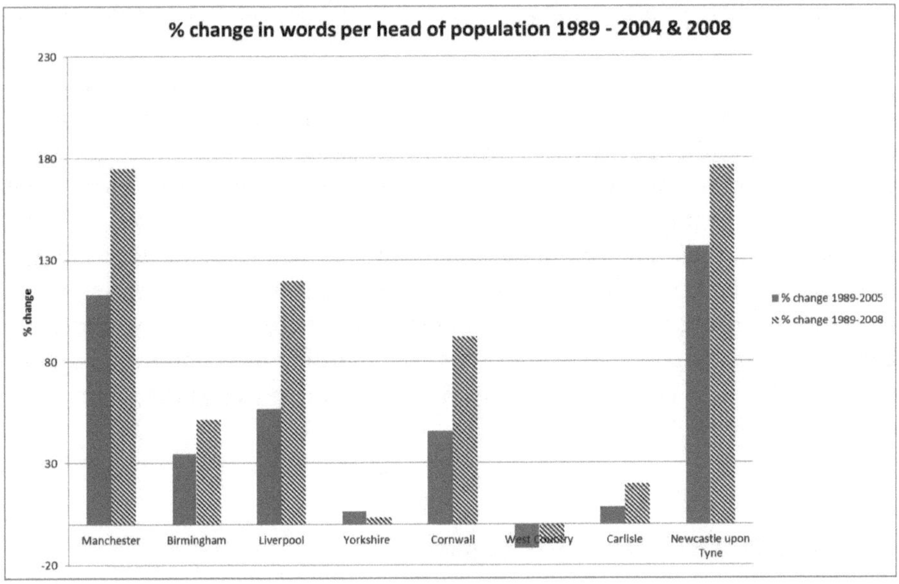

Fig. 10.8: Percentage change in words per head of population: 1989–2005; 1989–2008.

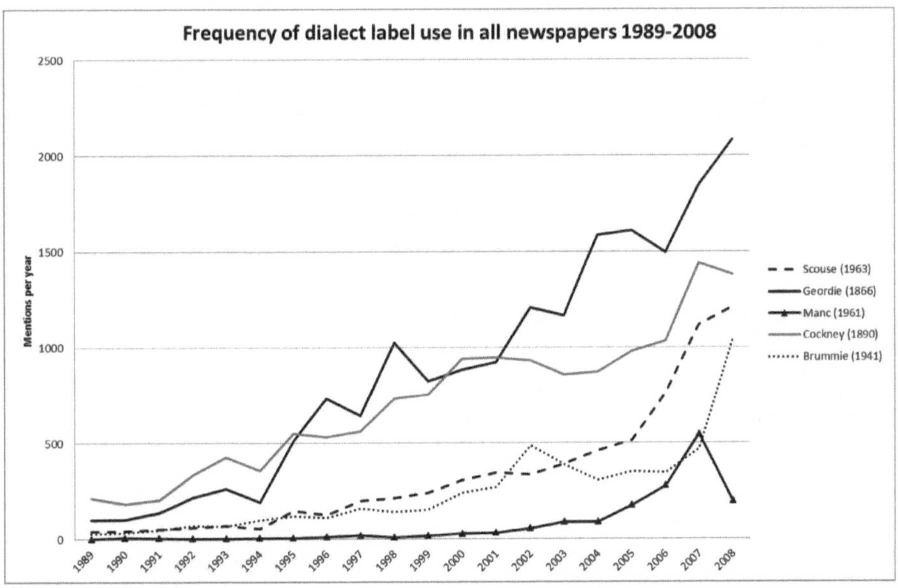

Fig. 10.9: Use of dialect area labels in all newspapers, 1989–2008.

which in turn have led to the *Manc* dialect area being added to perceptual map for the first time.

Of course, one could argue that despite the increased recognition of the area, that respondents were only really responding to an increasing awareness of the city, and this is what is seen on their maps. In order to assess this, I performed another search using LexisNexis in which dialect area names that were given by respondents were the target. Figure 10.9 shows the result of this search, and reveals that over time all dialect area labels are more frequently used. Some labels increased from around 100 uses in 1989 (*Cockney* and *Geordie*), and others such as *Scouse* from a much lower base. By contrast, the *Manc* label was not used at all in 1989 and is only used from 1998 onwards, suggesting that this might be the time at which *Manc* emerges as a variety with some national currency.

7 Summary and conclusions

In this chapter, I have demonstrated the various ways in which linguists have attempted to classify the dialect areas of England, using various data sources, and with varying results. The disagreement between many of the boundaries arrived at by different scholars merely seems to underline the difficulty in defining dialect areas using purely linguistic criteria. That "dialect areas" are a fictional construct is not a new observation, but the willingness of non-linguists to talk about them, and add them to maps, means that the concept is still one worth studying. In addition to this, the line density analysis performed on the variety of boundaries produced by others did show some agreement about the placement of some dialect areas.

Perceptual dialectology is one way to investigate the concept of dialect areas, and Inoue's results show that in some respects his respondents produced maps that demonstrated coincidence with the line density analysis of older dialect area maps. This means that to some extent, Inoue's respondents were "accurate" in their perceptions. Notable features of Inoue's results were a lack of city-based dialect areas (except in his *Northern* region). Particularly notable, especially in light of my later research, was the absence of a *Manc* area.

The *Manc* dialect area is prominent amongst my results, and was consistently added to maps by respondents from around England and Wales. This is a stark contrast to Inoue's results, which have no place at all for a Manchester-based dialect area. Not only this, but the dialect typologies discussed above also find no dialect area centered around Manchester, with the line density analysis finding a mass of lines in the area of the city. The recognition of the *Manc* area by respondents drawing maps 15 years after Inoue points towards a change in the perception of dialect landscape of England.

This changed perception finds a prominent Manchester-based dialect area brought about, I would contend, by the increased cultural prominence of the city of Manchester and the Greater Manchester area. This increased cultural prominence, demonstrated via the increased mentions per head of population, has resulted in the recognition of the area and its “dialect”. This new recognition could in turn bring about further cultural prominence, and ultimately result in the enregisterment of specific linguistic features as *Manc*, in effect giving “birth” to a new dialect area in the national consciousness.

Natalie Braber

Chapter 11
Dialect perception and identification in Nottingham

1 Introduction

The focus of this study is an examination of dialect perception and identification within the city of Nottingham. The city is situated in the East Midlands region of the United Kingdom. Although Wales (2000: 5) comments that the East Midlands may be the geographical centre of England, it is not in any sense the *perceived* centre of England. It is an area which can be hard to locate and is referred to by Wales as "neither here nor there" (Wales 2000: 7/8), by Montgomery as a "no-man's land" (Montgomery 2007: 352) and as an area which may be intermediate between north and south (Upton 2012: 260). Although the East Midlands does not have an official regional capital, Nottingham is treated by many as such (such as the 'Meet England' website[1]) and is referred to as *Queen of the Midlands* by John Beckett in his centenary history of Nottingham (Beckett 2006).

Linguistic variation in the city of Nottingham has long been neglected. There has been little survey of dialect in the East Midlands since the *Survey of English Dialects* in the 1950s. It is, moreover, striking that publications, which aggregate the findings of earlier surveys and more recent localised studies to present an overview of regional speech in the UK, either lack up-to-date research data from this area or simply ignore the region completely. There are a few studies which deal with specific regions of the East Midlands, for example Derby in Docherty and Foulkes (1999) and Milroy (1996) and Nottingham in Flynn (2007, 2012) and Braber and Flynn (2015). Despite this relative scarcity of empirical evidence, anecdotally it appears that language in Nottingham remains distinctive and there is difference, for instance, between speech in the major urban centres of the East Midlands – Nottingham, Derby and Leicester (see for example Scollins and Titford 2000; see also the linguistic descriptions given by the British Academy funded project carried out by Diane Davies and

---

1 See: http://www.meetengland.com/Incentives/Nottingham-and-the-East-Midlands.aspx where it states that "Nottingham is the capital of the East Midlands".

Natalie Braber from around the East Midlands: http://www.le.ac.uk/emoha/community/dialect/home.html).

However, it appears that local accents are not easily recognised by young people who live there (Braber, 2015). This seems to contradict theories about people being more accurate at recognising their own local accents (Wells 1982: 33; Williams, Garrett & Coupland 1999). This chapter examines perception of local language varieties with 200 sixth-form students, male and female, in the city of Nottingham. The question being asked here is whether lack of recognition could be related to the apparent lack of cultural salience of this particular city (for notions of cultural salience, see for example Preston 2010b: 88). Furthermore, this study looks at the possible relations between the very negative attitudes these speakers have about their own linguistic variety and this perceived lack of salience or the poor reputation of Nottingham. In other words, are young Nottingham speakers misidentifying local accents as part of a 'denial' phenomenon (see Montgomery & Beal 2011; Williams et al. 1999)?

2 Nottingham: Geography and linguistics

Nottingham is a relatively large city in the heart of England situated in the central part of the East Midlands, England's fourth largest region (see Figure 11.1 for Nottinghamshire's location in the United Kingdom and Figure 11.2 for Nottingham within Nottinghamshire). More than a quarter of a million people live in the city, many more work here and visit the area (Beckett 2006: 1). Before the ninth century there was very little of note in relation to the city and only during the last hundred years has it grown into urban city it is today (Beckett 2006: 21). Although there are officially defined government regions, boundaries of the East Midlands are not always clear (for a full discussion, see Braber 2014). Nottingham is clearly part of the East Midlands and the main centre of one of the three shires which are most frequently named in relation to the region – Nottinghamshire, Derbyshire and Leicestershire.

As mentioned in the introduction to this chapter, there are few linguistic studies carried out in the area (Braber & Flynn 2015; Flynn 2007; 2012; Wright 1986). However, it is an interesting area, particularly as it falls near some important isoglosses, namely the short /a/ of BATH and lack of a FOOT/STRUT split, both of which are important markers of northern vs. southern varieties of British English (for a full discussion, see Wells 1986). When considering dialectal variation in the UK, linguists have frequently considered the linguistic markers separating the two regions in the North/South divide (see for example Trudgill 1999; Wells 1982). However, it has been noted that this is not a straightforward issue (e.g., Beal 2008; Goodey, Gold, Duffett & Spencer 1971;

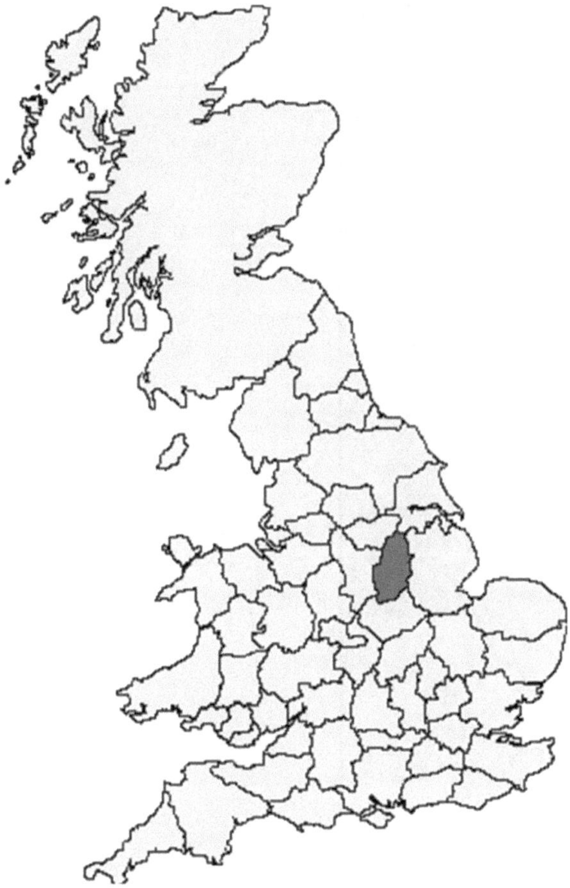

Fig. 11.1: Nottinghamshire within the UK[2].

Montgomery 2007; Wales 2002). The dialect of the region is said to belong to the northern dialect area as it shares similar features with those of the North (Wells, 1982; Beal, 2008) such as short /a/ and the unsplit FOOT-STRUT vowel, but there are also shared features with varieties found in southern England (Hughes, Trudgill and Watt, 2012) which include features such as l-vocalisation and th-fronting.

2 Map from: http://www.pavetechnology.co.uk/countries-and-counties-serviced-by-prl-pavement-rejuvenation-limited/Nottinghamshire.html

Fig. 11.2: Nottingham within Nottinghamshire[3].

3 Project and methods

As this volume considers perceptual dialectology, it is important not solely to look at actual language usage, but also at what people think about language, because these issues are inseparable (Williams et al. 1999: 348). This chapter

3 Map from: http://www.google.co.uk/imgres?um=1&hl=en&biw=1350&bih=867&tbm=isch&tbnid=pWU1i0Rktc9gRM:&imgrefurl=http://www.touristnetuk.com/em/nottinghamshire/towns/nottingham.htm&docid=OHibG1f2MdvETM&imgurl=http://www.touristnetuk.com/em/nottinghamshire/images/map.gif&w=250&h=411&ei=BmWXUpbtOs2LhQeU_4HoCA&zoom=1&iact=hc&vpx=1127&vpy=148&dur=8920&hovh=288&hovw=175&tx=132&ty=148&page=1&tbnh=151&tbnw=92&start=0&ndsp=45&ved=1t:429,r:27,s:0,i:167

considers where non-linguists think linguistic boundaries are (Preston 2010a: 179) – in this case, a group of Nottingham sixth-formers (17–18 year olds sitting their final exams in schools and colleges). This is an important age group to consider as we know that children already making judgements about language varieties and young adults are aware of linguistic norms and are considering their own identities (Williams et al. 1999: 347). Furthermore, dialects are not just abstract entities, they have emotional aspects (Inoue 1999a: 147) and such aspects should be considered.

Previous studies language attitudes including Nottingham have shown different results. Coupland and Bishop (2007: 79) report Nottingham as coming 25th out of 34 in ratings scored by social attractiveness in a large online, nationwide survey on different varieties of English. Flynn (2007) shows that the Nottingham accent was not perceived particularly negatively by outsiders and insiders – something which will contradict findings from this chapter. Generally, there is the assumption that people are accurate at recognising accents (Williams et al 1999: 345), but this chapter will evaluate whether this is actually the case.

The aim of the overall project discussed in this chapter was to investigate where a group of East Midlands adolescents believe the North/South boundary to fall in the UK, how they position themselves within this divide, i.e., do they think of themselves as 'Northern', or 'Southern' or as something else, and to get them thinking about (and listening to) voices around the UK. It combines methods from studies by scholars such as Long (1999a and 1999b), Fought (2002), Diercks (2002), Lance (1999), and Williams et al. (1999). Some of the tasks mentioned below are not fully discussed in this chapter but can be found in Braber (2014; 2015).

I invited state secondary schools from across the East Midlands to take part in a study investigating language variation in the UK. For this chapter only the participating schools in Nottingham have been selected. I attended an hour-long class with students in their final year to engage with students' opinions on language variation in the UK and have them listen to some recorded voices to try to place particular accents. The schools were not told in advance about considering the North/South divide or feelings of identification to try to avoid prior discussion and students influencing one another. Some of the schools specifically asked for A Level English classes to be involved, whereas other schools were happy for all A Level students to take part. This meant that some schools only had small groups of students participating, while others had larger groups. In all, 191 students in Nottingham were involved in this study (the overall East Midlands project contained 327 student participants).

Participants were given several maps of the UK and were informed that we would be carrying out a number of tasks to do with language variation in the

UK and attitudes towards accents, as well as listening to some speakers. Students were told there were no right or wrong answers and they should try to answer reflecting their own ideas and feelings, because it was their opinions that I was interested in. Students were encouraged to carry out this work alone, without consulting others in the class as others' opinions may be different to their own. For the first two tasks some students received a map on which some cities were indicated and some received blank maps. The cities on the labelled map were chosen to give a geographical spread around the UK, and contained Nottingham. This spread was given to allow students to orient themselves geographically, and one city in the East Midlands was chosen to ensure they would know the location of the area if required. Some perceptual dialectologists, such as Preston (personal communication) have questioned whether having a map with cities marked on it will influence what participants do with them. So having these two maps allowed for the comparison between the two groups, to see whether different results would be given by the two different groups. When comparing the boundaries on the maps for dialect areas it appeared that the labels did not influence the outcome because the results for both groups were identical in their rank ordering. The naming of dialect areas was counted and geographical accuracy was not judged for this task. For the identification task all students received the map with the city dots, although they were informed that the recordings did not necessarily come from these cities, but were there to guide them geographically.

In total, the participants were asked to carry out four main tasks (some of the results have already been discussed in Braber 2014; 2015):

1. Draw the North-South boundary (if you think it exists) on the map given and then write next to the map whether you feel yourself to be Northern/Southern/Neither;
2. Draw boundaries on maps where you believe accent areas can be found and label them with comments about any special characteristics from the region;
3. Listen to 14 short extracts and put a cross on the map where you think the speakers come from (these included locations from around the UK, including a Northern Irish, Scottish and Welsh accent, as well as one from Plymouth, Ashford (Kent), London, Newcastle, Birmingham, Burnley, Withernsea (East Yorkshire) and Liverpool and three “local” dialects – Nottingham, Leicester and Derby. The voices the participants were heard were all female, working-class and young to middle-aged. Most of the recordings come from the British Library archives (http://sounds.bl.uk/Accents-and-dialects/Millenium-memory-bank), although the voices from Nottingham and Derby had to be recorded additionally as there were no voices on the

British Library sound archives which matched the other voices and did not state where the listener was from. These recordings were examples of natural speech with the women talking about neutral topics).
4. With smaller groups, students were given a "mind map" of the East Midlands and asked to think about variation of language in the region and give any examples they could think of.

This chapter will focus on tasks 2, 3 and 4 and how reactions towards Nottingham may have influenced the results. The participants were also asked to complete information about themselves and their accent. Some of the students also took part in a mind map exercise, where they worked in small groups on maps of the East Midlands making comments about where accents differed and any thoughts they had about them. Students were asked to think about where language started to sound different, how it was different, and give examples of words and/or pronunciations of different accents in the region.

4 Results

4.1 Dialect divisions

Students were asked to circle areas on the map where they thought distinct linguistic varieties could be found, to label them with names if possible and to give any opinions they had about these varieties (what they sounded like, typical words or pronunciations, famous people who spoke with these accents, and any thoughts or feelings about the varieties).

Beal comments that patterns of mobility and contact lead to "talk about talk" (Beal 2009: 141) and an awareness of regional varieties being signals of authentic local identity. I was interested to see whether these students specifically recognized their own area as having an accent. This task allowed examination of respondents' "actual mental map of regional speech areas" (see Preston, 1999c: 361) to see where they think varieties are different, which is important as maps give powerful clues to people's perception of space (Preston 1989: 16). The responses given by these students were very variable, some included just one or two circles (e.g., Scottish, Geordie etc,) as can be seen in Figure 11.3 whereas others drew very detailed analysis of different words used in areas and included many circles covering larger areas of the UK, see Figure 11.4. Most students fell in between these two extremes.

The rank ordering of the ten most frequently named/circled dialect areas can be seen in Table 11.1 below. The interpretation of such maps is not always

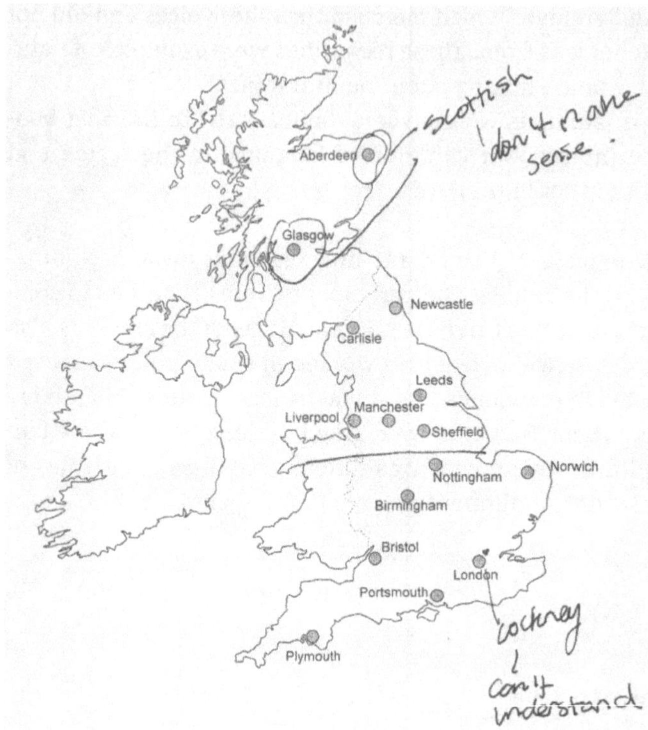

Fig. 11.3: Map with few labels.

straightforward. Some students circle city names whereas others circle larger areas. This was reflected in my analysis: where students circled a city only, this is how it was classified in my analysis, and where students circled larger areas this was also signified during analysis (for example, where some students circled Sheffield and Leeds together and named it Yorkshire). Most students provided labels for areas and this was also taken into account during the analysis stage.

This table shows that the four most commented on accents (Liverpool, Newcastle, Birmingham and London) had relatively similar scores and were recognised by over 144 students (out of 191) as being distinctive in the UK. After that, the numbers drop rapidly – Scotland was still seen by 120 students as being distinctive, but this declined to 84 and 82 for Ireland and Wales respectively, with Manchester and Glasgow coming behind them. The tenth most common accent named was Nottingham, but only 56 students (29 % of the total of 191) named or circled this variety. In a similar task carried out by Montgom-

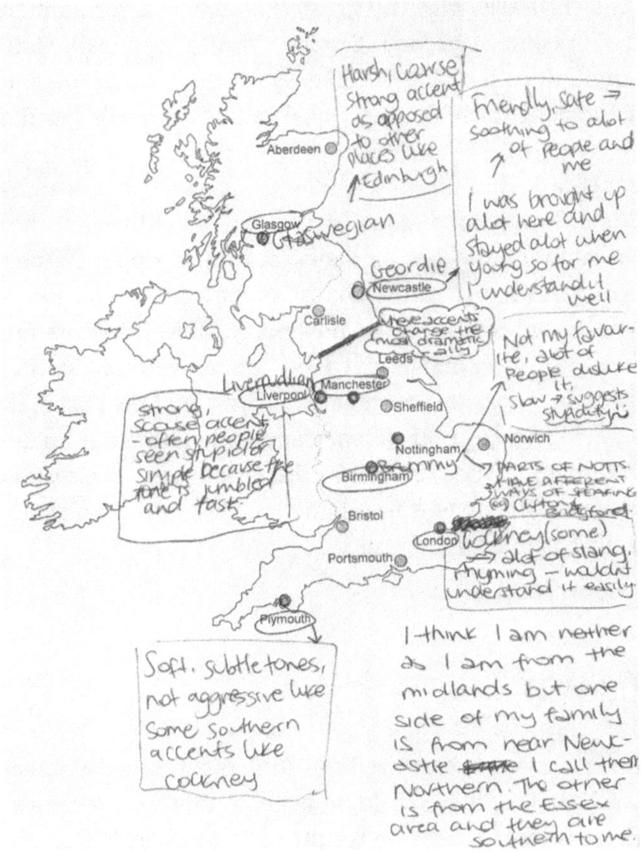

Fig. 11.4: Map with detailed labels.

Tab. 11.1: Rank ordering of dialect areas.

Area circled/named	Number of times circled/named
Liverpool	181 (95 %)
Newcastle	160 (84 %)
Birmingham	158 (83 %)
London	144 (75 %)
Scotland	120 (63 %)
Ireland	84 (44 %)
Wales	82 (43 %)
Manchester	74 (39 %)
Glasgow	68 (36 %)
Nottingham	56 (29 %)

ery (2007: 65) the five dialect areas with the greatest number of recognition were Newcastle/Geordie, Yorkshire, Liverpool/Scouse, Cornish/Cornwall, Manchester showing some similarities but also differences in which areas the students named. It seems for both these studies, local regions did not get the highest ratings.

So it appears from this task that the East Midlands accent is not seen as being an obvious candidate for having regional variation. This differs to other studies, which state that self-identification is likely (see for example Preston 1989: 118). This may be due to the age of the cohort as they are not as aware of having an accent themselves, but as stated previously, these students are aware of local and regional variation due to extensive contact with media that frequently contain non-standard varieties. It can also suggest that the East Midlands may be an area which lacks cultural saliency and is therefore not immediately recognised as an individual area, as will be discussed later. The results of the next task will bring out how well these participants can recognise accents, including that of their own city.

4.2 Dialect recognition

This study is particularly interested in how well the students were able to recognise the local voices and whether they would be better at this than recognising voices which come from further away. The results seem to indicate that the dialects commonly named in the previous task were also accurately labelled in this task (particularly Liverpool, Newcastle, Birmingham and Wales), which may be those with the highest sense of cultural salience, i.e., they are well represented on the media and therefore well recognised (see also Montgomery 2007: 331). However, London, Glasgow and Lissummon (Northern Ireland) as well as the local voices were generally labelled less accurately.

For example, the most frequently named language variety in the previous task was that of Liverpool and this was very accurately recognised by these students. The map for Newcastle was similarly accurate to this one and can be seen in Figure 11.5.

Unlike Williams et al. (1999) and Wells (1982), I discovered that participants were unable to situate local voices accurately (see also Clopper and Pisoni 2004: 113). Not only were participants unable to distinguish different dialects within the East Midlands (which many state in the mind maps task as being distinctive), but they are unable to recognise that the voices from Derby,

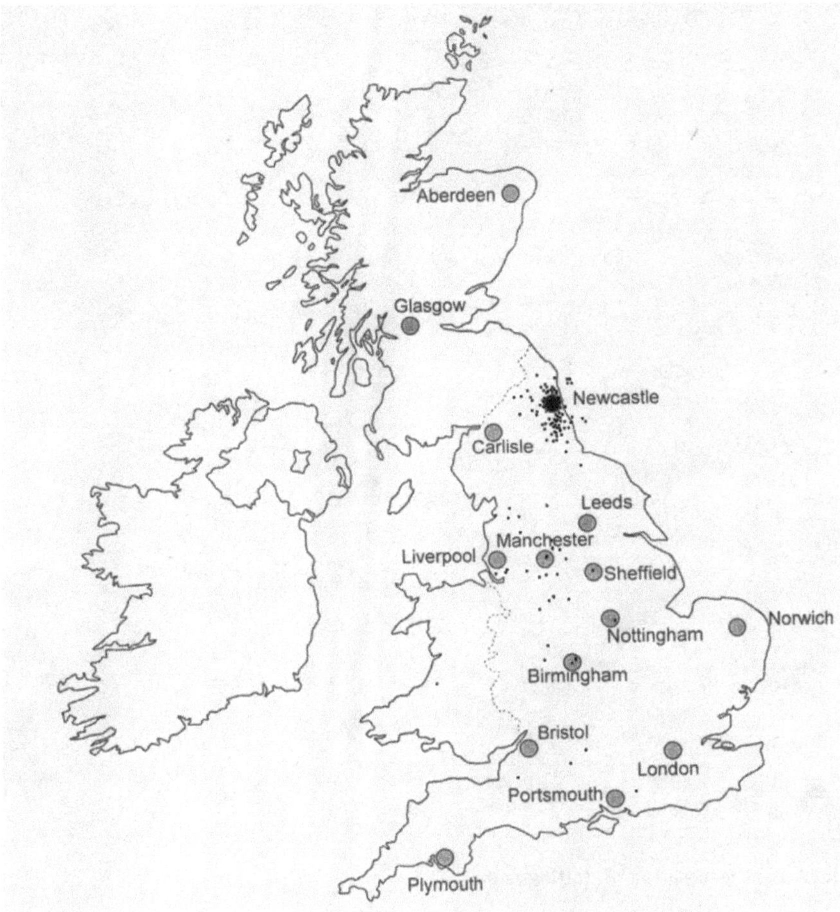

Fig. 11.5: Accent recognition of Newcastle.

Nottingham and Leicester are from the East Midlands. Instead they place them all around the country, with locations as far north as Manchester and Leeds, and as far south as Plymouth and London, see Figure 11.6 below. We can see from this map that it is not that these students think the Nottingham voice is from a particular region, because the spread ranges around England without any particular draw to another area. Could this be down to misrecognition? Williams et al. (1999: 352) question whether such misrecognition could simply be due to insufficient experience and awareness of varieties. I question whether this is the case and suggest it can be related to the lack of cultural salience of Nottingham as will be discussed this in the following section.

Fig. 11.6: Accent recognition of Nottingham.

4.3 Mind maps

A final activity was only carried out in three schools (with a total of six mind maps overall). The students in these schools were divided into smaller groups and were given a map of the East Midlands and a set of felt-tip pens. They were asked to talk about (and write down) language in their local and surrounding areas. Students were asked to think about where language started sounding different, how it was different and to give examples of words and/or pronunciations of different accents in the region. These conversations were not recorded as by their very nature they would have been almost impossible to analyse in this way, but students were encouraged to make extensive notes on the map and these were discussed in greater detail alongside some field notes made by the researcher. Students were also encouraged to feed back their dis-

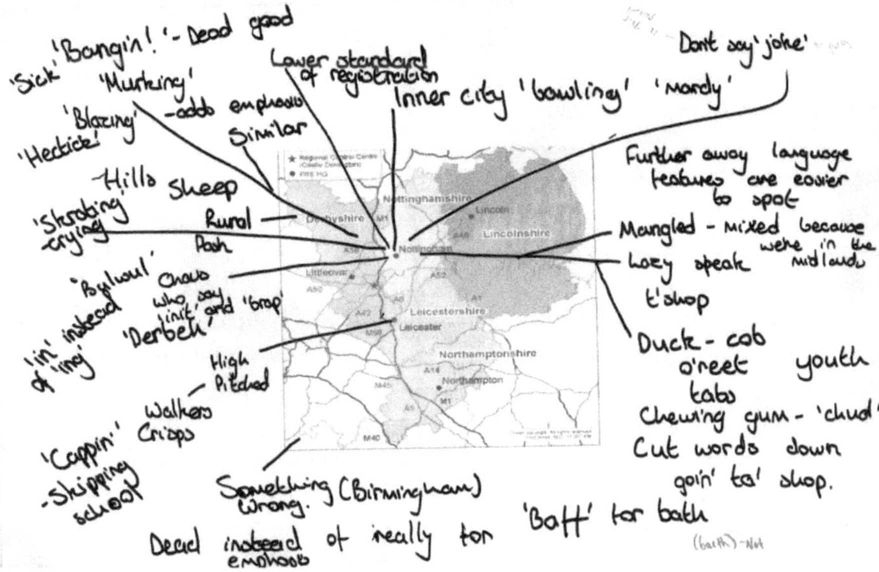

Fig. 11.7: Example of a mind map.

cussions to others within the group to examine whether students agreed or disagreed on the issues raised.

I had expected that some of the students would feel uncertain about expressing such opinions during some of the earlier tasks, and had thought during this task that students might be more willing to discuss local varieties on a more personal level. This turned out to be a very rich manner of finding out about perceptual dialectology and the task generated animated discussions, much of which centred on ideas about the "ugliness" and "negativity" surrounding their local accents (see Figure 11.7 for an example of one of these maps).

As was to be expected in such an exercise, there were certain issues which were constant for many of the participants and several which differed among the groups. Language variation within the East Midlands was a contentious issue. Some students commented that there was no variation in the East Midlands, but the majority of students agreed that there were differences within the East Midlands. They thought that Derbyshire was more rural and therefore sounded more "country" than the other counties. Others commented that Nottingham and Leicester (the cities, not the counties) were heavily influenced by being multi-cultural as well as being more urban and therefore more "rough". One interesting comment made (and then discussed further within the group) was that someone from London would not be able to tell the difference between someone from Nottingham, Derby and Leicester. Thereby the implication was that those from the East Midlands would be able to do so. However, as we can see from the previous task, this was certainly not the case for the majority of these participants.

With regard to typical features of the East Midlands, these students were very insightful and commented on features which I think are typical features of the East Midlands, dealing with phonological, lexical and morpho-syntactic variation. Some of these were lexical items (particularly "cob" for bread roll, "mardy" for grumpy, "mi duck" as a term of endearment, "reet" for right, "sen" for self as well as other items which may be more typical of youth language throughout the UK, such as "peng" for good/attractive and "innit" for isn't it). They also mentioned definite article reduction and reduction of prepositions. There were several phonetic features which were also frequently mentioned as being typical for the East Midlands, or particular areas within the East Midlands, such as the short /a/ in BATH, the non-split FOOT/STRUT in words like "cup," the happY vowel approaching the DRESS vowel, h-dropping, t-glottalisation, and the distinctive final shwa-vowel in Leicester being pronounced as "Lestah". There was much valuable information about these mind maps dealing with regional variation, which is currently being analysed and will form the basis of a future publication.

Much of the discussion centred on the perceived ugliness of the dialect. Students mentioned that it was *lazy, slurred, chavvy* ("chav" is a derogatory term used to describe loud and brash working-class youths), *Vicky Pollard* (a parody of a delinquent teenager in the popular TV sketch show *Little Britain*), *not nice, common, slang, boring, harsh, rough, horrible, not proper, nothing unique, fast* and some said that they did not like it or were trying to get rid of it. They also commented that local varieties miss out letters and that it is too colloquial. There were also mentions of *Shottingham* by the students (due to the links with high gun crime in the city). These mentions were frequent and widespread. Only one or two students said that it was relatively easy to understand and one student that it was friendly. The more neutral and positive comments only came up once or twice, whereas some of the negative evaluations were common. This is very different to other research which suggests that participants rate their own varieties as most pleasant, even if not seen as standard (Long 1999b: 202, 204, 211; Preston 1989: 82) Many of these descriptions imply a range of emotional consequences for these participants (see Garrett, Coupland, and Williams 2003: 119) and most of them are negative.

I was really interested to see that there were so many negative comments – a few of them were made by students who stated that they specifically did not speak like this, but most of them were made about the way they spoke themselves. This seems to go against research which states (e.g., Coupland and Bishop 2007) that local accents tend to be rated more positively for social attractiveness and prestige than many other accents. I wanted to learn more about why there are these negative perceptions about the Nottingham accent and what could be causing them. I started examining statistics about Nottingham and its representation in the media.

5 Nottingham: Reputation and media representation

Historically associated with Robin Hood, in recent years Nottingham has not found favour in the national press. During the last decennium a stigma has blemished the city. This covers aspects such as education, employment, crime figures, obesity and social deprivation. The indices of deprivation[4] put together by Nottingham City Council state that Nottingham ranked 17th worst (out of 326 districts) using the Average Rank measure. This is, however, an improvement from the 2004 index, where it was ranked 7th worst. Nottingham ranked fourth of the Core Cities behind Liverpool (1st), Manchester (4th) and Birmingham (9th). Half of the city areas rank within the 10 % most deprived in the country but crime is the domain in which Nottingham performs worst in media coverage. An article in *The Times* in 2006 stated "Nottingham has maintained its fearsome reputation in the latest survey of levels of serious crime committed in the biggest towns and cities" (2006: 6). The article said that residents are four times more likely to be victims of burglars, muggers and rapists than those living in the safest places. Nottingham had the highest number of murders, burglaries and vehicle crimes in the UK per head of the population. The figures for this article were taken from a study entitled *Urban Crime Rankings* which is based on national crime figures collected in 2005. These figures were also published on a website called SkyscraperCity, which stated that Nottingham tops Britain's crime league (the website also claimed residents of Nottingham were actually five times more likely to be victims of burglary, mugging or rape rather than the four times mentioned by *The Times*) [5].

This type of media attention has resulted in the city being dubbed as *Shottingham*, reflecting to the high rates of gun crime in the city (see Figure 11.8

4 See: The Indices of Deprivation, 2010. Compendium of Results for Nottingham City – April 2011. Produced by Nottingham City Council.

5 See: http://www.skyscrapercity.com/showthread.php?t=354112

Fig. 11.8: Shottingham (from notinham.wordpress.com).

for just one example) and led to widespread negative views about the city. A picture appeared on Facebook in 2013 (which has since disappeared) which looked like a postcard with different people representing the different areas of Nottingham, many of which reflect violence and crime in the city. The term also appears in the Urban Dictionary, stating that it is "a word used by many to describe Nottingham. A play on words due to the extreme amount of gun-crime in Nottingham" (Urban Dictionary, last accessed 7 July 2015)[6].

A recent BBC on-line news article discussing the riots which took place around the UK in the summer of 2012, mentioned five cities: London, Birmingham, Bristol, Liverpool and Nottingham[7]. Earlier, a 2006 television show labelled Nottingham as the 7th worst place in the UK to live in (*Best and Worst Homes*, Channel 4). More recent figures – from 2012 – from the *University Guide* stated the city has the second highest crime rate of university towns (after Manchester) according to official police data. The guide said that the city has been at the bottom of the table in three out of the four last rankings[8].

Figures from the Strategic Accessibility Assessment of Greater Nottingham[9] also show Nottingham in a poor light. Using data from the 2001 census, it states that although the city of Nottingham has two universities and is one of the UK's six science cities, there are problems with some of the educational developments. It claims that the prosperity is not shared out equally. Despite improvements in education, only 63 % of Nottingham's city residents are in employment (compared to 75 % nationally). Furthermore, in Nottingham city 74 % of the population lives in areas experiencing social need. The Greater

6 See: http://www.urbandictionary.com/define.php?term=shottingham

7 See: http://www.bbc.co.uk/news/uk-england-14457535

8 See: The Complete University Guide Website: http://www.thecompleteuniversityguide.co.uk/news/the-best-and-worst-cities-for-students

9 See: Chapter 7 of the Strategic Accessibility Assessment of Greater Nottingham 2006/7 – 2010/11.

Nottingham Development Strategy (2004) states that Nottingham has a large number of residents with low skills and no qualifications, particularly in the inner city and outer-estates; 34 % of Nottingham city residents have no qualifications compared to the national rate of 29 % (in 2001). The city suffers from a low value added economy, producing low quality products, using low skills and paying low wages.

Nottingham performs badly on “Education, Skills and Training” (Nottingham Insight). In the city, almost 10 % of 16–18 year olds are not in education, employment or training (compared with 2.8 % nationally). Absence levels in city schools are above the national average. Nottingham has a higher level of mental illness compared to neighbouring regions. The difference of life expectancy between the poorest and most affluent parts of the city is approximately 13 years. A recent article in MailOnline shows the East Midlands to be the worst region for obesity related diseases (7 May 2013).

Over recent years there has been a steady improvement in school performance in the Nottingham City area (5 % increase in attainment of 5 GCSE’s A-C since 1998). However, results still remain well below the national average. Although there has been an increase in post-16 participation, retention rates remain poor, resulting in city employers recruiting extensively from outside the city. It is estimated that over half of the jobs in the city are taken by people living elsewhere[10]. Even today, the city has the lowest percentage of 19-year-olds with five or more GCSEs in the country. A newspaper article published in the *Nottingham Post* on 29 November 2013 says that “results in Nottingham aren’t as high as they are in other parts of the country” (front page story).

Another interesting article which examines negative attitudes towards Nottingham can be found on *uncyclopedia*, a tongue-in-cheek Wikipedia which has a section on Nottingham written by a local resident[11]. He writes the following (all formatting of the original has been retained):

> Nestled languidly in a former colliery just north of South, the charming high-class resort plays host to some of the most dramatic scenery to be found anywhere south of North. Unfortunately for Nottingham, it is considered too **northern** by ~~pansies poofs shandy drinkers~~ *southerners* to warrant any interest; but also way, way, too far *down south* for the ~~pie-munching flat-cap-wearing whippet-fucking~~ Northerners to worry their heads about.
>
> Similarly, neither East nor West seem too bothered about claiming Nottingham for their own, leaving it in a place that doesn’t actually exist at all.

10 Taken from Nottingham Insight, a website containing statistics and policites relating to Nottingham and Nottinghamshire. *Http://www.nottinghaminsight.org.uk*

11 See: http://uncyclopedia.wikia.com/wiki/Nottingham

Fig. 11.9: Negative image of Nottingham (from Facebook).

> Nottingham is so remote from North, South, East and West that it doesn't even get a mention on it's *local* weather forecast, never mind the national one. (And it fucking *never* gets a mention on the national weather, trust me.)

The *uncyclopedia* article continues to talk about deprivation, poor educational achievement, low expectations, poor healthcare, gun crime, poverty, injustice, decades of underfunding and urban decay. The article highlights the negative feelings surrounding the city and also mentions the uncertainty of belonging of the region discussed in Braber (2014).

The negative publicity about the city leads to stereotypes (see Figure 11.9 – this photo appears on a Facebook group called *I live in Nottingham and I haven't been shot yet*). I was interested to see what ideas were triggered when Nottingham sixth-formers hear a local accent, and to examine any attitudes (positive or negative) that they discussed as part of the mind map tasks. As Edwards comments "listening to a given variety is generally considered to act as trigger or stimulus that evokes attitudes (or prejudices, or stereotypes) about the relevant speech community" (Edwards 1999: 102) although we do not know which linguistic features act as positive or negative triggers. Williams et al comment that "'recognizing' a dialect is inseparable from that same cluster of affective and evaluative processes" (Williams et al. 1999: 348). These views can also be found in Shuy and Williams (1973: 85) and Williams (1973: 113).

There is relatively little positive cultural salience in Nottingham. Other than Robin Hood and D. H. Lawrence, there are few famous people from the city and few TV programmes based in the city. There is some sporting prowess – the city has two football teams: Nottingham Forest (currently playing in the Championship League, which is the second highest division in England) and Notts County (this team is currently bottom of the League One division, the third highest division). Nottingham also boasts the Trent Bridge Cricket ground, home to Nottinghamshire County Cricket Club.

The question I wanted to raise after discovering these findings and considering the responses made by the students in this project was whether the lack of cultural salience, added to the negative prestige of the city, influenced the recognition and judgement of local voices. Could it be that the presumed negative social attributes of Nottingham are being transferred to the linguistic features associated with it (see for example in Preston 1999c: 360)? We know that attitudes towards language are tied in with attitudes towards people who use such varieties (Preston 2003: 40). Do these students have such negative evaluations of their own linguistic variety due to the negative association of Nottingham?

Interestingly, since the data collection took place for this project, there have been two recent appearances by Nottingham in popular culture. The first of these is the singer Jake Bugg, born in 1994. He was nominated for the 2013 Brit Award[12] for *British Breakthrough Act* and for *Best British Male* in 2014. The singer is from Clifton, a council estate in Nottingham, which at one point was Europe's biggest council estate. For Nottingham residents, Bugg represents a strongly 'local' character. The second is a TV drama production which appeared on the BBC in 2013 called *Truckers*. It is about a group of truck drivers and set in Nottingham. However, at the point of data collection, neither of these influences had come out yet.

Even though Jake Bugg can be regarded as a positive representation of the city, the link to Nottingham is not always positive. Lyrics of one of the songs,

12 The Brit Awards (stylised as the BRIT Awards; often simply called the Brits) are the British Phonographic Industry's annual pop music awards, and the British equivalent of the American Grammy Awards.

entitled *Trouble Town*, on his first album refer to problems in the city and the video is clearly shot in Clifton, Nottingham.

> Stuck in speed bump city
> Where the only thing that's pretty
> Is the thought of getting out
>
> There's a tower block overhead
> All you've got's your benefits
> And you're barely scraping by [...]
>
> Sitting on the pavement
> Boy you've missed your payment
> And they're gonna find you soon

Many of Bugg's other songs refer to excessive drinking, drug-taking and violence which he links to his own experiences in Nottingham. On the second album, released November 2013, one of the songs is entitled *Slumville Sunrise* the video to which is also set in Nottingham and features a robbery.

6 Discussion

The results section showed that students in this study were unlikely to name the East Midlands when naming dialect areas around the UK. Furthermore, they were also inaccurate when recognising the East Midlands voices in the dialect recognition tasks. I was interested to examine why the East Midlands rarely feature as one of the dialects named by students. Previous studies have shown that there are trends towards self-identification (Preston, 1989: 118) – so the question could be asked whether these empty spaces in the East Midlands reflect something else. Long (1999a: 186) has found that many participants in his study did not treat their home area as having dialect boundaries which goes against other perceptual dialectology studies which show people making more fine-grained distinctions near to home, also known as the proximity effect (Inoue 1999b: 164). Is the East Midlands an area which has no characteristic linguistic features or no popular cultural notoriety (Preston, 1989: 121)? It is certainly not the first statement as the students felt there were differences within the East Midlands and they expressed this in the mind map task. Many of the students commented that there are differences between the three main cities of the East Midlands. It is also worth considering the second statement regarding cultural notoriety. Kerswill and Williams suggest that it could be to do with the influence of the broadcast media (Kerswill and Willliams 2002: 200). This was also shown in my study as many participants named programmes and TV celebrities linked to particular accents, and as mentioned in the section on Nottingham, there are relatively few celebrities or programmes linked to Nottingham or even the East Midlands more generally. This is supported by other studies which have shown that participants were only able to identify dialects that are often heard in the media (see Bayard 1990: 92; Inoue, 1999b: 162) or those that are the focus of a particular activity, such as football (Wales 2000: 14–5). The East Midlands may not have much cultural salience and this could result in students making errors in recognising and categorising the recorded voices. We may also need to consider whether own community recognition will be better among those with strong local ties and consider the geographical mobility of individual participants.

Another aspect of the issue of the inability to label local varieties is brought out by students rating them negatively when they discuss the mind maps. This can be linked to the concepts of "claiming" and "denial", where voices are not recognised as local if they are not perceived positively (see Long 1999b: 220; Montgomery 2007: 336; Montgomery and Beal 2011: 138; Williams et al. 1999: 356) or seen as being embarrassing, in the sense that Bayard talks of "cultural cringe" of specific varieties (Bayard 1990: 68). When asked about their own linguistic varieties, some students in my study stated that they do not have an accent, although they then continue to discuss the negative features of the way they themselves talk and of the East Midlands region which were discussed in the results of the mind maps. We know that people who use non-standard varieties can see them as markers of solidarity, identity and local values (Preston (1999c: 367). But what happens when the local area is not seen as being positive – does this lead to accents being misrecognised, denied or simply not discussed? Kerswill and Williams (2002: 174) ask whether in-group membership has the power to make a voice attractive. Maybe this will change for Nottingham if it is perceived as a more positive place. We need to consider not only what people know about varieties, but also how they use this to refine their own group memberships (Williams et al. 1999: 358). We also need to consider how language helps construct social identities (E. Moore 2011: 222). By understanding more about these attitudes we can learn more about the perception of language in Nottingham.

I was interested to see to what extent these two factors are related: the lack of ability to recognise the Nottingham accent (in both tasks) and the negative comments surrounding it. As mentioned in the previous section, there are many negative associations surrounding Nottingham – poor educational attainment, lower than average rates of employment, high levels of violence. It could be that this is related to the 'denial' of this accent. It is not necessarily the case that these students are consciously aware of these statistics, and local

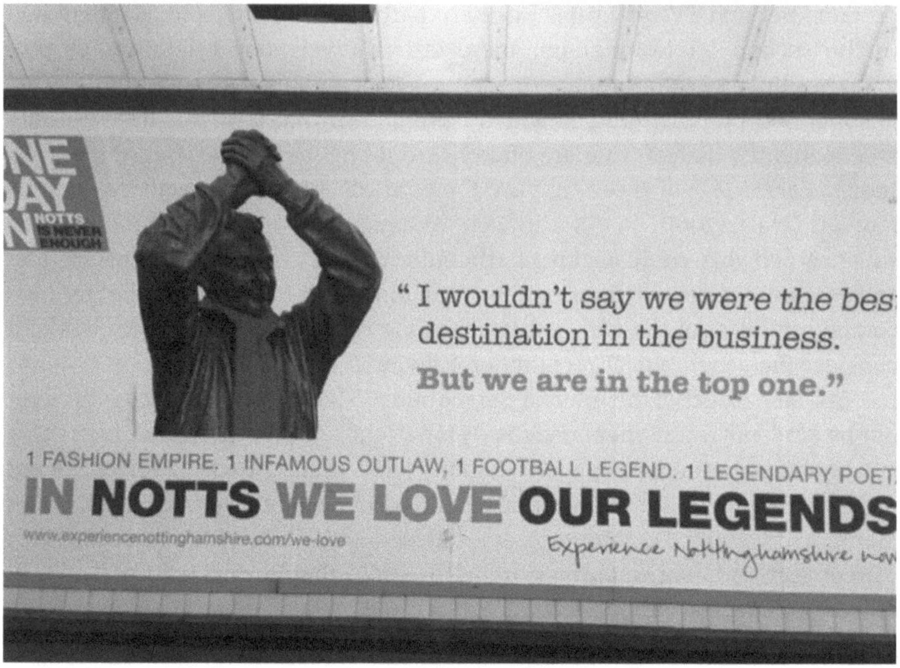

Fig. 11.10: Positive image of Nottingham (taken by Kathryn Harvey, 2013).

government is certainly doing its best to combat it, but there are certainly negative features surrounding Nottingham. Language is one of the ways in which people identify themselves (Johnstone 2004: 70) and language can be used as a commodity to represent a region (Beal 2009). Perhaps a more positive representation of Nottingham would change how the language used there is viewed by adolescents. As mentioned in the section on Nottingham, there are now two visible cultural representations of Nottingham in the TV and music world and businesses and local government in the region are working hard to promote the positive aspects of the city (see Figure 11.10, a photo taken on the underground in London to promote Nottingham).

7 Conclusion

Johnstone comments that “A sense of place comes both from immediate experience and more abstract modes of knowledge” (Johnstone 2010: 10). She adds that the subjective experience will always mediate the physical aspects of a place. This chapter has discussed the opinions of non-linguists, specifically sixth-form students in Nottingham, about language used in the city and understand how this language is recognised and viewed by them. It was found that these students were unlikely to name Nottingham when asked about varieties of language in the UK, and were unlikely to recognise local accents, even though they had strong views on language used in the region – many of which were negative. The question is whether this not naming of Nottingham, the inability to recognise local accents and the negative views on its regional language are linked to a lack of salience and Nottingham's poor reputation. Based on the opinions voiced by the students, this chapter suggests there is such a link.

Where we come from is an important aspect of our identity. We use language to classify people – ourselves as well as others (Alford and Strother 1999: 480). Such attitudes towards language and its users can predispose people to act in certain ways and may influence language change (Pearce 2009: 162). Therefore it is important to understand how a linguistic variety is viewed by its users. The negative prestige of a city or region could predispose attitudes of speakers towards the language used there. We saw in this chapter how Nottingham has been represented by the media and the many negative associations that are tied to the city. This could influence the attitudes of young people towards perception of the language used there. So perhaps the absence of positive cultural qualities or features along with a negative reputation of the city caused these students to misrecognise or deny that a Nottingham accent belongs to them.

Further research is needed to firmly establish the existence of this connection. This research can concentrate on issues such as the media representation on Nottingham as well as examining how the city is viewed by locals. Alternatively, new studies on Nottingham can compare language use and attitudes towards local language in different age groups as well as placing these local and regional findings in a wider UK context.

If current work on behalf of local councils and changes in media representation, in the shape of popular culture and music, can create a positive salience and improve Nottingham's reputation, it would be interesting to find whether the young people of this city will look more positively at their own language varieties.

John Lonergan

Chapter 12 Real and perceived variation in Dublin English

1 Introduction

Dublin is one of Ireland's oldest English speaking communities (Kallen 1994) and by far the largest city (Central Statistics Office, 2006; Simms 2001). Like many cities it is linguistically diverse, with the existing literature distinguishing vernacular and non-vernacular varieties of Dublin English (Bertz 1975, 1987; Hickey 2005; Wells 1982). It is sharply stratified by social class (Haase and Pratschke 2005) and some areas are historically quite distinct, including long-established inner city communities and relatively new suburban working class areas (Horner 1992). Bertz (1975, 1987) describes several of these historically distinct areas as linguistically distinct, although he argues that these differences were diminishing as a result of increased mobility. More recent work by Hickey (2005) does not describe local linguistic differences, although it does note that Dubliners often perceive a linguistic difference between the city's north and south. Hickey has argued that Dublin has long provided an unrecognised linguistic standard for Southern Irish English and is the point of origin for a series of vowel changes that are affecting Southern Irish English. Hickey argues that these vowel changes began as a shift away from the open vowel realisations of stigmatised, vernacular forms of Dublin English. In the case of the LOT lexical set, for example, Hickey describes a shift from /ɑ/ to /ɔ/ as increasing the contrast with the realisation used by speakers of the vernacular variety: /a/. This new pronunciation became known as a "Dublin 4 accent", after the prosperous south Dublin postcode. It became the subject of intense negative media attention (R. Moore, 2011), which Hickey (2005) argues led to the abandonment of some of the more salient vowel realisations, while other changes spread around Ireland. Given the importance attached by Hickey to the stigma and prestige of different varieties of Dublin English, a full understanding of these folk linguistic beliefs is of great importance.

This chapter investigates the number of varieties of Dublin English that Dubliners perceive to exist, as well as the geographical areas associated with these perceived varieties. It investigates Dubliners' folk linguistic perceptions in greater detail than has previously been attempted, attempting to determine whether these beliefs are more complex and multifaceted than good/bad or standard/non-standard opposition. A detailed investigation of these beliefs has the potential to provide valuable insights into the motivations of change in progress, as well as into the retention of stigmatised varieties of Dublin English. Finally, this chapter investigates the extent to which perceived linguistic varieties equate to phonetically distinct varieties of Dublin English.

An analysis of subjects' hand-drawn maps suggested that a clear linguistic difference was perceived between a varieties associated with the southeast and the north and west. A less clear difference was perceived between the inner city and the north and west. These three perceived varieties were most frequently described as "D4", "northside" and "inner city" accents. The "D4" perceived variety was found to be high in status, with a smaller advantage in perceived pleasantness. Subjects described the "D4" perceived variety as "well-spoken," "educated" and "easy to understand," but also "contrived" and "American". The "northside" and "inner city" perceived varieties are described as "rough," "broad," "strong," "common" and "unpleasant". A series of semi-structured interviews revealed a more complex set of associations, however. Inner city subjects were found to value their speech as an ingroup marker and as warm and direct. While southsiders were broadly critical of vernacular forms of Dublin English, they did associate these varieties with a "dry," "earthy" sense of humour. The interviews also suggested that American speech norms are a stigmatised characteristic of the "D4" perceived variety. Dubliners' perception of a north/south linguistic divide was not supported by the analysis of the vowel systems of the subjects in this study. Inner city Dubliners were found to have a vowel system that differs clearly from that of northsiders and southsiders, whose vowel systems differ little from each other. The popularity of the north/south dichotomy is argued to reflect historical and socio-economic differences, which are seldom discussed explicitly in Ireland.

Dubliners' perceptions about the number and location of varieties of Dublin English were investigated using hand-drawn maps, which were digitised and combined, highlighting frequently marked areas which are discussed in Section 4.1. Subjects also completed questionnaires about the varieties they marked. These results are discussed in Section 4.2. Geographical areas of interest were identified based on these tasks, combined with a social and historical study of Dublin, which is discussed in detail by Lonergan (2013). Subjects were recruited for interviews in each of these areas, with discussion focussing on subjects' perceptions of Dublin English. The results of these interviews are discussed in Section 4.3. Finally, subjects' vowel spaces were subjected to acoustic analysis. The results of this analysis are contrasted with subjects' folk linguistic perceptions in Section 4.4.

2 Perceptions of Irish English

An early investigation of the linguistic perceptions of Dubliners was conducted by Edwards (1977). Subjects were played a series of recordings, in a series of Irish accents, under the false impression that each was of a different speaker. Subjects rated each "speaker" on a series of scales. The (non-vernacular) Dublin English guise was rated as the least "important," "trustworthy," "ambitious," "industrious," and "intelligent" variety tested, but received the highest marks for "sense of humour" and the second highest for "friendly". In Hickey's (2005) more recent research subjects were played the same script read by speakers of several varieties of English and rated that speech on several scales. The 41 subjects from Dublin rated the "Dublin 4" variety as "important, intelligent" and "well-educated," while "Local Dublin" was rated much more poorly in both respects. On the other hand, there was little difference between the perceived "pleasant, friendly" ratings of the two varieties. Hickey (2005) conducted a second experiment that used hand-drawn maps to investigate respondents' perceptions of the number and location of varieties of Irish English. Hickey analysed 207 maps, of which 138 were drawn by Dubliners. 65 % of Dublin-based subjects divided the city into two or more varieties of English, usually making a north/south split. The annotations written on subjects' maps suggest that the perceived variety associated with North Dublin was seen as "strong" and "hard" as well as "normal" or "common". This is contrasted with a variety associated with south Dublin that was described as "posh" or "snobbish" but "clear" and "distinct".

Hickey's (2005) perceptual data suggests that Dubliners perceive a linguistic difference to exist between the city's north and south, but the national scale of his "draw-a-map" tasks precludes the detection of a greater number of perceived local differences (if any exist). In contrast to the widespread perception of a north/south linguistic difference, Hickey does not identify local differences in Dublin English. This is suggestive of a mismatch between perceptions of difference and actual linguistic behaviour that is worth further study. This chapter therefore further investigates perceptions of local differences in Dublin English and contrasts these beliefs with actual phonetic data from subjects recruited in the identified areas. The existing literature provides support for the characterisation of vernacular forms of Dublin English as stigmatised, while the "Dublin 4" variety is perceived as high in status. Further insights can also be gleaned from the annotations found on subjects' maps, but the number of comments made is comparatively small. This chapter therefore uses questionnaires to elicit more data than is found in previous work. Semi-structured interviews are also used to provide more a more nuanced view of the possibly complex and multifaceted views that Dubliners have about Dublin English.

These views may, in turn, allow a better understanding of the social significance of variation and change in Dublin English.

3 Methods and data

Groups of Dublin residents used maps to identify areas of Dublin that they believe are linguistically distinct and used questionnaires to describe those differences. Interview subjects were recruited in each area identified as linguistically distinct. Subjects discussed their beliefs about Dublin English and completed a task that elicited tokens of their vowels. A linguistic description of subjects from each area was then compared with subjects' perceptions of local linguistic differences. Undergraduate students at University College Dublin, as well as staff at the National Institute of Cell Biology at Dublin City University completed the draw-a-map task. Minimal demographic information was gathered about these subjects, but most were female and highly educated. 52 subjects, working mostly in groups, were provided with a blank map of Dublin, along with a reference map in which suburbs were named as a navigation aid. Subjects were asked to circle areas of Dublin where they believed most people spoke differently. Subjects were asked to number each area that was different, but to re-use a number if they believed most people spoke the same way in two different places. Each subject also completed a questionnaire. Subjects were asked to name each variety they marked and asked "What does this accent sound like?" and "What kind of people speak with this accent?" Finally, subjects were asked to rate on a scale of 1 to 5 how "pleasant" and "good English" each variety they indicated was. In order to provide an overview of the contrasts made by most Dubliners, subjects' hand-drawn maps were digitised and combined. Clusters were identified based on the names assigned to these perceived varieties, as well as the degree to which they overlapped. Based on these findings, three areas were selected for further data collection. Additional subjects were recruited and interviewed in each of these three areas. Most of these subjects also completed draw-a-map tasks and questionnaires and their results were added to the earlier data. The total number of draw-a-map tasks rose to 97 in this fashion.

73 subjects were recruited in the city's north (around Whitehall, Beaumont, Santry, Ballymun and Coolock), southeast (along the coast from Sandymount to Dún Laoghaire), and inner city (between the Royal and Grand canals). The majority of subjects were recruited using the friend of a friend methodology, although a number of older subjects were recruited through a historical society and a retirement group. Community groups such as a creche, a community centre, a youth project and a support group were crucial to recruitment in the

inner city. All subjects were interviewed in pairs, and in almost all cases subjects were very well known to each other. Subjects were allowed to select any quiet venue for the interview, with the majority of interviews taking place in subjects' homes, work places or community centres. The interviews were recorded using a Marantz PMD670/W1B solid-state recorder, with two Shure SM10A dynamic headset microphones.

In each interview the draw-a-map task was used to initiate a conversation about Dublin English. Subjects were allowed to raise any issues they wished, with the interviewer simply expressing interest, and occasionally asking questions to clarify statements. As well as recording large amounts of natural, animated speech, this component of the interview provided insights that could not have been obtained from a questionnaire and allowed the interviewer to further investigate subjects' views with additional questions. Finally, subjects completed a series of map leading tasks, in which they helped each other navigate maps using a series of landmarks designed to prompt a range of linguistic variables. By prompting the subject to focus on giving directions they distract him or her from their use of language and generate natural informal speech (Anderson et al. 1991). Praat scripts were used to isolate each lexical item and mark the beginning and end of each vowel. F_1 and F_2 readings were taken at the 25 % time point in each vowel and subjected to Lobanov (1971) normalisation to control for the acoustic differences between men and women's speech. In order to accurately track subjects' vowel formants accurately it was also necessary to tailor the program's settings to each subject, as recommended by Wood (2005). These results were graphed using R and the non-parametric max-t test[1] (Herberich, Sikorski, and Hothorn, 2010) was used to check for differences in normalised F_1 and F_2 between older (40+) and younger subjects, as well as between men and women and between inner city, northside and southside subject groups.

4 Perceptions of Dublin English

4.1 The structure of Dublin English: Draw-a-map task results

As seen in Figure 12.1, most subjects marked between two and six perceived varieties of Dublin English on their maps, with the most common number being four.

---

1 This test was used instead of Tukey's HSD because of the non-normal distribution of F_1 and F_2 values in some tokens, combined with unequal subject numbers from different age/gender/area combinations.

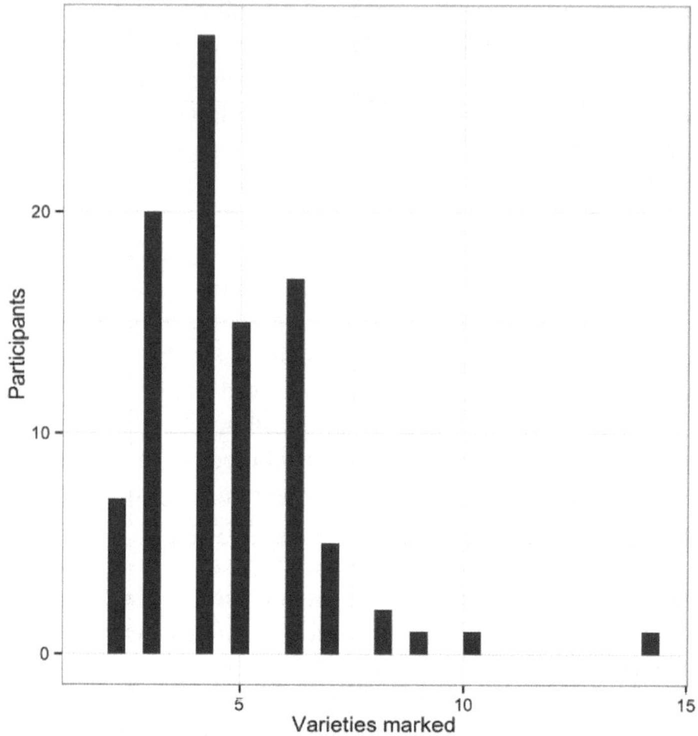

Fig. 12.1: Number of perceived varieties of Dublin English.

Subjects' maps were scanned and combined in order to determine which areas of the city (see Figure 12.2) were most frequently marked. In Figure 12.3, the most frequently marked pixels are darkest while the least frequently marked pixels are lightest. Four separate dark areas are discernible, in the north, inner city, southwest and a long band along the southeast coast.

The most frequently marked areas in Figure 12.3 correspond closely to historically and socio-economically distinct communities. Inner city Dublin contains the city's oldest working class communities (Prunty 2001) while suburban northern and southwestern suburbs were settled by late 20th century slum clearance programmes (Horner 1992). In contrast, the city's affluent southeast was developed in the 19th century as middle-class Dubliners moved to the suburbs (Brady 2001). These population movements are reflected in the distribution of social class in modern Dublin. The city's southeast remains predominantly affluent, while serious deprivation exists in parts of its inner city and northern and western suburbs (Haase and Pratschke 2005).

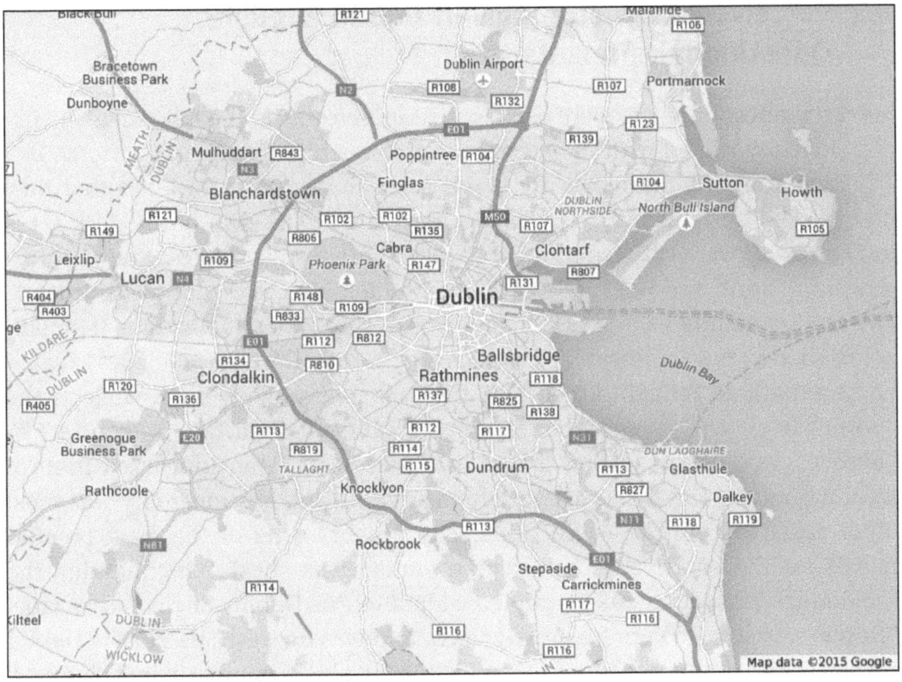

Fig. 12.2: Dublin city (Map credit: Google Maps).

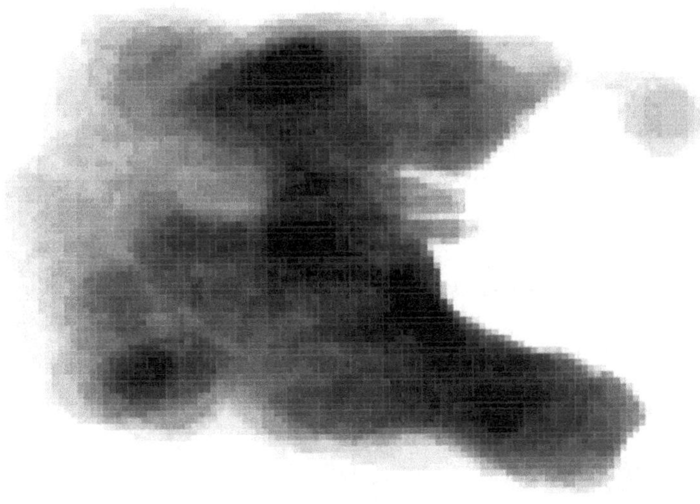

Fig. 12.3: The frequency with which each pixel was circled by respondents.

4.2 Perceived characteristics of Dublin English: Questionnaire results

Subjects' questionnaires were used to further investigate the folk linguistic impressions associated with the geographical areas highlighted in Figure 12.3.

Each subject's questionnaire asked them to name the varieties that they had marked on their map. The names associated with each of the dark areas in Figure 12.3 were collated into Table 12.1. These results suggest that the clusters in the north and west are typically both identified as a "northside accent". The predominantly working class southwestern suburb of Tallaght is frequently included in the "northside," although a minority of subjects did mark a separate a separate "Tallaght accent" on their maps. For many other subjects the "northside accent" is described simply a "Dublin accent," or a "neutral" or "normal" variety, suggesting that it is widely seen as an unmarked form of Dublin English. The perceived variety marked in Dublin's inner city was most often described simply as an "inner city accent". One of the most popular names for the perceived south-eastern variety was "posh," highlighting the importance of social class in any consideration of Dublin English. The label "Dublin 4" was also very frequently used, and was not restricted to the Dublin 4 postcode in southeast Dublin. "Southside" is also a popular name, suggesting a lack of agreement about whether the perceived Dublin 4 variety is separable from the Southside more broadly. A number of subjects described the English spoken in the southeast as "normal" or "neutral," suggesting that it represents an unmarked norm for them. The use of similar terminology in the north

Tab. 12.1: Names used to describe the three most commonly distinguished varieties of Dublin English.

North/West		Centre		Southeast	
(Strong) Northside	30 (17 %)	Inner city	39 (57 %)	D4	38 (24 %)
(Strong) Dublin	25 (15 %)	(Strong) Dublin	6 (9 %)	Posh	37 (24 %)
Neutral/normal	14 (8 %)	Old Dublin	3 (4 %)	Southside	25 (16 %)
Working class	12 (7 %)	Working class	3 (4 %)	Neutral/normal	12 (8 %)
Inner city	11 (6 %)			South County Dublin	7 (4 %)
Knacker/skanger/scumbag	11 (6 %)			Rural	5 (3 %)
Tallaght	10 (6 %)			Upper class	4 (3 %)
West	6 (3 %)				
Flat	5 (3 %)				
Rural	5 (3 %)				
Common	5 (3 %)				

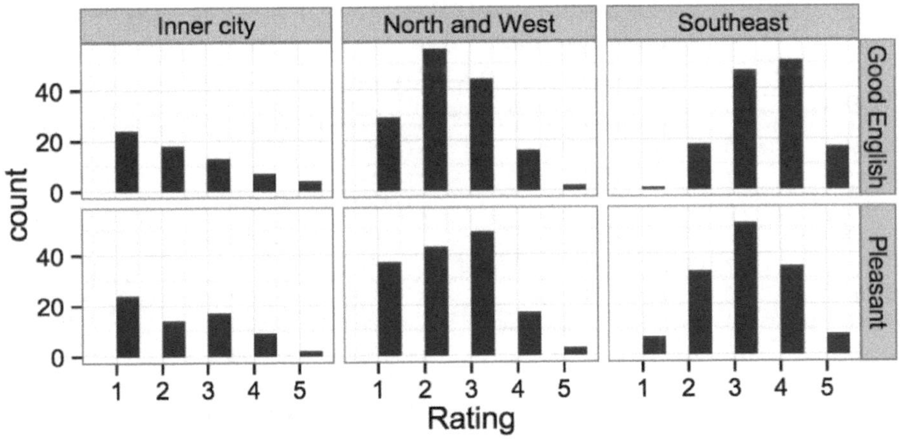

Fig. 12.4: "Good English" and "Pleasant ratings' for each of the three perceived varieties of Dublin English, from 1 ("Very unpleasant" / "Very bad English") to 5 ("Very pleasant" / "Very good English").

and west may suggest that the status of "normal" may be disputed. Finally, it is noteworthy that this perceived variety is never described as simply a "Dublin accent".

The scores for "good English" and "pleasant" assigned by subjects to the varieties they marked in the north or west, southeast and inner city are contrasted in Figure 12.4. This suggests that the "D4" south-eastern variety enjoys a relatively small advantage in perceived pleasantness relative to the "northside" and "inner city" perceived varieties. The chief difference between the two distributions is that very few subjects assign the "D4" perceived variety the lowest possible mark for pleasantness. A Mann-Whitney U test confirmed the presence of a significant ($p = 0.0002162$) difference between the "pleasantness' ratings of the "D4" and "northside" perceived varieties. Figure 12.4 suggests that the "D4" perceived variety enjoys a greater advantage in perceived correctness, suggesting that it is high in status. This is supported by the highly significant ($p = 1.93 \times 10^{-10}$) difference observed between the "good English" ratings of the "D4" and "northside" perceived varieties. The "inner city" perceived variety was not found to differ significantly from the "northside" perceived variety in either "good English" or "pleasant" ratings. Overall, these results suggest that Dubliners make a distinction between one variety that is good English and others that are not.

As seen in Table 12.2, the adjectives used by subjects to describe perceived varieties of Dublin English provide insights into the criteria that subjects use to evaluate them. The "posh" (used 84 times) D4 perceived variety is contrast-

Tab. 12.2: The most common adjectives used to describe "northside," "inner city" and "Dublin 4" perceived varieties of Dublin English.

Northside		**Inner city**		**D4**	
Rough	30 (17 %)	Rough	10 (13 %)	Posh	78 (29 %)
Broad	21 (12 %)	Broad	10 (13 %)	Clear	29 (11 %)
Common	15 (8 %)	Common	8 (10 %)	Contrived	27 (10 %)
Unpleasant	14 (8 %)	Flat	7 (9 %)	American	26 (10 %)
Unclear	12 (7 %)	Nasal	6 (8 %)	Slow	13 (5 %)
Clear	10 (6 %)	Unclear	5 (6 %)	Neutral	12 (4 %)
Neutral	10 (6 %)	Unpleasant	5 (6 %)	English	9 (3 %)
Slow	10 (6 %)	Slow	4 (5 %)	Educated	7 (3 %)
Flat	9 (5 %)			Nasal	7 (3 %)
Soft	7 (4 %)			Unpleasant	7 (3 %)
Rural	6 (3 %)			Pleasant	6 (2 %)
				Soft	5 (2 %)

ed with "common" northside and inner city perceived varieties (used 25 times), for example. This opposition is suggestive of concern for social class or social status. The disparity in "good English" ratings seen in section 4.2 is also mirrored by a concern for the clarity of these perceived varieties. Subjects frequently contrasted the "clear" or "easy to understand" D4 (50) variety with "unclear" or "hard to understand" (19) northside and inner city varieties[2]. Subjects also often contrasted a "soft," "mild" or "bland" (18) D4 variety with "strong" or "broad" (30) northside and inner city varieties. The D4 perceived variety was also frequently described as "contrived" (31), in contrast while the northside and inner city perceived varieties were sometimes described as "authentic" (2) or "traditional" (9). This is suggestive of ambivalent attitudes towards this high-status perceived variety, which is also regarded as "ridiculous' (5), "American" (26), and "English" or "British" (9). These adjectives are not paired with opposites. This rejection of a high-status variety as affected and influenced by foreign speech norms suggests strong parallels with Johnston's (1985) research on the Morningside/Kelvinside varieties of Scottish English. While there appears to be a consensus that speakers of the northside perceived variety are "rough" or "tough" (41), no subjects described speakers of the D4 perceived variety as "weak". Finally, it is noteworthy that both the northside and D4 perceived varieties are described as "normal" or neutral," which may suggest that this status is disputed.

2 Interestingly, the "Northside" perceived variety is described as "clear" and "unclear" with roughly equal frequency, suggesting a level of disagreement between subjects.

4.3 Perceived characteristics of Dublin English: Interview results

The questionnaire data and rating tasks discussed in sections 4.2 and 4.3 suggest a contrast between a prestigious "southside" variety and stigmatised "inner city" and "northside" varieties. Research on language attitudes has typically shown these attitudes to be more complex and multifaceted than this suggests, however. In particular, attitudes can be grouped into the categories of status and group solidarity (Ryan and Giles 1982). Those varieties that are low in status are often found to be strongly associated with high group solidarity. Conversely, high status is associated with lower group solidarity. Preston's (1999c: 368) research, for example, suggests that the English spoken in the American South is low-prestige, but also "casual," "friendly" and "down to earth".

Similarly, while the questionnaire data analysed in section 4.2 suggests that "northside" and "inner city" perceived varieties are seen as "common," this does not mean that they do not have other more positive attributes. In addition, the analysis of section 4.2 does not establish whether this perception of commonness (for example) is shared by inner city Dubliners and northsiders. These factors may be of substantial importance to explanations of variation and change in Dublin English. This section therefore analyses the semi-structured interviews conducted with southside and inner city subjects, contrasting the positive and negative descriptions of both in-group and out-group members. This analysis reveals that inner city subjects have adopted the negative term "common" to describe their own speech, as seen in interview 27 with two younger inner city men:

(1) 027A: No that's what we were brought up like that's what we were reared with that word like "common". I don't know like I don't even know what it means to be honest with you. I don't you know like like, com common as muck that's all you know and that's what I what I've heard like but
027B: Yeah. Inner city.
027A: Wuh wuh we are common compared to fucking [...] a Finglas, anywhere else in Dublin.

In general, "posh" appears to be used to indicate high status, while "common" indicates low status. Inner city subjects appear to regard their variety of Dublin English as the most "common" in the city, but they do not describe it as homogenous. Instead, a continuum of commonness is typically described. For example, two older women argued that while they speak with an inner city accent, drug dealers in the area speak a more common form.

(2)	003B:	And then there would be a real common accent within that again you know in the inner city, would there be? Yeah.
	003A:	There is, there's different accents.
	Interviewer:	Is there?
	003A:	Yeah, definitely.
	Interviewer:	And how can you tell one from the other?
	003A:	There's real, like if you heard the dealers selling on the street that's the real, Dublin Liberties accent, like do you know what I mean and then it gets kind of milder from that then doesn't it?
	003B:	Yeah.

In this excerpt the subjects refer to the Liberties area of inner city Dublin, an area strongly associated with vernacular Dublin speech, as discussed by Corrigan, Edge and Lonergan (2012). Inner city subjects routinely situated themselves and their family members on a continuum of commonness. Subject 003A, for example, argues that one of her daughters "is real soft spoken, me eldest daughter, but the youngest one is, she is as common as muck". Similarly, a younger inner city woman (001B) argues that "me ma wouldn't be as common as I would be now". While commonness is an apparently negative trait, it also appears to function as a powerful ingroup marker. This is demonstrated by the discussion between the two younger men in interview 27 about how their daughters speak. 27B describes his daughter as common, but claims that he has no wish to change it because "it's where we're from, like do you know what I mean and and we're puh I'm I'm proud of it like". In contrast, his friend expresses dismay that his daughter has begun to speak in a very different way to him after attending school in a relatively prosperous north Dublin suburb:

(3) 027A: My young one goes to a a snobby school in Glasnevin. You know and she's lives in the top ante in the flats like you know. And you can, and when she's at home she's as common as muck like you know, but when she goes to school it's "mammy" and "daddy" [...] and I do be taken back like saying where the fuck like. Where's she getting this? [...] I'd rather her be with her own people...

This man expresses the wish that his daughter "be with her own people," suggesting that these inner city Dubliners have a strong sense of belonging to a community that is separate to the city as a whole. This impression is strengthened by a young woman's (001A) description of moving to inner city flats as a child. She found that "people in flats don't usually take in outsiders. Like when you're new like you have to kind of settle in first until like, you'd be there months before they'd even acknowledge you like". For some inner city Dubliners their speech appears to be emblematic of their membership in their community, and outsiders are not entitled to use it, as seen below:

(4) 027B: I know people that try talk, now particularly girls, that try talk like people in the inner city [...]
[...]
027B: They be trying to talk with the real, you know the inner city tone, the *here* /hi:jə/, all like that you know. And you do say what the, where did they get that from like you know?
027A: <laughs> Yeah, yeah.
027B: And you know it's put on like you know. And like I'd I'd know it's put on because I'm, as I'm like <name deleted> we were born and reared in in the inner city, so we know, and as I said to you where you really hear that, the real *here* /hi:jə/ is Sherrif Street, or Sean Mac Dermott Street, around you know Summerhill, Ballybough. That's where you hear it. But to hear it in Darndale or Coolock or Finglas? Like you're saying where she's after buying that, the pound shop?

This man argues that residents of the north inner city areas of Summerhill and Ballybough have an exclusive right to the "inner city tone". In his view that right does not extend to Dublin's northern working class communities. This attitude is somewhat surprising, given the historical links between the communities and the fact that that they are socio-economically very similar (Haase and Pratschke 2005). Aside from "common," the draw-a-map task identified "rough" and "tough" as some of the most commonly used adjectives when describing inner city and northside speech. The semi-structured interviews showed that subjects were aware of this perception and somewhat bemused by it.

(5) 027B: You could be walking there, like I've often been probably on a bus or I've been walking down the road and you know if they hear you on your phone and you know they look behind you you can see the panic of them. You
Interviewer: OK, so they're afraid of you?

027B:	Yeah the snobbery, the snobs like you know can see the panic in them.
Interviewer:	Mm.
027B:	You can, like what the fuck?
027A:	I think that we're just we can be just more outspoken.
027B:	Outspoken, that's like that's the way
027A:	And we'll tell you how it is like and and we'll voice it like.

Here we see that the stereotype of their speech as threatening or rough is regarded as unfair. Many subjects did observe that inner city speech is loud and that this could be intimidating for people unused to it. However, no inner city subject described himself or herself as intimidated by any working class, inner city or northside form of Dublin English. In fact, while inner city Dubliners universally described their speech as "common" they were otherwise almost uniformly positive in how they described it, typically comparing it favourably to "posh" Dublin English. The inner city Dubliners interviewed regard their own speech as simple and concise where posh speech is pompous and long-winded. In addition, they believe it is warm where posh speech is cold, and welcoming where posh speech is dismissive and arrogant. Several subjects expressed strong disapproval of peers who had adopted "posh" pronunciations, but nonetheless acknowledged that it was sometimes necessary to modify one's speech depending on the social situation, as seen below:

(6)	028B:	[...] just say I met someone and that person could do something for me it was a person a manager or something like somewhere, that I didn't know.
	028A:	Mm.
	028B:	And he says hello, he'd be well spoken, I'm <name deleted> I'd say "hello I'm <name deleted>, very pleased to meet you". I'd define me words I would.
	028A:	But don't you, we do that anyway when we're talking on phones if we're ringing somewhere important and we're talking about
	028B:	Exactly.
	028A:	We try and put on that accent and we'll talk slower and
	028B:	Try and define our words in clear English. [...]
	028A:	These are important people I'm talking to so, let's talk important here you know make sound like

028B: Let's sound good.
028A: Good, yeah.

Implicitly, it would appear that while inner city speech is direct, warm and welcoming, it is neither "defined" nor "important". It therefore needs to be reduced when specific social goals must be obtained. Permanent linguistic change is, however, rejected as an almost dishonest act:

(7) 028B: [...] but I wouldn't hide me accent like I wouldn't, that's not saying that I'd try and be false.

Southsiders view of their own speech is, unsurprisingly, very different to the self-image of inner city Dubliners. While inner city Dubliners almost universally described their own accent as "strong" and "common," southsiders consistently described their own speech as "soft" or "neutral," as seen below.

(8) 35B: So my mum's family in Templeogue would have that almost neutrally kind of accent.

(9) 38A: So I think then there's kind of a neutral Dublin accent, which is what I think I have.

(10) 015A: I have found that anywhere I've gone around the world, that people don't know where I'm from.
Interviewer: Mm-hm.
015A: So, to me that kind of indicates that I've a bit of a neutral accent. But when I'm in Ireland, people know that it is, kind of they would say oh a posh Dublin accent.

This self-image is very different to the way these subjects are viewed by inner city Dubliners, who almost never use adjectives like "neutral" to refer to any variety of Dublin English. In addition, very few inner city subjects distinguish different varieties of "posh" Dublin English in their interviews, but southsiders overwhelmingly distinguish a "Dublin 4" variety that they don't speak and don't like. Two 29-year old southside women had the following exchange, for example, referring to students at University College Dublin (UCD), which is located in the city's affluent southeastern suburbs:

(11) 015B: I think people probably from around here had a bit of an accent but definitely the like even the girls a bit younger than us like if you look at the <15A: Oh God> first-year UCD

	Arts for example and they all have a particular look, with the hair back brushed and the makeup out to here and the Ugg boots and, they definitely their accents are much more from I think watching all the kind of programmes on TV like the Hills and all those American
Interviewer:	So sort of American?
015A:	Definitely. "Like".
015B:	Yeah so they've got that really California thing going on with their intonation and, definitely, I think there's a big change.
015A:	Yeah even sitting on the bus the other day listening to two girls behind me talking. Like I had to put in my earphones I couldn't bear to listen to them like. "Like" was every second word you know. Painful stuff.

These subjects' negative reaction to this perceived linguistic change is clearly evident, and this is typical of southsiders' comments. Their argument that American pronunciation norms are being adopted is also one of the most frequent comments heard in interviews, with television usually described as the cause. As in the example above, subjects typically mentioned the use of high rising terminals and discourse *like* (although this feature is present in their own speech). Almost all southsiders interviewed reported that a change had occurred in the form of English spoken in south Dublin, although almost none described themselves as participants in that change. Older subjects described it as affecting their children. Younger subjects described it as affecting even younger people, or people who attended different schools, as seen below, where subjects refer to several prestigious fee paying schools:

(12)	015B:	Like I think we have a south Dublin accent but then we're we're kind of somewhere on a continuum. But that continuum can go to really, really marbles in the mouth, like the whole rugby scene.
	015A:	Yeah, that's true. Yeah. Yeah.
	015B:	Like the Ross O'Carroll-Kelly characters the way he portrays them, but those people aren't a caricature type like they do talk like that. You know it's not, I don't think he's exaggerating.
	Interviewer:	He's just, that's not everybody in south Dublin?
	015A:	Yeah.
	015B:	Yeah, but that's only maybe, the stereotypically guys who went to Michael's and Belvedere and em Gonzaga, girls who

went to Mount Anvil, Foxrock. Just a few groups. And then not everybody in those groups [...]

Here we see that these subjects view the variety they condemn as an extreme version of their own, just as inner city Dubliners frequently describe themselves as intermediate in a continuum of commonness. It would seem that both extremes of the linguistic hierarchy are stigmatised, with subjects gravitating towards the linguistic centre in how they perceive themselves. The young women quoted above also make reference to the Ross O'Carroll-Kelly novels and newspaper columns. These publications satirise the affluence of residents of parts of Dublin. One book in the series is, for example, entitled *Ross O'Carroll-Kelly*'s *Guide to (South) Dublin: How To Get By On, Like, €10,000 A Day.* The publications focus in particular on the variety of English popularly associated with residents of these areas. They are written in eye dialect, with spellings suggestive of back vowel raising ("car" spelled as "cor") and the Irish English "soft t" ("right" spelled as "roysh"). This material was mentioned by most southsiders (but no inner city Dubliners), suggesting that it has been highly influential in forming opinions and stereotypes on the subject.

While inner city Dubliners were found to be quite positive about their speech, the questionnaires analysed above revealed clear stigma associated with the perceived varieties in the inner city and northside. These negative views are clearly seen in the southside interview data, where subjects used words like "flat," "careless," "indistinct," "lazy," "common" and "rough" very frequently. However, the interview data also revealed that southsiders' views of these varieties were not entirely negative. One younger southside woman argues, for example, argues that the coarseness of inner city speech is balanced by its wit:

(13) 015B: I think the Dublin just being, strong em, a bit coarse maybe, full of wit as well the way they speak em is brilliant. I actually quite like the Dublin accent the inner city Dublin accent though I think like what I think of Moore Street. I think of the street vendors. I think of their like really strong Dublin accent to the point that it nearly surprises me, you know I feel like wow, they speak like that and we're from the same, only a few miles apart. It's amazing.

Almost all interview subjects singled out the street traders of Moore Street as exemplars of the Dublin accent, as this woman has done. Subjects mentioned them explicitly or, when impersonating working class forms of Dublin English, frequently did so using the terminology of the street traders: "three for a pound" /tʃɹiː fəɹ ə pæʊənd/, being a particularly common catchphrase. For many interview subjects the speech of this group of Dubliners appears to be a valued part of Dublin life, despite its stigma. Southsiders' mixed feelings about marked forms of Dublin English are illustrated by two excerpts from an interview with two younger southsiders. In the first, we see that both agree that it has "character," but can be "grating":

(14) 038B: The Dublin accent I think is
038A: Character.
038B: Character, yeah. They and usually even if they're just talking normally or say something normally it's, just usually the way they express it is it can very witty to listen in on.
038A: Yeah. Yeah.
038B: Yeah.
Interviewer: So you like the sound of the strong Dublin accent?
038B: Yeah, it's just, constantly it grates on you after a while.
[...]
038A: There's a real humour to it.
038B: A real dry humour.

This subject's somewhat positive description of this variety of English transforms quite dramatically when he is asked about his own speech. After initially describing his own speech as "normal," he admits to a Dublin accent and reveals a conflicted attitude towards it:

(15) 038B: No in fairness I'd probably hedge a little bit more towards a, kind of a stronger Dublin accent. When I was younger, one stage it got really kind of Dublinish, because of I think, just the group where I was hanging around with. And then I lived in Kilkenny nearly to knock the Dublin-ness out of the voice, eh but it's starting to creep back in unfortunately.
Interviewer: And you wouldn't like to to think you were very Dublin.
038B: No I suppose I wouldn't like to be the real rough kind of, Dublin accent. I think it's great fun to listen to from time to time em but I think there's a certain association of real inner Dublin to it that that you'd be nearly judged on your accent alone on that one.
Interviewer: So if you had kids you wouldn't want them growing up speaking that?
038B: Ideally not, no, no I'd be knocking it out of them <laughs>.

Here the interview subject makes clear that he has struggled to modify his speech due to the stigma associated with that variety. This provides some support for Hickey's (2005) argument that speakers of non-vernacular varieties of Dublin English have sought to maximise their linguistic distance from vernacular varieties of Dublin English. The subject interviewed above makes it clear that the positive attributes of working class forms of Dublin English are clearly outweighed by its stigma. Like many southsiders, he appears to be glad to that this variety of Dublin English exists, but wouldn't want to speak it.

The interviews revealed that stop realisations of orthographic "th" are an extremely salient characteristic of vernacular forms of Dublin English for many subjects. Several subjects also commented on lenition, noting the prevalence of glottal stops in vernacular forms of Dublin English, as well as the use of a fricative "slit-t" that are well-known features of Irish English. Some subjects also noted the increasing prevalence of flapped realisations of intervocalic /t/ that have also been previously described in non-vernacular forms of Dublin English. Subjects made far fewer specific observations about the vowels Dublin English. Individual speakers did describe the D4 perceived variety as showing THOUGHT raising, central rhotic vowels in *fair hair* /fə˞ hə˞/ and centralised GOAT onsets /əʊ/. These are all realisations noted by Hickey (2005) as characteristic of recent change in Dublin English.

4.4 Summary

This chapter has investigated the number, geographical location and social significance of perceived varieties of Dublin English using hand-drawn maps, questionnaires and semi-structured interviews. Most subjects distinguished a "D4" variety in southeast Dublin from a "northside" variety in the city's north and west. A smaller number of Dubliners distinguished an "inner city" from the northside. Questionnaire data suggested that the D4 variety is high in status, receiving high marks for correctness and being described as "posh" and "clear" (but "contrived"). The northside and inner city varieties both appear to be low in status, with very low marks for correctness and pleasantness. These two perceived varieties were described in very similar terms, as "rough," "broad" and "common". A more complex set of attitudes was suggested by a series of semi-structured interviews. In particular, inner city subjects valued their speech an ingroup marker and as a warm and direct way of speaking. Interview data also suggested that southsiders believe that vernacular forms of Dublin have some positive attributes, but these are outweighed by its stigma. These views provide some support for Hickey's argument that ongoing change in Dublin English began as a shift away from stigmatised form of Dublin English. In addition, comments about linguistic change were limited to the high-status D4 perceived variety and the few linguistic variables specifically mentioned by subjects match well with the changes that Hickey describes in Dublin English.

5 Phonetic analysis

The analysis of the hand-drawn maps above suggests that Dubliners perceive a clear north/south difference to exist in Dublin English, with a less clear perceived difference between the inner city and northside communities. These perceived differences, combined with the socio-economic and historical differences between the areas, motivated the recruitment of subjects from each of these areas as part of a larger study on variation and change in Dublin English (Lonergan, 2013). This study examined subjects' entire vowel space, finding that the inner city subjects had a vowel system that differed dramatically from the other two subject groups. In contrast, subjects recruited in Dublin's north and southeast showed few differences, suggesting that the perceived northside/southside dichotomy is a highly inaccurate description of Dublin English. A clear example of the contrast between inner city subjects' vowel spaces and those of their northside/southside counterparts is seen in Figure 12.5, which contrasts realisations of the NURSE lexical set with realisations of TERM. The TERM lexical set has been used by Hickey (2005) to capture a contrast he identifies in vernacular forms of Dublin English between a back rounded realisation /ʊ/ and a front, unrounded one /ɛ/. This contrast is seen clearly in the speech of both older and younger inner city Dubliners, but not the northside or southside subjects.

The obvious differences in fronting between inner city Dubliners' realisations of NURSE and TERM were confirmed by max-t analysis ($p < 1 \times 10^{-5}$)[3]. Differences in the fronting of NURSE and TERM were much smaller for the other groups, suggesting that inner city subjects speak a vernacular variety of Dublin English, while northside and southside subjects speak a non-vernacular variety. The results also suggest a clear gender difference among northside and southside subjects. Male northsiders and southsiders show a small but signifi-

3 Male inner city subjects were also found to have a realisation of NURSE that was somewhat higher than their realisation of TERM ($p < 0.01$).

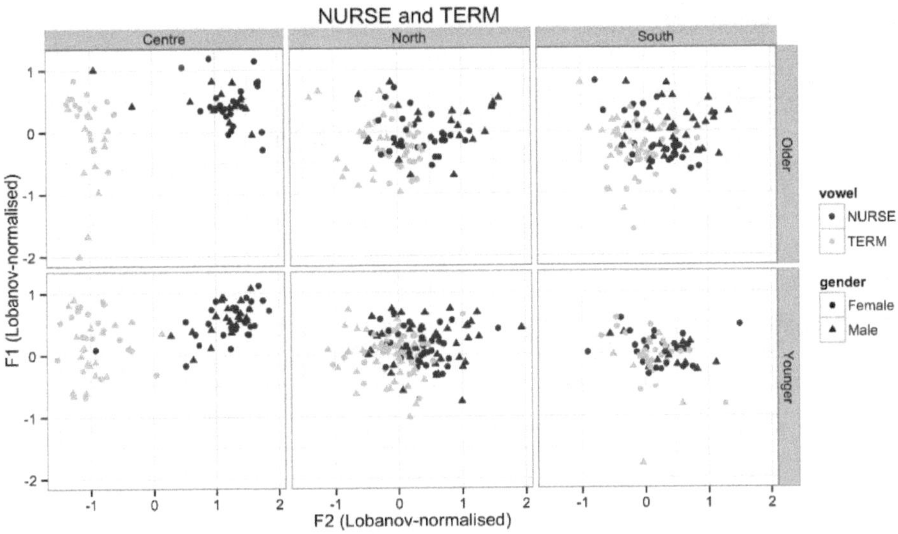

Fig. 12.5: The NURSE and TERM lexical sets in a two-dimensional vowel space, divided by age, gender and area.

cant ($p < 1 \times 10^{-5}$) difference in fronting between the two lexical sets, while almost all female groups do not[4].

A fundamental difference between the vowel spaces of the inner city subjects and their northside/southside counterparts is also suggested by Figure 12.6, which identifies several ongoing vowel changes. Two clear differences are apparent between older inner city Dubliners and their southside/northside counterparts. For inner city Dubliners FOOT and STRUT overlap almost completely, while for northside and southside subjects FOOT is clearly higher than STRUT. Secondly, among older northside and southside subjects GOOSE is only moderately fronted, while inner city subjects' realisation is near HIT at the front of the vowel space[5]. These two differences between inner city subjects and their northside/southside counterparts appear to be lost among younger subjects. Younger northsiders' and southsiders' STRUT is raised to the point where it overlaps with FOOT almost completely. In addition, their GOOSE is dramatically fronted to a realisation near HIT. Finally, younger northsider and

4 The difference in fronting is significant for older northside women ($p = 0.041$) and nearly significant for younger northside women ($p = 0.0547$).

5 There is evidence of diphthongisation in inner city men's realisations of GOOSE (Lonergan 2013).

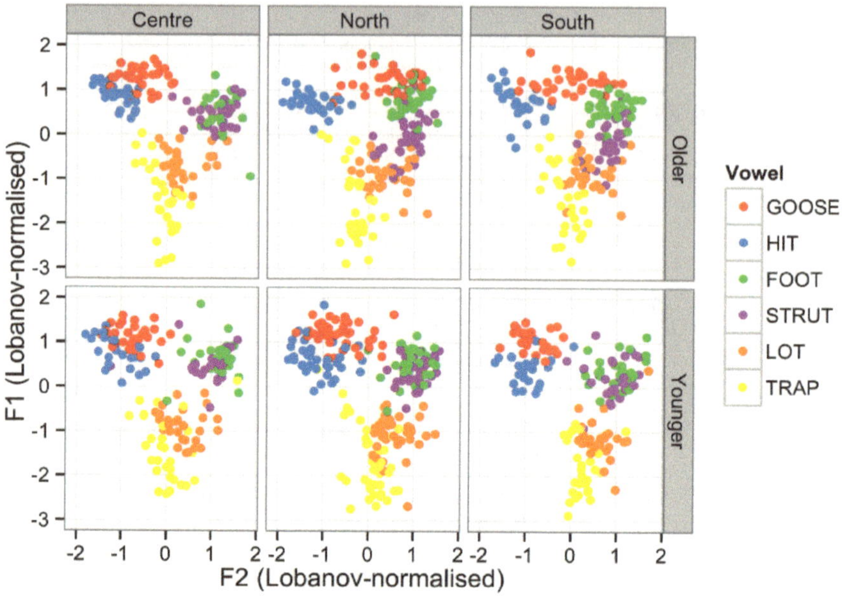

Fig. 12.6: A selection of female Dubliners' back vowels, along with TRAP and HIT.

southside subjects have more retracted realisations of TRAP than their older counterparts, a change that is not evident in the inner city subjects.

Overall, the results suggest a fundamental linguistic difference between the inner city subjects and the northside and southside subjects, not between the northside and southside subjects (as might have been expected based on the perceptual data).

6 Summary and conclusions

This chapter investigated the number and geographical location of perceived varieties of Dublin English using hand-drawn maps. This was combined with a more detailed investigation of the perceived characteristics of these varieties, using questionnaire and semi-structured interview data. Finally, Dubliners' folk linguistic perceptions were contrasted with the actual linguistic differences between the communities that were perceived to be linguistically distinct. Most subjects distinguished between a "D4" variety, associated with the city's southeast, and a "northside" variety, associated with much of the city's north and west. This mirrors the perceived north/south split suggested by Hickey's

(2005) research. A smaller number of subjects distinguished between the northside variety and one associated with the inner city. This suggests that the smaller scale of the hand-drawn maps used in this study were useful in documenting perceived differences that did not emerge with Hickey's national maps. Questionnaire data showed that the D4 perceived variety was high in status, while the northside and inner city perceived varieties were low in status (and described in very similar terms). Interview data did suggest more complex views, however, as well as important differences between ingroup and outgroup evaluations of these perceived varieties. Inner city subjects acknowledged the low status of their speech, but were otherwise highly positive about it. Southside subjects were also found to hold some positive views of the northside and inner city perceived varieties. These positive attributes did appear to be outweighed by the stigma of these perceived varieties. This provides some support for the argument that ongoing change in Dublin English may have begun as a response to the stigma of vernacular varieties of Dublin English.

While Dubliners perceive a north/south contrast to exist in Dublin English, the northside and southside subjects in this study differed little. In contrast, inner city Dubliners' realisations differed dramatically from those of the other two groups. One explanation for this apparent difference between perception and production is that labels like “northside” and “southside” should not be interpreted literally. Some subjects were found to describe residents of the south-western working class suburb of Tallaght as “northsiders,” for example, and one resident of the southwest inner city referred to locals as “northside people”. This suggests that the concept of a “northsider” is more complex than a simple matter of geographical location. Terms like “northside” and “D4” may refer more directly to social class. Broadly speaking, the south of the Dublin is more affluent than its north and this contrast is very long-established (Brady and Simms 2001; Haase and Pratschke 2005). Terms like “northside” and “southside” therefore offer a somewhat accurate means of referring to social class. There are, however, several working class communities in the city's south and several middle class communities in its north. The geographical distribution of social class has, furthermore been substantially complicated both by the development of suburban working class communities and the recent real estate boom (N. Moore 2008). The hand-drawn maps completed by subjects in this study suggest that Dubliners are aware of these complications, but terms like “northside” and “southside” appear to have survived. This may be a result of Irish people's discomfort with terminology like “working class' or “middle-class”. There is substantial evidence that Irish society is clearly stratified (Nolan and Maitre 2007) and that upward social mobility is low for working class Irish people (Breen and Whelan 1996; Whelan 1995). Nonetheless, many authors have argued that Irish people typically regard their society as classless (Ferriter 2004; Tovey and Share 2003) and that social class is little-discussed in Irish politics (Nolan and Maitre 2007). Whelan (2003), for example argues that many Irish people view social class as an English concept, abandoned with independence. This discomfort with openly discussing social class may, in part, explain the continued prevalence of terms like "northside" and "southside".

Nicole Palliwoda and Saskia Schröder

Chapter 13
Perceptual Dialectology, speech samples, and the concept of salience: Initial findings from the DFG-project "Lay Linguists' perspective on German regional varieties: Reconstructing lay linguistic conceptualizations of German in a perceptual dialectology approach"

1 Introduction: Aims and setting of the project

This chapter[1] is about the role of *salience* and salient features within the DFG[2]-project *Wahrnehmungsdialektologie*[3], and focuses on the key questions about the linguistic laypeople's mental maps, their labels for the varieties in the German language area, the subjective distance to standard language, the pleasantness of German dialects as well as the associated and perceived features. The project asks for the displayed knowledge of the linguistic laypeople concerning the German dialects and their speakers. The project aims are to conduct research into non-linguists speech perceptions and their linguistic knowledge. One major aim of the project is the elicitation of data concerning lay-linguistic conceptions of space and dialect.

The aims of the project are divided into the following categories:

1. Macro-level cartographic documentation of mental maps
2. Micro-level cartographic documentation of mental maps
3. Lay-linguistic concepts
4. Attitudes towards German dialects
5. Salience of single features and feature bundles
6. Lay-linguistic knowledge vs. expert knowledge.

---

1 The project is situated at the Christian-Albrechts-Universität in Kiel. It is funded for at least four years (April 2011 until March 2015).

2 *Deutsche Forschungsgemeinschaft* 'German Research Foundation'.

3 One of the translations of *perceptual dialectology*.

The first two aims deal with the mental maps including those of the whole German-speaking areas, as well as everyone's individual micromapping[4]. Moreover we want to find out more about associated and perceived dialect features. Thus, we will be able to understand the conceptualization of the linguistic laypeople concerning the German varieties. Further interests are *heterostereotypes*[5] and *autostereotypes*[6] and their evaluation as well as the salience of single features and feature clusters, especially their distance to the individual concept of standard language. The last aim is the comparison between linguistic lay's knowledge and that of professionals. Discovering convergences, divergences and investigate their functions are the major tasks here.

In order to reach these goals, the project uses quantitative and qualitative analyses based on non-linguists' assessments of speech samples. The results are aligned with control measures, i.e., the geographical knowledge and the spatial perception of the informants. By comparing ethnographic and cognitive lay-linguistic data with expert knowledge, it is possible to conclude the informants' attitudes and language behaviour. The project thus not only aims to fill important research gaps, but also combines theoretical and methodological approaches of empirical social research, perceptual dialectology, sociolinguistics, social and perceptual geography and cognitive sciences.

In order to fulfil these aims, we arranged interviews with three groups of informants covering different ages. The first group is aged between 16 and 20, the second one between 30 and 50 and the third one between 55 and 60. A common feature of the participants was the school: the interviews took place in schools, the youngest informants are pupils and the other two groups are teachers.

In the guided interviews, we firstly asked for demographic information such as age and hometown. This was followed by a thematic section divided into four branches. The first two sections are used to discover the mental maps as it was explained before. By using the draw-a-map-task, we are able to investigate the individual mental maps of our informants' home region on the one hand. The pilesort-method (cf. Tamasi 2003) on the other hand is used to visualize the dialect concepts of the whole German-speaking area. For the third part in the interview, we applied a quiz in order to investigate potential salient

4 Using this term, we mean the investigation of our informants' mental maps of their individual home region.

5 A *heterostereotype* means an image that a speaker or a group of speaker has/have of a different speaker or different speaker community.

6 We define *autostereotype* as the evaluation of everyone's own dialect (Hundt 2010: 183).

features. In the last section we ask questions, which focus on the status of standard language in Germany.

In the following chapter we will discuss results which are investigated by using the quiz. After a short introduction to the phenomenon of salience, we will focus on the potential of this quiz as a method which is useful to ascertain salient features in spoken language. Meta-communicative comments on the one hand and the evaluation of different speech samples we used on the other hand will be in the centre of this study.

2 Aspects of salience: A brief theoretical and methodological sketch

Schirmunski (1928/29 and 1930) firstly mentioned the *Auffälligkeit* "conspicuousness" of some language features in his work. He divided those features in two groups: primary features are less resistant to language change than secondary features. The linguistic term salience evolved with the work of Trudgill, who pointed out a subjective component which he claims to be very important on the evaluation of salient features (cf. Trudgill 1986; Lenz 2003, 2010). On the whole, he characterized salience to be containing "stigmatization, linguistic change, phonetic distance and phonological contrast" (Trudgill 1986: 11).

Other authors supposed the existence of standard language and its use as the variety to be aimed at as the requirement for salience (e.g., Mihm 1985: 184–185). A first methodological approach to phonetic conspicuousness was developed by Herrgen and Schmidt (1985), who ascertained that evaluations of dialect features are independent from the relevance of the underlying system. They do not depend either on the contrast within the language system or on the areal distribution (Herrgen/Schmidt 1985: 34–35).

In their newest work on the *Sprachdynamik-Theorie* "linguistic dynamics approach" (2011) Schmidt and Herrgen claimed dialect features to be salient, if they are acquired as a part of one's individual and intuitive knowledge (2011: 62)[7]. Following this, Purschke concluded that a test to classify salient features

7 Within a theory of knowledge and perception, Purschke (2011b) defined the *regional language knowledge* "as the entirety of acquired, cross-linked, weighted, implicit and explicit contents of memory about (regional) linguistic variation, which derive from everyday frameworks of action and circumstances and also serve as the basis of any (linguistic) action." (2011b: 93).

has to include aspectsof interactional acceptability as well as the significance depending on the situation (2011a: 109). As a consequence, he distinguishes salience and pertinence, which means the division of individual processes of perception and as a result the subjective relevance of salient features (2011a: 309).

Our project aims at eliciting salient features of regional speech on different ways. In the first instance, we apprehend salience to be perceptual distinction, which will be analyzed by the use of a quiz (cf. section 3). Moreover, the setting of the guided interviews allows us to investigate the evaluation of features as a result of their conspicuousness (or pertinence (Purschke 2011a)). Those evaluations cannot be examined in a highly systematic order, but it is perfectly possible to use meta-linguistic comments of the respondents in order to provide a first insight into the complex interaction of perception and evaluation of dialect features.

Within this article we will present results that derive from both, salience and pertinence. For this particular reason, we cooperated with the Deutsch-heute-project in Mannheim (cf. section 3) in order to refine and apply a quiz.

3 The quiz as a door opener to salient features

The basis to elicit potential salient features is the online version *Hör mal, wo der spricht*[8] of the *Deutsch heute-project*[9] from the *Institut für Deutsche Sprache* (IDS)[10] in Mannheim. We used the game as an offline version within our guideline-based interview.

Overall, the offline version contains nine speech samples and runs in three turns. Every sample has one corresponding city. The first turn contains two speech samples which the participants have to assign to one of three given cities situated on a map showing the whole German language area. In the next turn, the participants listen to three speech samples and choose between four cities shown on the German language area map. Finally they have to assign four samples to one of the five cities given on a German language area map.

The quiz itself starts for every participant with the same speech sample; namely the sample from Helmstedt (Eastphalian), which serves as our standard

8 'Listen where he is coming from' http://multimedia.ids-mannheim.de/hoermal/web/ [accessed 2 October 2013].
9 http://www1.ids-mannheim.de/prag/AusVar/Deutsch_heute.html [accessed 2 October 2013].
10 http://www1.ids-mannheim.de/index.php?id=1 [accessed 2 October 2013].

language variant. With this sample it shall be tested if the respondents identify this as standard variant. If this is the case, the sample could be used as actual reference for other parts within the interview (questions regarding the norm and standard German). Otherwise the samples are played randomly.

The speech samples[11] themselves are read text (the fable *Nordwind und Sonne* 'North wind and the sun'). The speakers were asked to read this text as close to standard German as possible. They are male and middle aged. Six of the speech samples were taken from the REDE-Project[12] located in Marburg and three speakers were ascertained by the research assistants.

The 61 cities, which are the basis for the shown cities to which the informants have to assign the speech samples, are taken from another part of our setting: the pilesort-method. These cities were chosen beforehand in an online investigation, where the people were asked to write down the cities and places which first come to their mind, when they see a map of the whole German language area. The most often mentioned cities provided the base for the pile sort method and for this game. These cities were randomized as well.

An additional feature of the quiz was a stimulus-response test in which participants were timed as they started to listen to the sample, with the timing stopped when the sample was assigned by the respondent. With this option we would like to find out which linguistic feature has led the participant to his or her assignment. Of course, we would not expect participants to be able to describe and name linguistic features exactly. Therefore, we compared the time they took to allocate the sample to a city with the time within the transcript of the speech sample in order to analyze which linguistic feature could be salient. We have not yet completed such analyses, which will be completed in the next period of the project.

Furthermore, the participants are asked why they allocated the sample to this city. They should justify their decision, so we get more background information about their assignment. Referring to Preston (1999b: xxiii–xl) the participants were asked to evaluate the speech samples (we chose scales of pleasantness and correctness). In this way it is possible to get meta-communicative comments and more information about salient features.

The picture in Figure 13.1 shows a screenshot of the quiz. It shows the first turn with two speech samples on the left and a short instruction on the top

11 The choice of the speech samples based on the results of the pilot study (cf. Hundt 2010: 179–219).

12 http://www.regionalsprache.de/ [accessed 2 October 2013].

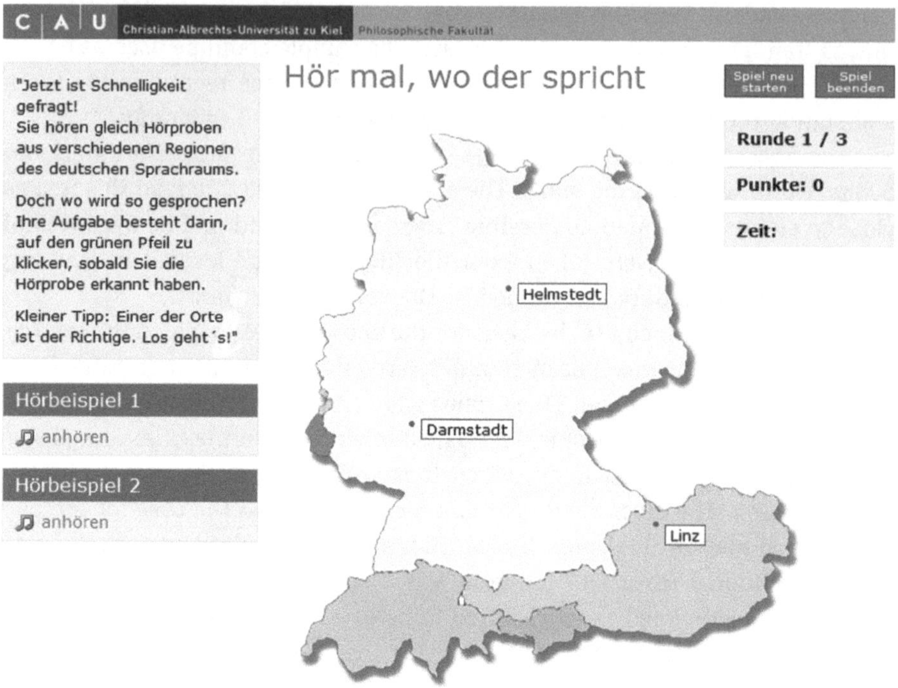

Fig. 13.1: Screenshot of the quiz.

left. In the middle, there is the German language area map with three cities to which the participants have to assign the speech samples. On the right you see the buttons showing the turns, the score and the time required until the speech sample was assigned. Every turn looks like this just with one more city and speech sample.

4 Initial findings

This section presents a short overview of the project's data. The previous analyses included 24 informants from four different places (Coburg in the north of Bavaria, Simmern in Rhineland-Palatinate, Vaduz in Liechtenstein and Meran in South Tirol) (see Figure 13.2).

Four participants aged between 16 and 20 years, eight aged from 30 to 50 years and 12 respondents aged from 55 to 60 years. The first table shows how often the speech samples were sorted correctly by their city (cf. Table 13.1).

Fig. 13.2: Map of the cities (Source: Google Maps).

Tab. 13.1: Allocation of the speech samples (*n* = 24).

	sorted incorrectly	sorted correctly
Bavarian	7	17
Brandenburgian	12	12
Hessian/Rhenish Franconia	12	12
High Alemannic	0	24
Central Austro-Bavarian	7	17
North Low German	9	15
Upper Saxonian	4	20
Swabian	7	17
Standard variety	12	12

Tab. 13.2: Pleasantness of the speech samples (*n* = 24).

Pleasantness (How pleasant is this speech sample for you, from −3 not at all till +3 very pleasant?)	**−3**	**−2**	**−1**	**0**	**1**	**2**	**3**	**Average**	**Standard deviation**
Bavarian	0	0	3	3	2	12	4	1,46	1,285
Brandenburgian	0	1	1	4	4	9	5	1,42	1,349
Hessian/Rhenish Franconia	1	2	2	5	7	7	0	0,50	1,445
High Alemannic	0	2	1	5	4	8	4	1,13	1,484
Central Austro-Bavarian	0	2	3	2	4	11	2	1,04	1,488
North Low German	0	2	3	4	3	10	2	0,92	1,501
Upper Saxonian	1	1	6	4	4	6	2	0,46	1,615
Swabian	0	0	4	5	9	5	1	0,75	1,113
Standard variety	0	0	1	3	4	6	10	1,87	1,227

If we take a look on the table, it can be seen that there are samples which could be assigned clearly like the High Alemannic or the Upper Saxonian sample. As it was found out in a preinvestigation of the project, both varieties are very present for the linguistic lays (cf. Hundt 2010). High Alemannic is situated in the very south of Germany and in the north of Switzerland. Previous studies suggest that there are specific features that are closely connected with this region: a very prominent lay-linguistic association is related to the articulation of High Alemannic speakers, which is meant to be much more retracted in the pharynx than other varieties. This association also comes up when linguistic lays are asked to describe Upper Saxonian speech samples. The remarkable difference is the evaluation of this variety, which is mostly very pejorative (cf. Anders, Palliwoda, and Schröder 2014).

On the other hand, there is insecurity in the assignment of some speech samples by the informants like the Brandenburgian, the Hessian/Rhenish-Franconian and the Standard variety samples. This could be related to the speech samples themselves – it is a read text – and also to the choice of given cities, because they were given randomized. Thus, it is possible that the interviewed person chose the city by using the principle of exclusion. Therefore the given cities have to be also analysed in the next period of the project.

The next table shows the evaluation (the pleasantness-scales) of the speech samples by the 24 participants (cf. Table 13.2).

The data in Table 13.2 show some interesting results: If we take a closer look at the values it can be seen that the Standard variety sample gets the highest values, closely followed by the Bavarian sample. The Upper Saxonian sample one takes the last place, followed by the Hessian/Rhenish Franconia

Tab. 13.3: Correctness of the speech samples (*n* = 24).

Correctness (How correct is this speech sample for you, from −3 not at all till +3 very correct?)	**−3**	**−2**	**−1**	**0**	**1**	**2**	**3**	**Average**	**Standard deviation**
Bavarian	0	1	5	4	5	5	4	0,83	1,523
Brandenburgian	0	1	0	0	5	12	6	1,87	1,076
Hessian/Rhenish Franconia	2	2	1	4	8	7	0	0,46	1,587
High Alemannic	0	3	6	4	4	6	1	0,29	1,517
Central Austro-Bavarian	0	2	4	4	7	5	2	0,63	1,439
North Low German	0	0	2	1	5	10	6	1,71	1,160
Upper Saxonian	0	2	4	4	4	8	2	0,75	1,511
Swabian	0	2	2	5	6	8	1	0,79	1,351
Standard variety	0	0	0	0	1	10	13	2,50	0,590

and the Swabian samples. The table shows also with the standard deviation the distribution of the evaluation of the samples given by the participants in order to examine the range of the results and the level disagreement within the evaluation of each speech sample. The smallest standard deviation can be found for the Swabian speech sample and the greatest for the Upper Saxonian sample. That means the participants were divided over the evaluation of the Upper Saxonian voice sample. Findings such as this demonstrate agreement with earlier investigations of the bipolarity of some speech samples depending on the question (cf. i.a. Anders 2011; Hundt 2010).

Table 13.3 shows the results for correctness ratings, and we see a similar picture to that in Table 13.2: the highest values obtain our Standard variety sample, our supposed-to-be standard language sample. All 24 informants agreed that this sample is correct or very correct[13].

The Standard variety results have also the minor distribution that means the participants are agreed that this is the correct variety of their concept of standard German. This sample is closely followed by the Brandenburgian and North Low German sample which could also be related to the read text and the choice of the given cities. The Bavarian and the Upper Saxonian sample were

13 It might be interesting to find a correlation between the location of our informants and their evaluation. This has not been done so far, but considering the high prestige of standard German within the whole German speaking population, the data suggest that the location of the linguistic lays only plays a minor part. The rating of standard language rather seems to be a question of prestige.

Tab. 13.4: Contingency table from pleasantness and correctness of the speech samples (n = 24).

Contingency table								
		Correctness (How correct is this speech sample for you, from −3 not at all till +3 very correct?)						
		−3	**−2**	**−1**	**0**	**1**	**2**	**3**
Pleasantness (How pleasant is this speech sample for you, from −3 not at all till +3 very pleasant?)	**−3**	1	1	0	0	0	0	0
	−2	1	4	5	0	0	0	0
	−1	0	3	7	5	4	3	2
	0	0	2	4	8	3	9	9
	1	0	1	3	7	15	12	3
	2	0	2	5	6	18	37	6
	3	0	0	0	0	5	10	15

perceived as relatively equal correct and the last place takes the High Alemannic sample at this time. It can be noticed that there are features within the High Alemannic speech sample that on the one hand trigger a positive concept of this sample and on the other hand these features are seen as incorrect by the informants.

The last preceding tables raise the question whether a correlation between pleasantness and correctness of the speech samples exists. Table 13.4 shows a contingency table which permits the investigation of possible correlations between the two scales.

Here it can be seen how often the pleasantness-values co-occur with the correctness-values of the speech samples. Obviously the correct and the rather correct values seem to correlate to the "pleasant" and "rather pleasant" values, which you can see in the grey shading showing the highest overlaps.

In order test the correlation between the pleasantness and correctness ratings for speech samples, we undertook a Pearson correlation test which confirmed the existence of a correlation, as demonstrated in Table 13.5.

The calculated correlation between pleasantness and correctness is 0.538 so the correlation coefficient is between 0.7 and 0.5. Thereby, we can verify our hypothesis that there is a middle correlation between these two factors which is also significant.

Of course, we have to treat these first results with caution, because the basis of these results is only a little sample and in addition we have to analyze

Tab. 13.5: Correlation between pleasantness and correctness.

		Pleasantness of the speech samples	Correctness of the speech samples
Pleasantness of the speech-samples	correlation by Pearson	1	.538**
	significance (2-tailed)		.000
	N	216	216
Correctness of the speech-samples	correlation by Pearson	.538**	1
	significance (2-tailed)	.000	
	N	216	216

** The correlation is at the level of 0.01 (2-tailed) significant.

other factors which could be influential on the informants' assignments. Therefore, it is important to take a closer look at the meta-communicative comments of the respondents which will be presented below.

Figure 13.3 shows the comments from Vaduz, Meran, Simmern and Coburg respondents categorized by the system proposed by Anders (2010). In all it includes 31 categories and thus offers a concise possibility to analyze meta-communicative data. Although Anders (2010) used categories for dealing with phonetic issues, we do not discuss these in relation to our data here[14].

Figure 13.3 shows the distribution of all comments that were given while the informants passed through the quiz. At the very first view, the chart depicts a very heterogeneous result in which no category seems to be extraordinary prominent. Nevertheless, there are some categories that do not emerge, such as *descriptions concerning word formation, lexical characteristics* and *phraseological features*. The only four categories with a frequency from 9 to 12% are *pronunciation, identifying descriptions, characterization with geographical or political character* and *horizontal variation*. In the following, two lay-linguistic concepts will be described in detail in order to explain the total result.

Figure 13.4 depicts the comments of the speech sample "Eastphalian". It has to be located in the very first round by each informant and serves as a control-sample: we suppose this speech sample as our standard variety[15]. Within the interviews, the participants are asked to evaluate the correctness and pleasantness of the other eight samples compared to the first one.

14 For further reading on salient features concerning the phonetic-phonological layer see Anders, Palliwoda, and Schroeder (2014).

15 It is a lay-linguistic myth that says "best German" (i.e., *standard language*) is spoken near Hanover, which is located in the Eastphalian dialect area.

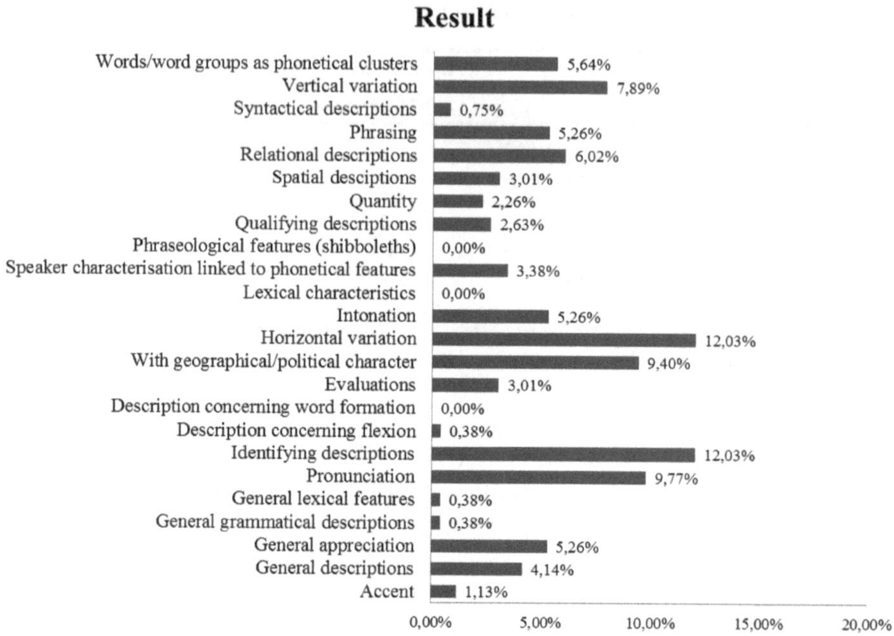

Fig. 13.3: Results of the meta-communicative comments within the quiz.

The chart implies four categories, which are very prominent. The first one contains descriptions concerning the *vertical variation* that means utterances, which are linked to the continuum between standard and dialect (e.g., *kind of standard language, no dialect*). The Eastphalian speech sample is typically commented with statements where the respondents assert that they cannot discern any dialect features.

This could correlate with the fact that this is the first speech sample in the quiz and the informants might be not so familiar with the procedure. On the contrary, the informants were not able to recognize any linguistic specifics even after several trials. This leads to the conclusion that the Eastphalian sample is rather unmarked. This result could be confirmed with the findings in Table 13.3. Furthermore, the evaluation of the Eastphalian speech sample concerning correctness showed that the respondents assumed this sample to be very correct, as discussion in section 3 depicts[16].

This assumption is supported by the second category *phrasing*, which contains descriptions like *neutral* and *clear*, as well as the descriptions with *gener-*

16 See also Anders, Palliwoda, and Schröder (2014).

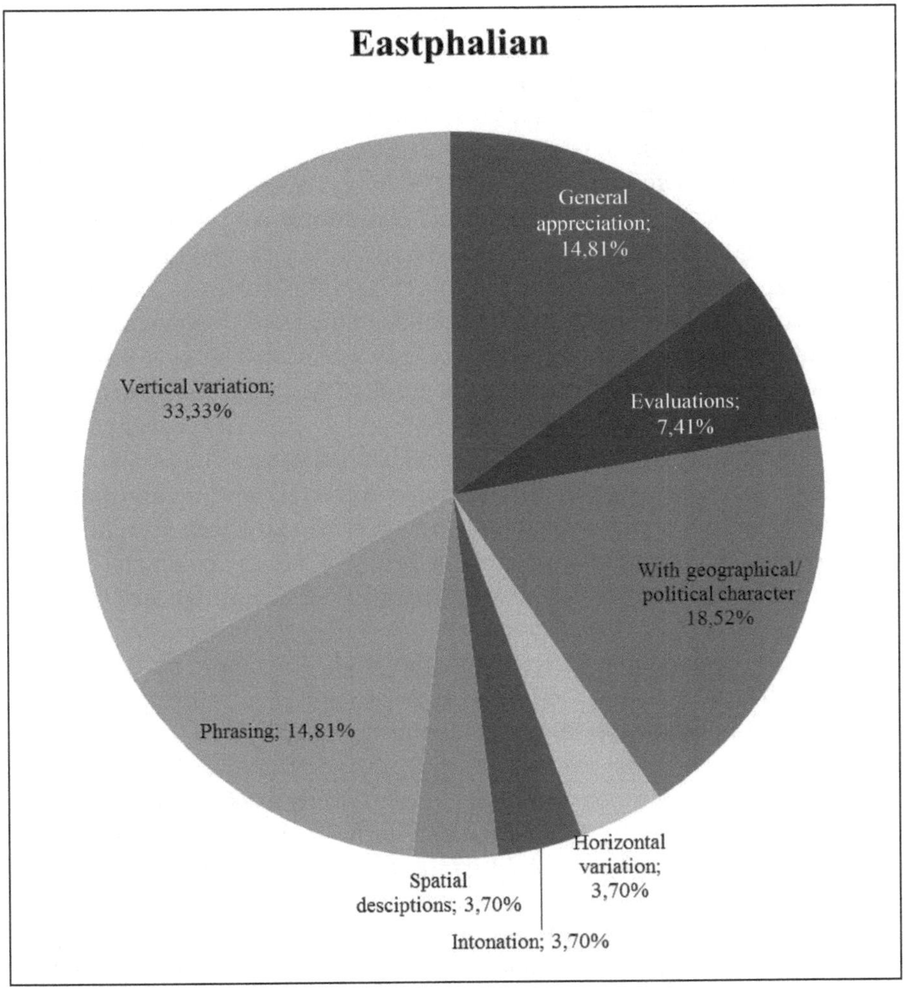

Fig. 13.4: Result of the Eastphalian speech sample.

al appreciations. Most of these comments in the Eastphalian speech sample estimate the language of the speaker to be *good* or *perfect*, linked to a *High-German* (i.e., standard language) concept. Initial analyses on the concept which arise when lay linguists are asked to describe their ideal of standard language[17] showed that those concepts consist of attributes like *no dialect features, easy to understand, clear pronunciation* etc.

---

17 Within the interview we used the term *gutes Deutsch* 'good German' instead of *standard language* in order to evoke the lay linguists' expression for the linguistic term.

The third dominating category contains descriptions *with geographical* or *political character*. These statements (e.g., "somewhere in the middle of Germany," "might be somewhere around Hamburg") often result due to an exclusion principle: many respondents tend to exclude one city due to the argument that it matches fewest at all mostly because of its geographic position (e.g., "Berlin is just not possible, so the speaker must be from Vienna!").

Within this category, the expressions concerning geographical issues are very often related to Central or North Germany. This might be the result of the setting itself: as described above, the speech samples' sequence as well as the choice of cities are randomized and cannot be influenced. That means the cities can be placed in a small-scaled region or they are located all over the map. In that case, most informants make a lucky guess without knowing exactly where the sample's speaker may come from.

In summary, it can be stated that the respondents from Vaduz, Meran, Simmern und Coburg characterized the sample as a clear and neutral language especially in its phrasing and its status in the German language area. Any comments on a special intonation, syntax or lexis are missing completely, so the Eastphalian speech sample can be described as unmarked with a lack of salient features.

This is a remarkable difference to the High Alemannic speech sample which results are shown in the following illustration (Figure 13.5).

The chart offers a more differentiated result compared to the Eastphalian speech sample. The most relevant category is the pronunciation. Its content mostly deals with consonants or rarely vocals, which are formed "at the back of the mouth" or sound like "they speak a lot in the larynx". The mainly discussed feature in here is the voiced uvular fricative [χ], which is meant to be "typical Swiss German".

Identifying descriptions played in the Eastphalian sample only a minor role, whereas the statements in here (e.g., "typically Zurich," "they can't deny their heritage") show descriptively, what the participants immediately associate, when they listen to the audio. Those meta-communicative assertions correspond to the correction of assignment: the High Alemannic speech sample was always located correctly within the first few seconds (cf. Table 13.1).

Apart from pronunciation, the intonation played an important part in the descriptions. The respondents mostly mentioned the *sing-song* and *a special sound* which they judged to be very unfamiliar with.

In this context, the High Alemannic speech sample was often evaluated, which is the fourth dominating category. Examples like this "they shouldn't read texts like these, it sounds lousy" are mostly provoked by the inquiring of the speech sample's correctness and pleasantness. Especially the comparison between the two presented samples is very productive for evaluations. The

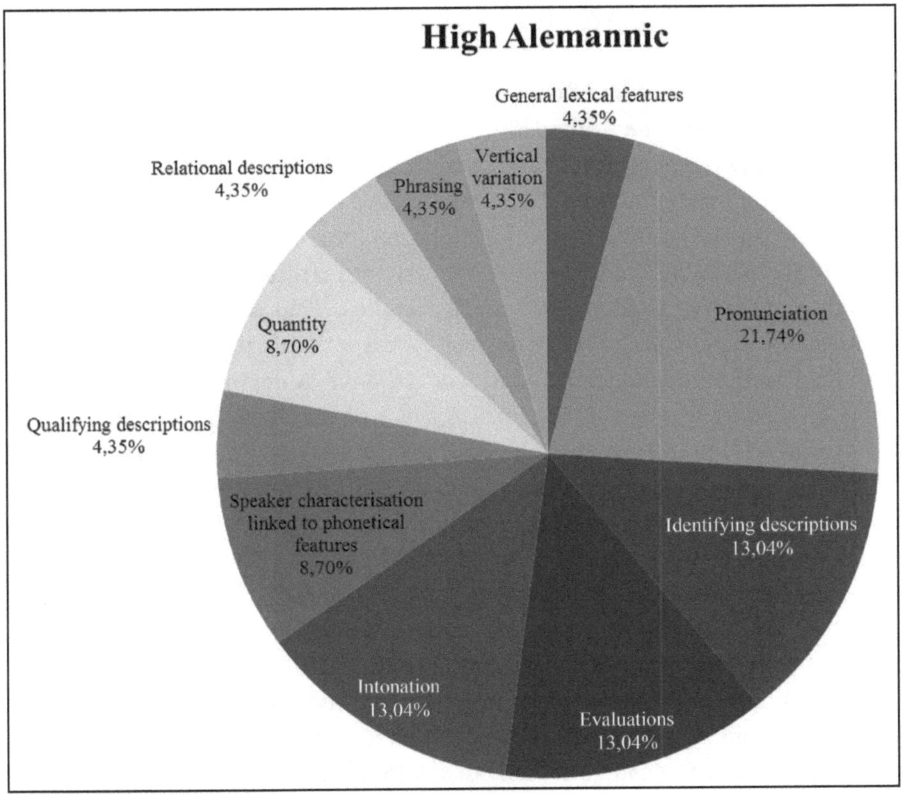

Fig. 13.5: Result of the High Alemannic speech sample.

most important subject of discuss in there is the difference between standard language use, which is the "desirable way of speaking", and a variety, which is markedly different to formal standard language. Our informants evaluate the High Alemannic sample to be something very different from "correct" language.

This reveals a bipolar view: either one is able to speak standard language ("High German"), which is strongly appreciated, or one cannot, which is depreciated. The fact that the High Alemannic speech sample is described as to be very different to standard language, leads to the conclusion that lay linguists seem to have a clear concept which they use to firstly describe and secondly evaluate speech samples. That correlates with a study by Hundt (2010), who operated via mapping tasks in order to depict all known dialect areas of linguistic lays. High Alemannic (mostly called *Schweizer Deutsch* 'Swiss German') was highly prominent within the corpus.

5 Discussion and outlook

The analysis of evaluation of the speech samples concerning pleasantness and correctness and the meta-linguistic comments within the quiz showed a wide and interesting range of results and presented two speech samples, which were described quite differently by the linguistic lays.

On the one hand the results show a correlation between pleasantness and correctness (cf. Table 13.5). On the other, salient features could trigger positive as well as negative concepts of speech samples. That means features within the speech sample could be seen as positive respectively pleasant but the same features could be evaluated as incorrect (cf. High Alemannic, Table 13.2 and Table 13.3). Also features which lead to a not pleasant evaluation are seen as incorrect (cf. Upper Saxonian, Table 13.2 and Table 13.3). Speech samples could be also evaluated, when the participants do not recognize any salient features like the Eastphalian example (cf. Standard variety, Table 13.2 and Table 13.3), which was perceived as a clear and neutral language. The respondents often struggled with naming and describing dialect features. This notice correlates with the results of the rating scales concerning the speech sample's correctness. However, the High Alemannic sample was always allocated correctly on the map (cf. Table 13.1) but its description fulfilled more categories than the Eastphalian did. It was much more evaluated and the lay-linguistic associations mostly aimed at categories that deal with articulation in general (pronunciation, intonation). As a second step, on the basis of the perceived features, the respondents associate extra-linguistic characteristics with the sample.

Again, the results of the Eastphalian speech sample reveal an old language myth about the inhabitants of Hanover who are supposed to speak the "best German" (in the sense of interference-free German standard language "High German"). This myth arose in the 18th century, when the Upper Saxon variety slowly lost its function as a role model (cf. Elmentaler 2012) Since then, this myth persists not only in Hanover but also in almost the whole German language area, despite the fact that the variety in and around Hanover does not exclude more dialect features than others, i.e., it is not a "more correct German" than it is spoken in other regions of Northern Germany (cf. Elmentaler 2012: 109). Nevertheless, the concept of a well-pronounced, easy-to-understand language is strongly connected to speakers, who are born and educated around Hanover. Even the Eastphalian speaker of the sample, who is from Helmstedt, a city about 97 km to the East of Hanover, was immediately associated with this language variety because of the close proximity of the two cities ("das ist *Hochdeutsch*" 'this is High German').

Connected with these results, there is a need to discuss the reliability of the setting, which might have revealed two weak spots:

1. The role of the given stimulus: It is a general problem of perceptual dialectology that lay-linguistic descriptions and evaluations are highly influenced by the design of the provided stimulus (cf. Schmidt and Herrgen 2011; Kehrein, Lameli, and Purschke 2010). The exclusion principle, which is due to the randomized choice of cities, is one result of this problem. On the one hand, there is a need to avoid a strict arrangement of their appearance in order to minimize the predetermination of the interviewers. On the other, this approach implies the difficulty of comparing the results with each other because they were provoked under different circumstances.
2. The problem of eliciting salient features in nearly standard language: A problematic sphere might be the elicitation of salient features within varieties that are almost standard language. Can we call these features salient? Don't we have to consider especially those features that are part of the variety in small-scaled regions because they clearly belong to the dialect and thus determine the lay-linguistic conceptualizations of the German language area? Those reproaches imply that the respondents have a determined concept of what is dialectal and what is standard language. But recent works shows that this is not the case. Kleiner (2011) proofed in his research on the use of German derivation suffixes that the pronunciation of /g/ in all forms of <ig>, <igste> etc. as a plosive or a fricative is not clear at all within the speaker-community of German (cf. Schröder 2013). They do not know the appropriate pronunciation of /g/ as a plosive or a fricative within different surroundings and thus are not able to decide which form is standard language use and which one is dialect use. Moreover, by the use of nearly standard language, we are in a position to create a basis that reflects reality: small-scaled dialects that differ from village to village are not representable for the German language area, that's why we use nearly standard language samples[18].

Apart from that, these results provide an interesting insight in the potential of the data. In order to fulfil the aims of the project (cf. section 1), it is planned to consider further research questions in upcoming papers:

---

18 On the dialect landscape of German see the newest work of Kehrein 2012 or Schmidt and Herrgen 2011 and Lenz 2003 (especially West-Middle-German).

- Correlation of the informants' conceptualisations and the rating scales
- Comparison of the informants within the Federal Republic of Germany with those of the neighbouring countries
- Differences between old and young participants
- Which phonetic/phonological features are salient[19]?

19 First results can be found in Anders, Palliwoda, and Schröder (2014).

References

AGAM (Agence d'Urbanisme de l'Agglomération Marseillaise). 2009. *Radioscopie des nouveaux Marseillais*. www.agam.org (accessed 29 July 2014).

Anders, Christina A. 2010. *Wahrnehmungsdialektologie: Das Obersächsische Im Alltagsverständnis von Laien*. Berlin: Mouton de Gruyter.

Agha, Asif. 2007. *Language and social relations*. Cambridge: Cambridge University Press.

Alford, Randall L. & Judith B. Strother. 1990. Attitudes of native and non-native speakers towards selected regional accents of US English. *TESOL Quarterly* 24, 479–495.

Anders, Christina A. 2010. *Wahrnehmungsdialektologie. Das Obersächsische im Alltagsverständnis von Laien*. Berlin, New York: De Gruyter.

Anders, Christina A. 2011. „Platt ist nicht uncool" – Zu den „coolsten" und „uncoolsten" Dialekten des Deutschen und ob das schon immer so war. In Lieselotte Anderwald (ed.), *Sprachmythen – Fiktion oder Wirklichkeit?* (Kieler Forschungen zur Sprachwissenschaft 3), 117–135. Frankfurt am Main: Peter Lang.

Anders, Christina A., Nicole Palliwoda, & Saskia Schröder. 2014. *"In dem moment wo ich es dann erkenne dann ist es auch gleich wieder weg". Salienzeffekte in der sprachperzeption.* https://bop.unibe.ch/linguistik-online/article/view/1572/2662 (accessed 9 July 2014)

Anderson, Anne. H., Miles Bader, Ellen G. Bard, Elizabeth Boyle, Gwyneth Doherty, Simon Garrod, Stephen Isard, Jacqueline Kowtko, Jan McAllister, Jim Miller, Catherine Sotillo, Henry S. Thompson and Regina Weinert. 1991. The HCRC map task corpus. *Language and Speech* 34(4), 351–366.

Andrade, Amália. 1999. On /l/ Velarization in European Portuguese. *Proceedings of the XIVth ICPhS*, San Francisco, 543–546. Berkley, CA: Linguistics Department, University of California.

Armstrong, Nigel & Tim Pooley. 2010. *Social and linguistic change in European French*. Palgrave Macmillan.

Asia Pop Project. 2013. Estimated population density in South Korea in 2010 and most populous urban areas. Wikimedia Commons. Retrieved from: http://schools-wikipedia.org/images/3254/325464.png.htm. (accessed 14 July 2015).

Ask a Korean. 2013. Your one stop guide to Korean dialects. Retrieved from: http://askakorean.blogspot.com/2013/06/your-one-stop-guide-to-korean-dialects.html. (accessed 29 July 2014).

Atwood, Bagby E. 1953. *A survey of verb forms in the Eastern United States*. Ann Arbor, MI: University of Michigan Press.

Auer, Peter. 1995. The pragmatics of code-switching: a sequential approach. In Lesley Milroy & Peter Muysken (eds.), *Social and linguistic Change in European French: One speaker, two languages*, 115–135. Cambridge University Press: Cambridge.

Auer, Peter. 2007. Introduction. In Peter Auer (ed.), *Style and social identities. Alternative approaches to linguistic heterogeneity*, 1–21. Berlin & New York: Mouton de Gruyter.

Auer, Peter. 2012. Sprachliche heterogenität im Deutschen. Linguistik zwischen variation, varietäten und stil. *Zeitschrift für literaturwissenschaft und linguistik* (Sonderheft "Verschwommene Dialekte") 166. 7–28.

Auer, Peter & Helmut Spiekermann. 2011. Demotisation of the standard variety or destandardisation? The changing status of German in late modernity (with special reference to south-western Germany). In Nikolas Coupland & Tore Kristiansen (eds.),

Standard languages and language standards in a changing Europe, 161–176. Oslo: Novus.

Auzanneau, Michelle & Caroline Juillard. 2012. Introduction. Jeunes et parlers jeunes : catégories et catégorisations. *Langage et société* 141. 5–20.

Asprey, Esther. 2006. The North-South divide: A view from the Midlands. Paper presented at the *Northern Englishes Workshop*, 1 (NEW 1), University of Lancaster, March.

Bailey, Guy, Tom Wikle, Jan Tillery, & Lori Sand. 1993. Some patterns of linguistic diffusion. *Language Variation and Change* 5. 359–390.

Bakhtin, Mikhail. 1981. *The dialogic imagination*. Austin: University of Texas Press.

Bayard, Donn. 1990. 'God help us if we all sound like this': attitudes to New Zealand and other English accents. In Allan Bell & Janet Holmes (eds.), *New Zealand ways of speaking English*, 67–96. Clevedon: Multilingual Matters.

Beal, Joan C. 2008. English dialects in the north of England: Morphology and syntax. In Bernd Kortmann & Clive Upton (eds.), *Varieties of English. Vol. 1: The British Isles*. 373–403. Berlin: Mouton de Gruyter.

Beal, Joan C. 2009. Enregisterment, commodification and historical context: "Geordie" vs. "Sheffieldish". *American Speech* 84(2). 138–156.

Beckett, John. 2006. *A centenary history of Nottingham*. West Sussex: Phillimore.

Benson, Erica J. 2003. Folk linguistic perceptions and the mapping of dialect boundaries. *American Speech* 78(3). 307–330.

Bertz, Siegfried. 1975. *Der Dubliner Stadtdialekt*. Freiburg: Albert Ludwigs Universitat.

Bertz, Siegfried. 1987. Variation in Dublin English. *Teanga* 7, 35–53.

Billings, Dwight, Gurney Norman, & Katherine Ledford (eds.), 1999. *Back talk from an American region: Confronting Appalachian stereotypes*. Lexington, KY: University Press of Kentucky.

Binisti, Nathalie & Médéric Gasquet-Cyrus. 2001. *Le français de Marseille. Description sociolinguistique*, report to the Délégation générale à la langue française, Ministry of Culture and Communication.

Binisti, Nathalie & Médéric Gasquet-Cyrus. 2003. Les accents de Marseille. *Cahiers du français contemporain* 8. 127–129.

Binisti, Nathalie & Médéric Gasquet-Cyrus. 2007. Imitation et diffusion de l'accent 'quartiers Nord' à Marseille. In Patricia Lambert, Agnès Millet, Marielle Rispail, & Cyril Trimaille, *Variations au cœur et aux marges de la sociolinguistique*, 107–118. Paris: L'Harmattan.

Bishop, Hywel, Nikolas Coupland, & Peter Garrett. 2005. Conceptual accent evaluation: Thirty years of accent prejudice in the UK. *Acta Linguistica Hafniensia*. 37. 131–154.

Blanchet, Philippe. 2004. L'identification sociolinguistique des langues et des variétés linguistiques : pour une analyse complexe du processus de catégorisation fonctionnelle. In *Actes du colloque Identification des langues et des variétés dialectales par les humains et par les machines*, 31–36. Paris: Ecole Nationale Supérieure des Télécommunications / CNRS.

Blommaert, Jan. 2005. *Discourse – a critical introduction*. Cambridge University Press, Cambridge.

Blommaert, Jan. 2010. *The sociolinguistics of globalization*. Cambridge: Cambridge University Press.

Board of Yakima County Commissioners. 2010. Ordinance 4–2010. https://yakimacounty.us/cmrs/ordinance/2010/4_2010.pdf (accessed 22 January 2014).

Bochmann, Klaus. 2003. Linguistique de conflit : considérations générales et plaidoyer pour une science critique. In Klaus Bochmann, Peter H. Nelde, & Wolfgang Wölck (eds.),

Méthodologie de la linguistique de conflit (Plurilingua XXIV), 3–12. St Agustin: Asgard/ Verlag.
Boersma, Paul & David Weenink. 2008. Praat: doing phonetics by computer (Version 5.0.35). http://www.fon.hum.uva.nl/praat/ (accessed 29 July 2014).
Bonaparte, Prince Louise Lucien. 1876. XIII.–On the dialects of Monmouthshire, Herefordshire, Worcestershire, Gloucestershire, Berkshire, Oxfordshire, South Warwickshire, South Northamptonshire, Buckinghamshire, Hertfordshire, Middlesex, and Surrey, with a new classification of the English dialects. *Transactions of the Philological Society* 16(1). 570–579.
Boutet, Josiane & Monica Heller. 2007. Enjeux sociaux de la sociolinguistique : pour une sociolinguistique critique. *Langage et société* 121–122. 305–318.
Boyer, Henri. 1990. Matériaux pour une approche des représentations sociolinguistiques. Eléments de définition et parcours documentaire en diglossie. *Langue française* 85. 102–124.
Braber, Natalie. 2014. The concept of identity in the East Midlands of England. *English Today* 30(2). 3–10.
Braber, Natalie. 2015. Language perception in the East Midlands. *English Today* 31(1). 16–26
Braber, Natalie & Nicholas Flynn. 2015. The East Midlands: Nottingham. In Raymond Hickey (ed.), *Researching Northern Englishes*, 369–391. Amsterdam: John Benjamins.
Brady, Joseph. 2001. Dublin in the nineteenth century: An introduction. In Joseph Brady & Anngret Simms (eds.), *Dublin through space and time (c. 900–1900)*, 159–165. Dublin: Four Courts Press.
Branca-Rosoff, Sonia. 1996. Les imaginaires des langues. In Henri Boyer (ed.), *Sociolinguistique: territoire et objets*, 79–114. Neuchâtel: Delachaux & Niestlé.
Brandstätter, Julia & Sylvia Moosmüller. 2015. Neutralisierung der hohen ungerundeten Vokale in der Wiener Standardsprache – A sound change in progress?. In Manfred Glauninger & Alexandra Lenz (eds.), *Standarddeutsch in Österreich – Theoretische und empirische Ansätze*, 183–203. Wien: Vandenhoeck & Ruprecht.
Breen, Richard & Christopher T. Whelan. 1996. *Social mobility and social class in Ireland.* Dublin: Gill and Macmillan.
Britain, David. 2002a. Space and spatial Diffusion. In J. K. Chambers, Peter Trudgill, & Natalie Schilling-Estes (eds.), *The Handbook of language variation and change*, 603–637. Oxford: Blackwell.
Britain, David. 2002b. Phoenix from the ashes?: The death, contact and birth of dialects in England. *Essex Research Reports in Linguistics*. 42–72.
Britain, David. 2010. Conceptualisations of geographic space in linguistics. In Alfred Lameli, Roland Kehrein, & Stefan Rabanus (eds.), *Language and space: An international handbook of linguistic variation, Volume 2: Language mapping*, 69–97. Berlin: Mouton de Gruyter.
Britain, David. 2012. Countering the urbanist agenda in variationist sociolinguistics: dialect contact, demographic change and the rural-urban dichotomy. In Sandra Hansen, Christian Schwarz, Phillip Stoekle, & Tobias Streck (eds). *Dialectological and folk dialectological concepts of space (linguae & litterae 17)*, 12–30. Berlin/Boston: De Gruyter.
Brook, George L. 1963. *English Dialects.* London: Deutsch.
Bulot, Thierry. 2002. La double articulation de la spatialité urbaine: « espaces urbanisés » et « lieux de ville » en sociolinguistique. *Marges linguistiques 3.* 91–105.

Bulot, Thierry. 2009, « Pour une gestion durable des rapports entre le local et le global (intervention et sociolinguistique urbaine) ». In Sabine Klaeger & Britta Thörle (eds.), *Sprache(n), Identität, Gesellschaft*, 63–72. Stuttgart: Ibidem.

Burbano-Elizondo, Lourdes. 2006. Regional Variation and Identity in Sunderland. In Tope Omoniyi & Goodith White (eds.), *The Sociolinguistics of Identity*, 113–128. London: Continuum.

Burbano-Elizondo, Lourdes & Joan Beal. 2006. "The Tyne and Wear derby": Analysing lexciality and identity in the north-east of England. Paper presented at the *Northern Englishes Workshop* 1 (NEW 1), University of Lancaster, March.

Bucholtz, Mary, Nancy Bermudez, Victor Fung, Lisa Edwards, & Rosalva Vargas. 2007. Hella nor Cal or totally so Cal?: The perceptual dialectology of California. *Journal of English Linguistics* 35(4). 325–352.

Bucholtz, Mary, Nancy Bermudez, Victor Fung, Rosalva Vargas, & Lisa Edwards. 2008. The normative North and the stigmatized South: Ideology and methodology in the perceptual dialectology of California. *Journal of English Linguistics* 36(1). 62–87.

Calvet, Louis-Jean. 1994. *Les voix de la ville. Introduction à la sociolinguistique urbaine.* Paris: Payot.

Calvet, Louis-Jean. 1999. *Pour une écologie des langues du monde.* Paris: Plon.

Campbell-Kibler, Kathryn. 2007. Accent, (ING), and the social logic of listener perceptions. *American Speech* 82(1). 32–64.

Canut, Cécile. 2000a. Subjectivité, imaginaires et fantasmes des langues : la mise en discours "épilinguistique". *Langage et société* 93. 71–97.

Canut, Cécile. 2000b. Le nom des langues ou les métaphores de la frontière. *Ethnologies comparées* 1. http://alor.univ-montp3fr.fr/cerce/revue.htm. (accessed 15 Jul 2014).

Canut, Cécile. 2001. À la frontière des langues. Figures de la démarcation. *Cahiers d'études africaines* 3, 163–164. 443–464.

Carter, Paul. 2003. Extrinsic phonetic interpretation: spectral variation in English liquids. In John Local, Richard Ogden, & Rosalind Temple (eds.), *Papers in Laboratory Phonology VI: Phonetic Interpretation*, 237–252. Cambridge: Cambridge University Press.

Carter, Paul & John Local. 2007. F2 variation in Newcastle and Leeds English liquid systems. *Journal of the International Phonetic Association* 37. 183–199.

Carver, Craig M. 1991. "In Pursuit of the Subjective Dialect Region". *American Speech* 66(4). 432–437.

Castellotti, Véronique & Danièle Moore. 2002. *Représentations sociales des langues et enseignements, Division des politiques linguistiques*, Direction de l'éducation scolaire, extra-scolaire et de l'enseignement supérieur, DGIV, Conseil de l'Europe, Strasbourg.

Central Statistics Office. 2006. Census of Ireland. Dublin.

Chambers, J. K. & Peter Trudgill. 1998. *Dialectology* (2nd ed.). Cambridge: Cambridge University Press.

Chen, Melinda Yuen-Ching. 1998. "I am an Animal!": Lexical reappropriation, performativity, and queer. In Suzanne Wertheim, Ashlee C. Bailey, & Monica Corston-Oliver (eds.), *Engendering communication: Proceedings from the fifth Berkeley women and language conference*, 128–140. Berkeley, CA: Berkeley Women and Language Group.

Chirrey, Deborah A. 2003. "I hereby come out": What sort of speech act is coming out? *Journal of Sociolinguistics* 7(1). 24–37.

Cloke, Paul. 2006. Conceptualizing rurality. In Paul Cloke, Terry Marsden, & Patrick Mooney (eds.), *Handbook of rural studies*. 18–29. London: Sage.

Cloke, Paul & Paul Milbourne. 1992. Deprivation and lifestyles in rural Wales-II: Rurality and the cultural dimension. *Journal of Rural Studies* 8(4). 359–371.

Clopper, Cynthia & David Pisoni. 2004. Some acoustic cues for the perceptual categorization of American English regional dialects. *Journal of Phonetics* 32. 111–140.

Collective. 2009. Marseille, *Z. Revue itinérante de critique sociale*, 2.

Corrigan, Karen, Richard Edge, & John Lonergan. 2012. Is Dublin English 'Alive Alive Oh' Or Is It 'With O'Leary In The Grave'? In Bittina Migge & Máire Ní Chiosáin (eds.), *New Perspectives in Irish English*, 1–28. Amsterdam: John Benjamins.

Coupland, Nikolas. 2007. *Style: Language variation and identity*. Cambridge University Press, Cambridge.

Coupland, Nikolas & Hywel Bishop. 2007. Ideologised values for British accents. *Journal of Sociolinguistics* 11(1), 74–93.

Coupland, Nikolas, Angie Williams, & Peter Garrett. 1999. "Welshness" and "Englishness" as attitudinal dimensions of English language varieties in Wales. In Dennis R. Preston (ed.), *Handbook of Perceptual dialectology*, Vol. 1, 333–343. Amsterdam: John Benjamins.

Cramer, Jennifer. 2009. Steady-state Patterns of /ai/ in Southern and Midland Dialects: The Case of Louisville. Paper presented at the *Annual Meeting of the Linguistic Society of America*, San Francisco, CA, 8–11 January.

Cramer, Jennifer. 2010. "The Effect of Borders on the Linguistic Production and Perception of Regional Identity in Louisville, Kentucky". Unpublished PhD dissertation, University of Illinois at Urbana-Champaign. https://www.ideals.illinois.edu/bitstream/handle/2142/18426/Cramer_Jennifer.pdf?sequence=1. (accessed 29 July 2014).

Cramer, Jennifer. 2012. What and Where is Appalachian English? The Perceptions of Outsiders. Paper presented at the *Annual Meeting of the Appalachian Studies Association*, Indiana, PA, 23–25 March.

Cramer, Jennifer. 2013. Styles, stereotypes, and the south: constructing identities at the linguistic Border. *American Speech*, 88(2). 144–167.

Cramer, Jennifer & Nathan Hardymon. 2014. Southern Twangs and Urban Brogues: Understanding dialect perceptions across Kentucky. Poster presented at the *Annual Meeting of the American Dialect Society*, Minneapolis, MN, 2–4 January.

Criekingen, Mathieu van. 2008. Comment la gentrification est devenue, de phénomène marginal, un projet global. *Agone* 38–39. 71–88.

Cukor-Avila, P. 1997. An Ethnolinguistic Approach to the Study of Rural Southern AAVE. In Cynthia Bernstein, Thomas Nunnally, & Robin Sabino (eds.), *Language Variety in the South Revisited*, 447–62. Tuscaloosa, AL: University of Alabama Press.

Cukor-Avila, Patricia. 2012. Some structural consequences of diffusion. *Language in Society* 41(5). 615–640.

Cukor-Avila, Patricia. Forthcoming. "Boots an' jeans and Mexican slang: Texans perceptions of dialect variation from the Panhandle to the Rio Grande Valley". *American Speech*.

Cukor-Avila, Patricia, Lisa Jeon, Patricia C. Rector, Chetan Tiwari, & Zac Shelton. 2012. "'Texas – it's like a whole nuther country': Mapping Texans' perceptions of dialect variation in the Lone Star State". In *Proceedings of the Twentieth Annual Symposium About Language and Society – Austin*, 55: 10–19. Austin, TX: Texas Linguistics Forum.

Dailey-O'Cain, Jennifer. 1999. The perception of post-unification German regional speech. In Dennis R. Preston (ed.), *Handbook of perceptual dialectology, vol. 1*, 227–242. Amsterdam: John Benjamins.

Darquennes, Jeroen. 2010. Language contact and language conflict in autochtonous language minority settings in the EU: A preliminary round-up of guiding principles and research desiderata. *Multilingua* 29(3–4). 337–351.

Darquennes, Jeroen. 2013. Language policy and planning in indigenous language minority settings in the EU, *Revue française de linguistique appliqué* XVIII(2). 103–119.

Demirci, Mahide & Brian Kleiner. 1999. The Perception of Turkish Dialects. In Dennis R. Preston (ed.), *Handbook of Perceptual dialectology, vol. 1*, 263–281. Amsterdam: John Benjamins.

Demirci, Mahide. 2002. Gender Differences in the Perception of Turkish Regional Dialects. In Daniel Long & Dennis R. Preston (eds.), *Handbook of Perceptual dialectology, vol. 2*, 41–50. Amsterdam: John Benjamins.

Diercks, Willy. 2002. Mental Maps: Linguistic-Geographic Concepts. In Daniel Long & Dennis R. Preston (eds.), *Handbook of perceptual dialectology, vol. 2*, 51–70. Philadelphia: John Benjamins.

Docherty, Gerard & Paul Foulkes. 1999. Derby and Newcastle: Instrumental phonetics and variationist studies. In Paul Foulkes & Gerard Docherty (eds.), *Urban voices: Accent studies in the British Isles*, 47–71. London: Arnold.

Dressler, Wolfgang U. & Ruth Wodak. 1982. Sociophonological methods in the study of sociolinguistic variation in Viennese German. *Language in Society* 11. 339–370.

Durand, Jacques. 2008. Essai de panorama phonologique: les accents du Midi. In Luc Baronian & France Martineau (eds.), *Le français d'un continent à l'autre*, 123–170. Montréal: Presses de l'Université de Laval.

Durand, Jacques, Bernard Laks, & Chantal Lyche (eds.), 2009. *Phonologie, variation et accents du français*. Paris: Hermès.

Durand, Jacques, Julien Eychenne, & Chantal Lyche. 2013. On levelling and counter-levelling in French: A phonological perspective. In Mari Jones & David Hornsby (eds), *Language and Social Structure in Urban France*, 58–68. Oxford: Legenda.

Durian, David. 2007. Getting [S]tronger Every Day?: Urbanization and the socio-geographic diffusion of (str) in Columbus, OH. *University of Pennsylvania Working Papers in Linguistics* 13(2). 65–79.

Eckert, Penelope. 2000. *Linguistic Variation as Social Practice*. Oxford: Blackwell

Eckert, Penelope. 2010. 'Who's there? Language and space in social anthropology and interactional sociolinguistics'. In Peter Auer & Jürgen E. Schmidt (eds.), *The Language and Space. Theories and methods. An International Handbook of Linguistic Variation*, 163–178. New York: Walter de Gruyter.

Eckert, Penelope & Sally McConnell-Ginet. 2003. *Language and Gender*. Cambridge: Cambridge University Press.

Edwards, John R. 1977. Students' Reactions to Irish Regional Accents. *Language and Speech* 20(3). 280–286.

Edwards, John R. 1982. Language attitudes and their implications among English speakers. In Ellen Bouchard Ryan & Howard Giles (eds.), *Attitudes towards language variation: Social and applied contexts*, 20–33. London: Edward Arnold.

Edwards, John R. 1999. Refining our understanding of language attitudes. *Journal of Language and Social Psychology* 18(1). 101–110.

Eichinger, Ludwig M., Anne-Kathrin Gärtig, Albrecht Plewnia, Janin Roessel, Astrid Rothe, Selma Rudert, Christiane Schoel, Dagmar Stahlberg, & Gerhard Stickel. 2009. *Aktuelle Spracheinstellungen in Deutschland*, Institut für Deutsche Sprache und Universität Mannheim, Mannheim.

Ellis, Alexander J. 1889. *On Early English pronunciation, Part V: Existing dialectal as compared with West Saxon pronunciation*. London: Trübner and Co.
Elmentaler, Michael. 2012. In Hannover wird das beste Hochdeutsch gesprochen. In Lieselotte Anderwald (ed.), *Sprachmythen – Fiktion oder Wirklichkeit?* (Kieler Forschungen zur Sprachwissenschaft 3), 101–116. Frankfurt am Main: Peter Lang.
ESRI. 2011. What is GIS? | GIS.com. http://www.esri.com/what-is-gis. (accessed 28 July 2014).
Evans, Betsy E. 2002a. Attitudes of Montreal students towards varieties of French. In Daniel Long & Dennis R. Preston (eds.), *Handbook of Perceptual dialectology, Vol. 2*, 71–93. Amsterdam: John Benjamins.
Evans, Betsy E. 2002b. An Acoustic and Perceptual Analysis of Imitation. In Daniel Long & Dennis R. Preston (eds.), *Handbook of Perceptual dialectology, vol. 2*, 97–116. Amsterdam: John Benjamins.
Evans, Betsy E. 2010. Aspects of the Acoustic Analysis of Imitation. In Dennis R. Preston & Nancy Niedzielski (eds.), *A Reader in Sociophonetics*, 379–391. Berlin & New York: Mouton de Gruyter.
Evans, Betsy E. 2011a. Seattletonian to Faux Hick: Mapping perceptions of English in WA. *American Speech* 86(4). 383–413.
Evans, Betsy E. 2011b. "Seattle to Spokane: Mapping English in Washington State". http://depts.washington.edu/folkling/ (accessed 28 July 2014).
Evans, Betsy E. 2013a. Seattle to Spokane: Mapping perceptions of English in WA. *Journal of English Linguistics* 41. 268–291.
Evans, Betsy E. 2013b. "Everybody sounds the same": Otherwise overlooked ideology in perceptual dialectology. *American Speech* 88(1). 63–80.
Fagyal, Zsuzsanna. 2010. *Accents de banlieue*. Paris: L'Harmattan.
Fairclough, Norman. 2001. *Language and Power* (2nd ed.), Longman, London.
Feagin, Crawford. 2000. Sound Change in the South. *American Speech* 75(4). 342–344.
Ferriter, Diarmaid. 2004. *The transformation of Ireland, 1900–2000*. London: Profile.
Fijalkow, Yankel & Edmond Préteceille. 2006. Gentrification : discours et politiques urbaines (France, Royaume-Uni, Canada). *Sociétés Contemporaines* 63, 5–13.
Fitzgerald, Peter. (2009. South Korea regions map. Wikimedia Commons. https://commons.wikimedia.org/wiki/File:South_Korea_regions_map.png (accessed 14 July 2015).
Flynn, Nicholas. 2007. *A sociophonetic comparison of adolescent speech in two areas of Nottingham*. Unpublished MA dissertation, University of Essex.
Flynn, Nicholas. 2012. *A sociophonetic study of Nottingham speakers*. Unpublished PhD dissertation, University of York.
Fought, Carmen. 2002. California students' perceptions of, you Know, regions and dialects? In Daniel Long & Dennis R. Preston (eds.), *Handbook of Perceptual dialectology, Vol. 2*, 113–134. Amsterdam: John Benjamins.
Fridland, Valerie, Kathryn Bartlett, & Roger Kreuz. 2004. Do you hear what I hear? Experimental measurement of the perceptual salience of acoustically manipulated vowel variants by Southern speakers in Memphis, TN. *Language Variation and Change* 16: 1–16.
Fridland, Valerie, Kathryn Bartlett, & Roger Kreuz. 2005. Making Sense of Variation: Pleasantness and Education Ratings of Southern Vowel Variants. *American Speech* 80(4). 366–387.

Fridland, Valerie & Kathryn Bartlett. 2006. Correctness, Pleasantness, and Degree of Difference Ratings across Regions. *American Speech* 81(4). 358–386.
Gordon, Matthew. 2010. The sociolinguistic construction of the urban/rural divide in Missouri. Paper presentation at *New Ways of Analyzing Variation* 39, San Antonio, TX, 4–6 November.
Gadet, Françoise. 2007. *La variation sociale en français*, 2nd edn. Paris: Ophrys.
Gadet, Françoise. 2008. Du bon usage du purisme quand on est linguiste. Preface of Marie-Anne Paveau & Laurence Rosier, *La langue française. Passions et polémiques*, 5–9. Paris: Vuibert.
Gadet, Françoise. 2012. La variation, des variétés à la variabilité. In Martine Dreyfus & Jean-Marie Prieur (eds.), *Hétérogénéité et variation. Perspectives sociolinguistiques, didactiques et anthropologiques*, 27–38. Paris: Michel Houdiard Editeur.
Gamboa, Erasmo. 1981. Mexican migration into Washington State: A history, 1940–1950. *Pacific Northwest Quarterly* 72(3). 121–131.
Garrett, Peter, Nikolas Coupland, & Angie Williams. 2003. *Investigating language attitudes*. Cardiff: University of Wales Press.
Garrett, Peter, Williams, Angie, & Evans, Betsy E. 2005. Accessing social meanings: values of keywords, values in keywords. *Acta Linguistica Hafniensia*. 37(1). 37–54.
Gartner, Theodor. 1900. Lautbestand der Wiener Mundart. *Zeitschrift für hochdeutsche Mundarten* 1. 141–147.
Gastil, John. July 18, 2006. Parting the cascade curtain: Rethinking the state's cultural fault line. *The Seattle Times*. http://seattletimes.com/html/opinion/2003129898_sungastil16.html (accessed 19 June 2013)
Gasquet-Cyrus, Médéric. 2002. Sociolinguistique urbaine ou urbanisation de la sociolinguistique? *Marges Linguistiques* 3. 54–67.
Gasquet-Cyrus, Médéric. 2004a. *Pratiques et représentations de l'humour verbal. Etude sociolinguistique du cas marseillais.* Aix-en-Provence: Université de Provence dissertation.
Gasquet-Cyrus, Médéric. 2004b. The sociolinguistics of Marseilles. *International Journal of the Sociology of Language*. 169. 107–123.
Gasquet-Cyrus, Médéric. 2009. Territorialisation, stigmatisation et diffusion. L'accent "quartiers Nord" à Marseille. In Thierry Bulot (ed.), *Formes & normes sociolinguistiques. Ségrégations et discriminations urbaines*, 209–222. Paris: L'Harmattan.
Gasquet-Cyrus, Médéric. 2012a. Hétérogénéité, catégorisation et description linguistique : de l'accent "du Midi" aux accents du sud de la France. In Martine Dreyfus & Jean-Marie Prieur (eds.), *Hétérogénéité et variation. Perspectives sociolinguistiques, didactiques et anthropologiques*, 50–65. Paris: Michel Houdiard Editeur.
Gasquet-Cyrus, Médéric. 2012b. La discrimination à l'accent en France : idéologies, discours et pratiques. In Cyril Trimaille & Jean-Michel Eloy (eds.), *Idéologies linguistiques et discriminations*. Carnets d'Atelier de Sociolinguistique 6, 227–245. Paris: L'Harmattan.
Gasquet-Cyrus, Médéric. 2012c. Variations, représentations et impact sociolinguistiques des dictionnaires : la lexicographie "identitaire" marseillaise. In Christophe Rey & Phillipe Reynés (eds.), *Dictionnaires, norme(s) et sociolinguistique, Cahiers d'Atelier de Sociolinguistique* (n° 5), 333–358. Paris: L'Harmattan.
Gasquet-Cyrus, Médéric. 2013. Perspectives dynamiques sur la ségrégation sociolinguistique en milieu urbain : le cas de Marseille. *Glottopol*. 21. 9–21.
Gasquet-Cyrus, Médéric. 2015. Changements urbains et conflits sociolinguistiques : l'impact de la gentrification sur le français de Marseille. *International Journal of the Sociology of*

Language 235, special issue "Language conflict", Jeroen Darquennes et Diarmait MacGiolla Chríost (eds), 77–101.

Gasquet-Cyrus, Médéric & J.-M. Géa (eds.). In press (to be published in 2016). Marseille: gentrification et ségrégation. *Langage et société.*

Gaudio, Rudolf P. 1994. Sounding gay: Pitch properties in the speech of gay and straight men. *American Speech* 69(1). 30–57.

Giles, Howard. 1970. Evaluative reactions to accents. *Educational Review* 22. 211–227.

Giles, Howard & Richard Y. Bourhis. 1975. Linguistic assimilation: West Indians in Cardiff. *Language Sciences.* 38. 9–12.

Giles, Howard & Richard Y. Bourhis. 1976. Methodological issues in dialect perception: Some social psychological perspectives. *Anthropological Linguistics* 18(7). 294–304.

Giles, Howard & Peter F. Powesland. 1975. *Speech style and social evaluation.* London: Academic Press.

Goebl, Hans. 1997. Some dendrographic classification of the data of CLAE 1 and CLAE 2. In Wolfgang Viereck & Heinrich Ramisch (eds.), *The computer developed linguistic atlas of England*, vol. 2, 23–32. Tübingen: Max Niemeyer.

Goebl, Hans & Guillaume Schiltz. 1997. A dialectometrical compilation of CLAE 1 and CLAE 2: Isoglosses and dialect integration. In Wolfgang Viereck & Heinrich Ramisch (eds.), *The computer developed linguistic atlas of England*, 13–21. Tubingen: Niemeger.

Goeman, Ton. 2002. Perception of dialect distance: Standard and dialect in relation to new data in Dutch varieties. In Daniel Long & Dennis R. Preston (eds.), *Handbook of perceptual dialectology*, Vol. 2, 135–149. Amsterdam: John Benjamins.

Gomarasca, Mario A. 2009. *Basics of geomatics.* London & New York: Springer Dordrecht Heidelberg.

Goodchild, Barry. 1974. Class differences in environmental perception: An exploratory study. *Urban Studies* 11. 157–169.

Goodey, Brian, John Gold, Alan Duffett, & David Spencer. 1971. *City scene. An exploration into the image of Central Birmingham as seen by area residents.* University of Birmingham: Centre for Urban and Regional Studies.

Gordon, Elizabeth. 1997. Sex, speech, and stereotypes: Why women use prestige speech forms more than men. *Language in Society* 26(1). 47–63.

Gordon, Matthew. 2001. *Small town values, big city vowels: A study of the Northern cities shift in Michigan.* Durham: Duke University Press.

Gould, Peter & Rodney White. 1986. *Mental maps.* 2nd edn. Boston: Allen & Unwin.

Grootaers, Willem. 1959. Origin and nature of the subjective boundaries of dialects. Orbis. 8(2). 355–384.

Grootaers, Willem. [1964] 1999. The discussion surrounding the subjecting boundaries of dialects. Reprinted in Dennis R. Preston (ed.), *Handbook of perceptual dialectology*, Vol. 1, 115–129. Amsterdam: John Benjamins.

Haase, Trutz & Jonathan Pratschke. 2005. *Deprivation and its spatial articulation in the Republic of Ireland.* Dublin: Pobal.

Halfacree, Keith. 1993. Locality and social representation: Space, discourse and alternativedefinitions of the rural. *Journal of Rural Studies.* 9(1). 23–37.

Halfacree, Keith. 2006. Rural space. In Paul Cloke, Terry Marsden, & Patrick Mooney (eds.), *Handbook of rural studies,* 44–62. London: SAGE.

Hall-Lew, Lauren. 2005. One shift, two groups: When fronting alone is not enough. *University of Pennsylvania Working Papers in Linguistics* 10(2). 105–116.

Hall-Lew, Lauren. 2012. Country talk. *Journal of English Linguistics* 40(3). 256–280.

Hartley, Laura C. 1999. A view from the West: Perceptions of U.S. dialects by Oregon residents. In Dennis R. Preston (ed.), *Handbook of perceptual dialectology*, Vol. 1, 315–332. Amsterdam: John Benjamins.
Hartley, Laura C. & Dennis R. Preston. 1999. The names of US English: Valley girl, cowboy, Yankee, normal, nasal, & ignorant. In Tony Bex & Richard J. Watts (eds), *Standard English*, 207–38. London: Routledge.
Hartley, Laura C. 2005. The consequences of conflicting stereotypes: Bostonian perceptions of U.S. dialects. *American Speech* 80(4). 388–405.
Herberich, Esther, Johannes Sikorski, & Torsten Hothorn. 2010. A robust procedure for comparing multiple means under heteroscedasticity in unbalanced designs. *PLoS ONE 5*(3). 1–8.
Herrgen, Joachim & Jürgen Erich Schmid. 1985. Systemkontrast und Hörerurteil. Zwei Dialektalitätsbegriffe und die ihnen entsprechenden Messverfahren. *Zeitschrift für Dialektologie und Linguistik* 52. 20–42.
Heyer, Julia Amalia. 2011. Marseille. La ville laboratoire. Courrier international, supplement to the n° 1096. www.courrierinternational.com/article/2011/11/03/la-ville-laboratoire. (accessed 27 July 2014)
Hickey, Raymond. 2005. *Dublin English: Evolution and change*. Amsterdam: John Benjamins.
Hoenigswald, Henry. 1966. A proposal for the study of folk-linguistics. In William Bright (ed.), Sociolinguistics, 16–26. The Hague: Mouton de Gruyter.
Hofer, Lorenz. 2004. Sprachliche und politische Grenzen im (ehemaligen) Dialektkontinuum des Alemannischen am Beispiel der trinationalen Region Basel (Schweiz) in Karten von SprecherInnen. *Linguistik Online* 20: 23–46.
Holliday, John, Clive Upton, Ann Thompson, Jonnie Robinson, Jon Herring, & Holly Gilbert. 2013. Geographical analysis of the vernacular. *Journal of Information Science*. 39(1). 1–12.
Honey, John. 1997. Sociophonology. In Florian Coulmas (ed.), *The handbook of sociolinguistics*, 92–106. Oxford: Blackwell.
Horner, Arnold. 1992. From city to city-region – Dublin from the 1930s to the 1990s. In F. H. A. Aalen & Kevin Whelan (eds.), *Dublin city and county: From prehistory to present*. Dublin: Geography Publications.
Hundt, Markus. 2010. Bericht über die Pilotstudie „Laienlinguistische Konzeptionen deutscher Dialekte“. In Christina A. Anders, Markus Hundt, & Alexander Lasch (eds.), *Perceptual dialectology: Neue Wege der Dialektologie*, 179–219. Berlin & New York: De Gruyter.
Hunt, Andrew J. & Allen W. Black. 1996. Unit selection in a concatenative speech synthesis system using a large speech database. In *Acoustics, Speech, and Signal Processing, 1996. ICASSP-96. Conference Proceedings., 1996 IEEE International Conference, Vol. 1*, 373–376.
Hughes, Arthur, Peter Trudgill, & Dominic Watt. 2012. *English accents and dialects: An introduction to social and regional varieties of English in the British Isles*. (5th edn.) London: Hodder Education.
Hymes, Dell. 1974. Foundations in sociolinguistics: An ethnographic approach. Philadelphia: University of Pennsylvania Press.
Inoue, Fumio. [1995] 1999a. Classification of dialects by image: English and Japanese. Reprinted in Dennis R. Preston (ed.), *Handbook of perceptual dialectology*, Vol. 1, 147–159. Amsterdam: John Benjamins.

Inoue, Fumio. [1996] 1999b. Subjective dialect division in Great Britain. Reprinted in Dennis R. Preston (ed.), *Handbook of perceptual dialectology*, Vol. 1, 161–176. Amsterdam: John Benjamins.

Irons, Terry Lynn. 2007. On the status of low back vowels in Kentucky English: More evidence of merger. *Language Variation and Change* 19. 137–180.

Irvine, Judith T. & Susan Gal. 2000. Language ideology and linguistic differentiation, In Paul V. Kroskrity (ed.), *Regimes of languages: Ideologies, politics, and identities*, 35–83. Sante Fe: School of American Research Press.

Ito, Rika. 1999. *Diffusion of urban sound change in rural Michigan: A case of the northern cities shift*. Unpublished MA dissertation, Michigan State University.

Jamin, Mikaël, Cyril Trimaille, & Médéric Gasquet-Cyrus. 2006. De la convergence dans la divergence: le cas des quartiers pluri-ethniques en France. *Journal of French Language Studies*. 16(3). 335–356.

Jeon, Lisa. 2011. Drawing Boundaries and Revealing Language Attitudes: The Relationship of Language and Place in Korea. Poster presented at *New Ways of Analyzing Variation* 40, Georgetown University, Washington, DC, 28–30 October.

Jeon, Lisa & Cukor-Avila, Patricia. 2015. 'One country, one language?': Mapping perceptions of dialects in South Korea. In Maria-Pilar Parea & Gotzon Aurrekoetxea (eds.), *Dialectologia* 14. 17–46.

Johnstone, Barbara. 2000. *Qualitative methods in sociolinguistics*. New York: Oxford University Press.

Johnstone, Barbara. 2004. Place, globalization, and linguistic variation. In Carmen Fought (ed.), *Sociolinguistic variation. critical reflections*, 65–83. Oxford: Oxford University Press.

Johnstone, Barbara. 2009. Pittsburghese Shirts: Commodification and the enregisterment of an urban dialect. *American Speech* 84(2). 157–175.

Johnstone, Barbara. 2010. Language and geographical space. In Peter Auer & Jürgen Erich Schmidt (eds.), *Language and space. An international handbook of linguistic variation, Vol. 1*, 1–18. Berlin: Walter de Gruyter.

Johnstone, Barbara. 2011. Making Pittsburghese: Communication technology, expertise, and the discursive construction of a regional dialect. *Language and Communication* 31(1). 3–15.

Johnston, P. A. 1985. The rise and fall of the Morningside/Kelvinside accent. In Manfried Gorlach (ed.), *Focus on Scotland*, 37–56. Amsterdam: John Benjamins.

Jørgensen, Jens Norman. 2010. *Languaging – Nine years of poly-lingual development of young Turkish-Danish grade school students*, Copenhagen Studies in Bilingualism, The Køge Series, Vol. K16. Copenhagen: University of Copenhagen, Faculty of Humanities.

Jourdan, Silvère. 2008. Un cas aporétique de gentrification: la ville de Marseille, *Méditerranée* 111. 85–90.

Jones, Owain. 1995. Lay discourses of the rural: Developments and implications for rural studies. *Journal of Rural Studies* 2(1). 35–49.

Kallen, Jeffrey. 1994. English in Ireland. In Robert Burchfield (ed.), *The Cambridge History of the English Language*, Vol. 5, 148–196. Cambridge: Cambridge University Press.

Kehrein, Roland. 2012. *Regionalsprachliche Spektren im Raum. Zur linguistischen Struktur der Vertikale*. Stuttgart: Steiner.

Kehrein, Roland, Alfred Lameli, & Christoph Purschke. 2010. Stimuluseffekte und Sprachraumkonzepte. In Christina Anders, Marcus Hundt, & Alexander Lasch (eds.),

Perceptual dialectology: Neue Wege der Dialektologie. 351–384. Berlin: Mouton de Gruyter.
Kerswill, Paul. 2001. Mobility, meritocracy and dialect levelling: the fading (and phasing) out of Received Pronunciation. In Pilvi Rajamäe & Krista Vogelberg (eds.), *British studies in the new millennium: the challenge of the grassroots*, 45–58. Tartu: University of Tartu.
Kerswill, Paul. 2003. Dialect levelling and geographical diffusion in British English. In David Britain & Jenny Cheshire (eds.), *Social dialectology. In honour of Peter Trudgill*, 223–243. Amsterdam: John Benjamins.
Kerswill, Peter & Angie Williams. 2002. Dialect recognition and speech community focusing in new and old towns in England. In Daniel Long & Dennis R. Preston (eds.), *Handbook of perceptual dialectology*, Vol. 2, 173–204. Philadelphia: John Benjamins.
Kim, Pyeongje. 1988. *Chosŏn Ŏnŏjirihak Sigo (An Essay in Korean Dialect Geography)*. P'yŏngyang: Kwahak Paekkwa Sajŏn Chonghap Ch'ulp'ansa.
Kim, Wan-Jin. 1983. Phonological Structure of the Korean language. In Korean National Commision for UNESCO (ed.), *The Korean Language*, 157–170. Seoul: Si-sa-yong-o-sa.
King, Ross. 2006. Dialectal variation in Korean. In Ho-Min Sohn, (ed.), *Korean Language in Culture and Society*, 264–280. University of Hawaii Press.
Kleiner, Stefan. 2011. Zur Aussprache von nebentonigem *-ig* im deutschen Gebrauchsstandard. *Zeitschrift für Dialektologie und Linguistik* 77(3). 259–304.
Kolb, Eduard, Beat Glauser, Willy Elmer, & Renate Stamm. 1979. *Atlas of English sounds*. Bern: Franke.
Koops, Christian. 2010. /u/-fronting is not monolithic: Two types of fronted /u/ in Houston Anglos. *University of Pennsylvania Working Papers in Linguistics* 16(2). 113–122.
Korean Wiki Project. 2009. Korean Wiki Project. Retrieved from: http://www.koreanwikiproject.com/wiki/Category:Dialects. (accessed 14 July 2015).
Kranzmayer, Eberhard. 1953. Lautwandlungen und Lautverschiebungen im gegenwärtigen Wienerischen. *Zeitschrift für Mundartforschung* 21. 197–239.
Kristiansen, Gitte. 2003. How to do things with allophones: Linguistic stereotypes as cognitive reference points in social cognition. In René Dirven, Roslyn Frank, & Martin Pütz (eds.), *Cognitive models in language and thought: Ideology, metaphors and meanings*, 69–120. Berlin & New York: Mouton de Gruyter.
Kristiansen, Tore. 2009. The macro-level meanings of late-modern Danish accents. *Acta Linguistica Hafniensia* 41: 167–192.
Kristiansen, Tore & Stefan Grondelaers (eds). 2013. *Language (de)standardisation in Late Modern Europe: Experimental studies*. Oslo: Novus Press.
Kristiansen, Tore & Nikolas Coupland. 2011. SLICE: Critical perspectives on language (de)standardisation. In Tore Kristiansen & Nikolas Coupland (eds.), *Standard Languages and Language Standards in a Changing Europe*, 11–35, Oslo: Novus.
Krosnick, Jon A., Charles M. Judd, & Bernd Wittenbrink. 2005. The measurement of attitudes. In Dolores Albarracín, Blair T. Johnson, & Mark P. Zanna (eds.), *The Handbook of Attitudes*, 21–76. New York & London: Psychology Press.
Kruse, Jan. 2008. *Reader – Einführung in die qualitative Interviewforschung*. http://www.soziologie.uni-freiburg.de/kruse (accessed 29 July 2014).
Kuiper, Lawrence. 1999. Variation and the norm: Parisian perceptions of regional France. In Dennis R. Preston (ed.), *Handbook of perceptual dialectology*, Vol. 1, 243–262. Amsterdam: John Benjamins.
Kuiper, Lawrence. 2005. Perception is reality: Parisian and provençal perceptions of regional varieties of French. *Journal of Sociolinguistics* 9(1). 28–52.

Kurath, Hans. 1931. The geography of speech: Plans for a linguistic atlas of the United States and Canada. *Geographical Review* 21(3). 483–486.
Kurath, Hans. 1949. *Word geography of the Eastern United States*. Ann Arbor, MI: University of Michigan Press.
Kurath, Hans, Miles L. Hanley, Bernard Bloch, Guy S. Lowman, Jr., & Marcus L. Hansen. 1939–1943. *Linguistic atlas of New England*. 3 vols. Providence, RI: Brown University and American Council of Learned Societies.
Kurath, Hans & Raven I. McDavid Jr. 1961. *The pronunciation of English in the Atlantic states*. Ann Arbor, MI: University of Michigan Press.
Kvale, Steinar. 2005. *InterView – En introduktion til det kvalitative forskningsinterview*. Copenhagen: Hans Reitzel.
Labov, William. 1966. *The social stratification of English in New York City*. Arlington VA: The Center for Applied Linguistics.
Labov, William. 1972. *Sociolinguistic patterns*. Philadelphia: University of Pennsylvania Press.
Labov, William. 2001. *Principles of linguistic change, Vol. 2: Social Factors*. Oxford: Blackwell.
Labov, William, Sharon Ash, & Charles Boberg. 2006. *The atlas of North American English: Phonetics, phonology and sound change, a multimedia reference tool*. Berlin: Mouton de Gruyter.
Ladd, Florence C. 1970. Black Youths View Their Environment. *Environment and Behaviour* 2(2). 74–99.
Lambert, W. E. 1967. A social psychology of bilingualism. *Journal of Social Issues* 23(2). 91–109.
Lambert, W. E., R. C. Hodgson, R. C. Gardner, & S. Fillenbaum. 1960. Evaluational reactions to spoken languages. *Journal of Abnormal and Social Psychology* 60(1). 44–51.
Lameli, Alfred, Christoph Purschke, & Roland Kehrein. 2008. Stimulus und kognition: Zur aktivierung mentaler raumbilder. *Linguistik Online* 35. 55–86.
Lance, Donald M. 1999. Regional variation in subjective dialect divisions in the United States. In Dennis Preston (ed.), *Handbook of perceptual dialectology*, Volume 1, 283–314. Philadelphia: John Benjamins.
Leap, William L. 1996. *Word's out: Gay men's English*. Minneapolis: University of Minnesota Press.
Lee, Sang-Oak. 1983. Remarks on tone in Middle Korean. In the Korean National Commission for UNESCO (ed.), *The Korean Language*, 171–189. Seoul: Si-sa-yong-o-sa.
Lehiste, Ilse & Gordon E. Peterson. 1961. Transitions, glides, and diphthongs. *Journal of the Acoustical Society of America*. 33. 268–277.
Lenz, Alexandra N. 2003. *Struktur und Dynamik des Substandards. Eine Studie zum Westmitteldeutschen*. Stuttgart: Steiner.
Lenz, Alexandra N. 2010. Zum Salienzbegriff und zum Nachweis salienter Merkmale. In Christina A. Anders, Markus Hundt, & Alexander Lasch (eds.), *Perceptual dialectology: Neue Wege der Dialektologie*, 89–110. Berlin & New York: De Gruyter.
Levon, Erez. 2006. Hearing "gay": Prosody, interpretation, and the affective judgments of men's speech. *American Speech* 81(1). 56–78.
LexisNexis. 2011. LexisNexis library. http://www.lexisnexis.com/uk/legal/ (accessed 17 August, 2012).
Le Douaron, Michelle. 1983. *Contribution à l'étude sociolinguistique et phonétique des parlers méridionaux*. Unpublished Université de Provence dissertation.

Le Page, Robert & Andrée Tabouret-Keller. 1985. *Acts of identity: Creole-based approaches to language and ethnicity*. Cambridge: Cambridge University Press.

Liang, A. C. 1997. The creation of coherence in coming-out stories. In Anna Livia & Kira Hall (eds.), *Queerly phrased: Language, gender, and sexuality*, 287–309. New York: Oxford University Press.

Lichter, Daniel T. & David L. Brown. 2011. Rural America in an urban society: Changing spatial and social boundaries. *Annual Review of Sociology* 37. 565–592.

Lindsey, Geoff & Allen Hirson. 1999. Variable robustness of nonstandard /r/ in English: evidence from accent disguise. *The International Journal of Speech, Language and the Law* 6(2). 278–288.

Lippi-Green, Rosina. 2012. *English with an accent: Language, ideology, and discrimination in the United States*, 2nd edn. London: Routledge.

Llamas, Carmen. 1999. A new methodology: Data elicitation for social and regional language variation studies. *Leeds Working Papers in Linguistics and Phonetics* 7. 95–118.

Llamas, Carmen. 2000. Middlesbrough English: Convergent and divergent trends in a "Part of Britain with no identity". *Leeds Working Papers in Linguistics and Phonetics* 8. 123–148.

Lobanov, B. M. 1971. Classification of Russian vowels spoken by different listeners. *Journal of the Acoustical Society of America* 49. 606–608.

Lodge, Anthony. 1996. *Le français. Histoire d'un dialecte devenu langue*. Paris: Fayard.

Lonergan, John. 2013. *An acoustic and perceptual study of Dublin English phonology*. Unpublished PhD dissertation, University College Dublin, Dublin.

Long, Daniel. 1999a. Geographical perceptions of Japanese dialect regions. In Dennis Preston (ed.), *Handbook of perceptual dialectology*, volume 1, 177–198. Philadelphia: John Benjamins.

Long, Daniel. 1999b. Mapping nonlinguists' evaluations of Japanese language variation. In Dennis Preston (ed.), *Handbook of perceptual dialectology*, volume 1, 199–226. Philadelphia: John Benjamins.

Long, Daniel & Dennis R. Preston (eds.), 2002. *Handbook of perceptual dialectology, Vol. 2*. Amsterdam: John Benjamins.

Long, Daniel & Young-Cheol Yim. 2002. Regional differences in the perception of Korean dialects. In Daniel Long & Dennis R. Preston (eds.), *Handbook of perceptual dialectology, Vol. 2*, 249–75. Amsterdam: John Benjamins.

Luick, Karl. 1904. *Deutsche Lautlehre. Mit besonderer Berücksichtigung der Sprechweise Wiens und der österreichischen Alpenländer*. Leipzig & Wien: Franz Deuticke.

Lynch, Kevin. 1960. *The image of the city*. Cambridge, Mass: MIT Press.

Maguire, Warren. 2012. Mapping the existing phonology of English dialects. *Dialectologia et Geolinguistica* 20. 84–107.

MailOnline. 2013. http://www.dailymail.co.uk/news/article-2320561/Fat-map-Britain-revealed-East-Midlands-worst-place-obesity-related-hospital-admissions.html (accessed 30 July 2014)

Manessy, Gabriel. 1997. Norme endogène. In Marie-Louise Moreau (ed.), *Sociolinguistique. Concepts de base*, 223–225. Sprimont: Mardaga.

Markham, Duncan. 1999. Listeners and disguised voices: the imitation and perception of dialect accent. *The International Journal of Speech, Language and the Law* 6(2). 289–299.

Marotta, Giovanna & Nadia Nocchi. 2003. A phonetic conspiracy. Low vowels and velarized lateral in Leghorn Italian. In Daniel Recasens & Maria Josep Solé (eds.), *Proceedings of the XVth ICPhS*, 909–912. Barcelona: Universitat Autònoma de Barcelona.

Martel, Claude. 1999. Le parler de Marseille vu par les Provençaux (provençal-français régional). *Skholê. Cahiers de la Recherche et du Développement* 10, 25.
McDavid, Raven I. 1983. Retrospect. *Journal of English Linguistics* 16. 47–54.
McDonald, John F. 2013. Urban Areas in the Transformation of the South: A Review of Modern History. *Urban Studies Research* 2013. 14 pages. doi:10.1155/2013/376529 (accessed 30 July 2014).
Mihm, Arend. 1985. Prestige und Stigma des Substandards. Zur Bewertung des Ruhrdeutschen im Ruhrgebiet. In Mihm, Arend (ed.), *Sprache an Rhein und Ruhr*, 163–193. Stuttgart: Steiner.
Meseck, Donald. 2013. Chelan and Douglas counties profile. Employment Security Department, Washington State. https://fortress.wa.gov/esd/employmentdata/reports-publications/regional-reports/county-profiles/chelan-and-douglas-counties-profile (accessed 30 July 2014).
Meuter, Anne 2008. *The emergence of structured variation in the speech of children and adolescents in Bonn, Germany*. Unpublished PhD dissertation, University of Essex.
Milroy, James. 1996. A current change in British English: Variation in (th) in Derby. *Newcastle and Durham Working Papers in Linguistics* 4. 213–222.
Milroy, Lesley. 2002. Introduction: Mobility, contact and language change working with contemporary speech communities. *Journal of Sociolinguistics* 6. 3–15.
Milroy, Lesley & Dennis R. Preston. 1999. Introduction. *Journal of Language and Social Psychology* 18(1). 4–9.
Montgomery, Chris. 2014. Perceptual ideology across the Scottish-English border. In Dominic Watt & Carmen Llamas (eds.), Language, Borders and Identities, 118–136. Edinburgh: Edinburgh University Press.
Montgomery, Chris. 2007. Northern English dialects: A Perceptual Approach. Unpublished PhD dissertation, University of Sheffield. http://etheses.whiterose.ac.uk/1203/ (accessed 30 July 2014).
Montgomery, Chris. 2011a. Sprachraum and its perception. In Alfred Lameli, Roland Kehrein, & Stefan Rabanus (eds.), *Language and Space – An International Handbook of Linguistic Variation, Volume 2: Language Mapping*, 586–605. Berlin: Mouton de Gruyter.
Montgomery, Chris. 2011b. Perceptual dialectology, bespoke methods and the future for data processing. Paper presented at the *8th United Kingdom Language Variation and Change* conference, Edge Hill University, 12–14 September.
Montgomery, Chris. 2012. The effect of proximity in perceptual dialectology. *Journal of Sociolinguistics* 16(5). 638–668.
Montgomery, Chris & Joan C. Beal. 2011. Perceptual dialectology. In Warren Maguire & April McMahon (eds.), *Analysing variation in English*, 121–148. Cambridge: Cambridge University Press.
Montgomery, Chris & Philipp Stoeckle. 2013. Geographic information systems and perceptual dialectology: a method for processing draw-a-map data. *Journal of Linguistic Geography* 1(1). 52–85.
Moon, Kyuwon. 2010. Doing cuteness in Korea: The social meaning of rising-falling tone. MS. Stanford University.
Moore, Emma. 2011. Variation and identity. In: Warren Maguire & April MacMahon (eds.), *Analysing variation in English*, 219–236. Cambridge: Cambridge University Press.
Moore, Niamh. 2008. *Dublin docklands reinvented: The post-industrial regeneration of a European city quarter*. Dublin: Four Courts Press.

Moore, Robert. 2011. The unmentionable: Verbal taboo and the moral life of language. "If I Actually Talked Like That, I'd Pull a Gun on Myself": Accent, Avoidance, and Moral Panic in Irish English. *Anthropological Quarterly* 84(1). 41–64.

Moosmüller, Sylvia. 1987. Soziophonologische Variation im gegenwärtigen Wiener Deutsch. Wiesbaden: Steiner.

Moosmüller, Sylvia. 1991. *Hochsprache und Dialekt in Österreich*. Wien: Böhlau.

Moosmüller, Sylvia. 2007. *Vowels in standard Austrian German. An acoustic-phonetic and phonological Analysis*. Habilitationsschrift: Wien.

Moosmüller, Sylvia. 2010. Stereotyping the Viennese dialect. In Rudolf de Cillia, Helmut Gruber, Michal Krzyżanowski, & Florian Menz (eds.), *Diskurs – Politik – Identität. Festschrift für Ruth Wodak*, 401–408. Tübingen: Stauffenberg.

Moosmüller, Sylvia. 2015. Methodisches zur Bestimmung der Standardaussprache in Österreich. In Manfred Glauninger & Alexandra Lenz (eds.), *Standarddeutsch in Österreich – Theoretische und empirische Ansätze*, 163–182. Wien: Vandenhoeck & Ruprecht.

Moosmüller, Sylvia & Julia Brandstätter. 2014. Phonotactic information in the temporal organization of Standard Austrian German and the Viennese dialect. *Language Sciences* 46. 84–95.

Moreno Fernández, Juliana & Francisco Moreno Fernández. 2002. Madrid Perceptions of regional varieties in Spain. In Daniel Long & Dennis R. Preston (eds.), *Handbook of perceptual dialectology*, Vol. 2, 295–320. Amsterdam: John Benjamins.

Moscovici, Serge. 1998. The History and Actuality of Social Representations. In Flick, Ewe (ed.), *The Psychology of the Social*, 209–247. New York: Cambridge University Press.

Munck, Ronaldo. 2007. Social class and inequality. In Sara O'Sullivan (ed.), *Contemporary Ireland: A sociological map*, 301–317. Dublin: University College Dublin Press.

Nakamura, Tamotsu, Tong-Jun Kim, & Hiroyuki Umeda. 1991. Kankukugo Seoul hogen no sedaisa nit suite, *Gakujutsu Geppo* 44(4). 348–354.

Narayanan Shrikanth S., Abeer A. Alwan, & Katherine Haker. 1997. Toward articulatory-acoustic models for liquid approximants based on MRI and EPG data, Part I: The laterals. *Journal of the Acoustical Society of America* 101(2). 1064–1077.

National Language Research Institute (NLRI). 1992. *P'yojun Hwapop Haesol* (Explanations on Standard Ways of Speaking). Seoul: Kungnip Kugo Yon'guwon.

National Youth Gang Center 2009. National Youth Gang Survey Analysis. http://www.nationalgangcenter.gov/Survey-Analysis (accessed 30 July 2014).

Nelde, Peter. 1997. Language conflict. In Florian Coulmas (ed.), *The handbook of sociolinguistics*, 285–300. Oxford: Blackwell.

Nelde, Peter. 2006. La linguistique de contact, la recherche sur le conflit linguistique et l'aménagement linguistique au XXe siècle. In Sylvain Auroux, E. F. K. Körner, Hans Niederehe, & Kees Versteegh (eds.), *History of the language sciences*, Vol. 3, 2413–2429. Berlin & New York: De Gruyter.

Nerbonne, John. 2007. Crosstalk in humanities Computing. *International Journal of Humanities and Arts Computing* 1(2). 85–96.

Nerbonne, John & William A. Kretzschmar. 2013. Dialectometry++. *Literary and Linguistic Computing* 28(1). 2–12.

Neuhauser, Sara. 2008. Voice disguise using a foreign accent: phonetic and linguistic variation. *The International Journal of Speech, Language and the Law* 15(2). 131–159.

Nguyen, Emily. 2011. The urban/rural distinction and so much more: Monophthongal (ow) in Minnesota. Paper presented at *New Ways of Analyzing Variation* 40, Georgetown University, Washington, D.C., 27–30 October.

Niedzielski, Nancy. 2002. Attitudes Toward Midwestern American English. In Daniel Long & Dennis R. Preston (eds.), *Handbook of perceptual dialectology*, Vol. 2, 321–327. Amsterdam: John Benjamins.
Niedzielski, Nancy & Preston, Dennis. 2003. *Folk Linguistics*. New York: Mouton.
Nolan, Brian & Bertrand Maitre. 2007. Economic Growth and Income Inequality: Setting the Context. In Tony, Fahey, Helen Russell, & Christopher Whelan (eds.), *Best of Times? The Social Impact of the Celtic Tiger*. Dublin: Institute of Public Administration.
Nottingham Post. 29 November 2013. Ofsted warning for six schools. Front page story.
Ogura, Shimpei. 1940. The outline of the Korean dialects. Memoirs of the Research Department of The Toyo Bunko 12. Tokyo: Toyo Bunko.
Onishi, Isao & Daniel Long. 1997. *Perceptual dialectology Quantifier (PDQ) for Windows*. http://nihongo.hum.tmu.ac.jp/~long/maps/perceptmaps.htm (accessed 2 July 2012).
Orleans, Peter. 1973. Differential cognition of urban residents: Effects of social scale on mapping. In Roger Downs & David Stea (eds.) *Image and environment: Cognitive mapping and spatial behaviour*, 115–130. Chicago: Aldine Publishing Company.
Orton, Harold & Michael V. Barry (eds.), 1969. *The survey of English dialects (B): The basic material*, Vol. 2: The West Midland counties. Leeds: E. J. Arnold.
Orton, Harold & Eugen Dieth. 1962. *Survey of English dialects: Introduction*. Leeds: E. J. Arnold.
Orton, Harold, Stewart Sanderson, & J. D. A. Widdowson (eds.). 1978. *The linguistic atlas of England*. London: Croom Helm.
Pavlovskaya, Marianna. 2009. Non-quantitative GIS. In Meghan Cope & Sarah Elwood (eds.), *Qualitative GIS*, 13–38. London: Sage.
Pearce, Michael. 2009. A perceptual dialect map of North East England. *Journal of English Linguistics* 37. 162–192.
Peraldi, Michel & Michel Samson. 2006. *Gouverner Marseille. Enquête sur les mondes politiques marseillais*. Paris: La Découverte.
Petitjean, Cécile. 2008. Représentations linguistiques et accents régionaux du français, *Journal of Language contact*. Varia 1. 29–51.
Plichta, Bartłomiej & Dennis R. Preston. 2005. The /ay/s have it: The perception of /ay/ as a north-south stereotype in United States English. In Tore Kristiansen, Peter Garrett, & Nikolas Coupland (eds), *Acta Linguistica Hafniensia* 37 (*Subjective Processes in Language Variation and Change*): 107–30.
Preston, Dennis R. 1982. Perceptual dialectology: Mental maps of United States dialects from a Hawaiian perspective. *Hawaii Working Papers in Linguistics* 14(2). 5–49.
Preston, Dennis R. 1986. Five Visions of America. *Language in Society* 15. 221–240.
Preston, Dennis R. 1986. Introduction. *International Journal of the Sociology of Language* 57. 5–8.
Preston, Dennis Richard. 1988. Change in the perception of regional varieties. In Jacek Fisiak (ed.), *Historical dialectology*. 475–504. Berlin: Mouton de Gruyter.
Preston, Dennis R. 1989. *Perceptual dialectology: Non-linguists' view of aerial linguistics*. Dordrecht: Foris.
Preston, Dennis R. 1993. The uses of folk linguistics. *International Journal of Applied Linguistics* 3(2). 181–259.
Preston, Dennis R. 1996. Where the worst English is spoken. In Edgar Schneider (ed.), *Focus on the USA*, 297–360. Amsterdam: John Benjamins.
Preston, Dennis R. 1998. Why we need to know what real people think about language. *The Centennial Review* 42(2). 255–284.

Preston, Dennis R. 1999a. *Handbook of perceptual dialectology*, vol 1. Philadelphia: John Benjamins.

Preston, Dennis R. 1999b. Introduction. In Dennis R. Preston (ed.), *Handbook of perceptual dialectology*, xxiii–xxxix. Amsterdam: John Benjamins.

Preston, Dennis R. 1999c. A language attitude approach to the perception of regional variety. In Dennis Preston (ed.), *Handbook of perceptual dialectology*, Vol 1, 359–373. Philadelphia: John Benjamins.

Preston, Dennis R. 2003. Language with an attitude. In J. K. Chambers, Peter Trudgill, & Natalie Schilling-Estes (eds.), *The handbook of language variation and change*, 40–66. Oxford: Blackwell.

Preston, Dennis R. 2004. Folk metalanguage. In Adam Jaworski, Nikolas Coupland, & Dariusz Galasinski (eds.), *Metalanguage: Social and ideological perspectives*, 75–101. Berlin: Mouton de Gruyter.

Preston, Dennis R. 2010a. Language, space and folk. In Peter Auer & Jürgen Erich Schmidt (eds.), *Language and Space. An international handbook of linguistic variation, vol 1: Theories and methods*, 179–201. Berlin: Walter de Gruyter.

Preston, Dennis R. 2010b. Language, people, salience, space: Perceptual dialectology and language regard. *Dialectologia* 5. 87–131.

Preston, Dennis R. To appear. Does language regard vary? In Alexei Prikhodkine & Dennis R. Preston (eds). *Language attitudes: Variability, processes, and outcomes*. Amsterdam/ Philadelphia: John Benjamins.

Preston, Dennis R. & George M. Howe. 1987. Computerized studies of mental dialect maps. In Keith M. Dennning, Sharon Inkelas, Faye McNair-Knox, & John Rickford (eds.), *Variation in language: NWAV-XV at Stanford (Proceedings of the fifteenth annual conference on new ways of analyzing variation)*, 361–78. Stanford CA: Department of Linguistics, Stanford University.

Prunty, Jancita. 2001. Improving the urban environment. In Joseph Brady & Anngret Simms (eds.), *Dublin through space and time (c. 900–1900)*, 166–220. Dublin: Four Courts Press.

Pucher Michael, Friedrich Neubarth, Volker Strom, Sylvia Moosmüller, Gregor Hofer, Christian Kranzler, Gudrun Schuchmann, & Dietmar Schabus. 2010. Resources for speech synthesis of Viennese varieties. *Proceedings of the 7th International conference on Language Resources and Evaluation*, Valletta, Malta, May 2010. 105–108.

Purschke, Christoph. 2011a. *Regionalsprache und Hörerurteil. Grundzüge einer perzeptiven Variationslinguistik.* Stuttgart: Steiner.

Purschke, Christoph. 2011b. "Regional linguistic knowledge and perception – On the conceptualization of Hessian". *Dialectologia* (Special Issue 2). 91–118.

Rampton, Ben. 2009. Interaction ritual and not just artful performance in crossing and stylization, in *Language in Society* 38. 149–176.

Recasens, Daniel. 2012. A cross-language acoustic study of initial and final allophones of /l/. *Speech Communication* 54. 368–383.

Recasens Daniel & Aina Espinosa. 2005. Articulatory, positional and coarticulatory characteristics for clear /l/ and dark /l/: evidence from two Catalan dialects. *Journal of the International Phonetic Association* 35. 1–25.

Rensink, W.G. [1955] 1999. Informant classification of dialects. Reprinted in Dennis R. Preston (ed.), *Handbook of perceptual dialectology*, Vol. 1, 3–7. Amsterdam: John Benjamins.

Robillard, Didier de. 2003. Français, variation, représentations : quelques éléments de réflexion. *Cahiers du Français contemporain* 8. 35–61.

Romanello, Maria Teresa. 2002. The Perception of Urban Varieties: Preliminary Studies from the South of Italy. In Daniel Long & Dennis R. Preston (eds.), *Handbook of perceptual dialectology*, Vol. 2, 329–349. Amsterdam: John Benjamins.

Ruffin François. 2010. Au bonheur des riches, *Manière de voir. Le Monde diplomatique* 114.

Ruoff, Arno. 1997. Sprachvarietäten in Süddeutschland. In Gerhard Stickel (ed.), *Varietäten des Deutschen. Regional- und Umgangssprachen*, 142–154. Berlin/New York: de Gruyter.

Ryan, Ellen Bouchard & Howard Giles. 1982. *Attitudes towards language variation: Social and applied contexts*. London: Hodder Arnold.

Sayer, Andrew. 1989. The 'new' regional geography and the problems of narrative. *Society and Space* 7. 253–276.

Schirmunski, Victor. 1928/1929. Die schwäbischen Mundarten in Transkaukasien und Südukraine. *Teuthonista* 5. 38–60 and 157–171.

Schirmunski, Viktor. 1930. Sprachgeschichte und Siedlungsmundarten. *Germanisch-Romanische Monatsschrift* 18. 113–122 and 171–188.

Schmid, Carolin & Sylvia Moosmüller. Forthcoming. Geschlechtsspezifische Realisierung des Laterals im Wiener Dialekt. In Sylvia Moosmüller, Manfred Sellner, & Carolin Schmid (eds.), *Phonetik in und über Österreich*. Wien: Österreichische Akademie der Wissenschaften.

Schmidt, Jürgen Erich & Joachim Herrgen. 2011. *Sprachdynamik. Eine Einführung in die moderne Regionalsprachenforschung*. Berlin: Erich Schmidt Verlag.

Schröder, Saskia. 2013. Zur Dynamik der /g/-Spirantisierung in den deutschen Regionalsprachen. In Yvonne Hettler, Carolin Jürgens, Robert Langhanke, & Christoph Purschke (eds.), *Variation, Wandel, Wissen. Studien zum Hochdeutschen und Niederdeutschen*. (Sprache in der Gesellschaft 32), 37–60. Frankfurt am Main: Peter Lang.

Scollins, Richard & John Titford. 2000. *Ey up mi duck! Dialect of Derbyshire and the East Midlands*. Newbury: Countryside Books.

Sannada, Shinji & Young-Cheol Yim. 1993. Sahoi Ŏŏn' ŏ-hak ui Chŏn'gae (Developments in Sociolinguistics). Seoul: Japanese Language Promotion Center.

Shuy, Roger & Frederick Williams. 1973. Stereotyped attitudes of selected English dialect communities. In Roger Shuy & Ralph Fasold (eds.), *Language attitudes: Current trends and Prospects*, 85–96. Washington: Georgetown University Press.

Sibata, Takesi. [1959] 1999. Consciousness of Dialect Boundaries. Reprinted in Dennis R. Preston (ed.), *Handbook of perceptual dialectology*, Vol. 1, 39–62. Amsterdam: John Benjamins.

Šimáčková, Šárka. 2009. Variable quality of the Czech lateral liquid: A perception experiment with young Czech listeners. In Frank Kügler, Caroline Féry, & Ruben van der Vijver (eds.), *Variation and gradience in phonetics and phonology*, 125–139. Berlin & New York: Mouton de Gruyter.

Simms, Anngret. 2001. Origins and early growth. In In Joseph Brady & Anngret Simms (eds.), *Dublin through space and time (c. 900–1900)*, 15–65. Dublin: Four Courts Press.

Soukup, Barbara. 2009. *Dialect use as interaction strategy*. Wien: Braumüller.

Soukup, Barbara & Sylvia Moosmüller. 2011. Standard language in Austria. In Tore Kristiansen & Nikolas Coupland (eds.), *Standard languages and language standards in a changing Europe*, 39–46. Oslo: Novus.

Spiekermann, Helmut. 2004. Regionale Standardisierung, nationale Destandardisierung. In Ludwig M. Eichinger & Werner Callmeyer (eds.), *Standardvariation – Wie viel Variation verträgt die deutsche Sprache?*, 100–125. Berlin/New York: Mouton de Gruyter.

Spini, Mathilde. 2013. *Analyse sociolinguistique de la variation phonétique à Marseille: dynamiques et complexité*. Unpublished MA dissertation, Aix-Marseille University.
Spolsky, Bernard. 2004. *Language policy*. Cambridge: Cambridge University Press.
Sproat, Richard & Osamu Fujimura. 1993. Allophonic variation in English /l/ and its implications for phonetic implementation. *Journal of Phonetics* 21. 291–311.
Steinhauser, Walter. 1953. 250 Jahre Wienerisch. Zur Geschichte einer Stadtmundart. *Zeitschrift für Mundartforschung* XXI(3). 159–190.
Stoeckle, Philipp. 2012. The folk linguistic construction of local dialect areas: Linguistic and extra-linguistic factors. In Sandra Hansen, Christian Schwarz, Philipp Stoeckle, & Tobias Streck *(eds.), Dialectological and folk dialectological concepts of space: Current methods and perspectives in sociolinguistic research on dialect change*, 142–163. Berlin: De Gruyter Mouton.
Sundgren, Eva. 2001. Men and women in language change. A Swedish case study. *NORA – Nordic Journal of Feminist and Gender Research* 9(2). 113–123.
Svenstrup, Christoph. H. 2014. Language attitudes in south-west Germany. In Tore Kristiansen & Stefan Grondelaers (eds.), *Language (De)standardisation in Late Modern Europe: Experimental Studies*, 55–70, Oslo: Novus.
Tabouret-Keller, Andrée (ed.), *Le nom des langues I. Les enjeux de la nomination des langues*, BCILL 95, Louvain-la-Neuve: Peeters.
Tamasi, Susan. 2003. Cognitive patterns of linguistic perceptions. Unpublished PhD dissertation, University of Georgia. https://getd.libs.uga.edu/pdfs/tamasi_susan_l_200305_phd.pdf (accessed 30 July 2014).
The Times. 23 May 2006. City of outlaws leads the league of crime hotspots, page 6. May 23.
Thomas, Eric. 1997. A rural/metropolitan split in the speech of Texas Anglos. *Language Variation and Change* 9. 309–332.
Thomas, Erik R., Norman E. Lass, & Jeannine Carpenter. 2010. Identification of African American Speech. In Dennis R. Preston & Nancy Niedzielski (eds.), *A Reader in Sociophonetics*, 265–285. Berlin & New York: Mouton de Gruyter.
Tillery, Jan & Guy Bailey. 2003. Urbanization and the evolution of Southern American English. In Stephen J. Nagle & Sara L. Sanders (eds.), *English in the Southern United States*, 159–172. Cambridge: Cambridge University Press.
Tobler, Waldo. 1970. A computer movie simulating urban growth in the Detroit region. *Economic Geography* 46(2). 234–240.
Torstensson Niklas, Erik J. Eriksson, & Kirk P. H. Sullivan. 2004. Mimicked accents. Do speakers have similar cognitive prototypes? *Proceedings of the 10th International Australian Conference on Speech Science & Technololgy*. Macquarie University, Sidney: Australian Speech Science & Technology Association Inc. 271–276.
Tovey, Hilary & Perry Share. 2003. *A sociology of Ireland, 2nd Edn*. Dublin: Gill and Macmillan.
Trimaille, Cyril. 2005. Spatialité vécue, dite et (inter)agie par des adolescents dans un quartier péricentral en mutation. *Signalétiques et signalisations linguistiques et langagières des espaces de ville (configurations et enjeux sociolinguistiques)* 36(1). 61–96.
Trimaille, Cyril & Médéric Gasquet-Cyrus. 2013. Sociolinguistic change in the city: Gentrification and its linguistic correlates in Marseille. In Mari Jones & David Hornsby (eds), *Language and social structure in urban France*, 133–150. Oxford: Legenda.
Trudgill, Peter. 1972. Sex, covert prestige and linguistic change in urban British English. *Language in society* 1. 179–195.
Trudgill, Peter. 1986. *Dialects in contact*. Oxford: Blackwell.
Trudgill, Peter. 1990. *The dialects of England*, 1st edn. Oxford: Blackwell.

Trudgill, Peter. 1999. The *dialects of England*, 2nd edn. Oxford: Blackwell.
Upton, Clive. 2012. The importance of being Janus. Midland speakers and the 'North-South divide'. In Manfred Markus, Yoko Iyeiri, Reinhard Hemberger, & Emil Chamson (eds.), *Middle and Modern English Corpus*, 257–268. Amsterdam: John Benjamins.
Upton, Clive, David Parry, & J. D. A. Widdowson. 1994. *Survey of English dialects: The dictionary and grammar*. London: Routledge.
Upton, Clive, Stewart Sanderson, & John Widdowson. 1987. *Word maps: A dialect atlas of England*. London: Croom Helm.
US Census Bureau. 2010a. American FactFinder. Profile of general population and housing characteristics: 2010, Seattle, Washington. http://factfinder2.census.gov/faces/tableservices/jsf/pages/productview.xhtml?src=bkmk (accessed 30 July 2014).
US Census Bureau. 2010b. American FactFinder. Profile of general population and housing characteristics: 2010, Wenatchee, Washington. http://factfinder2.census.gov/faces/tableservices/jsf/pages/productview.xhtml?src=bkmk (accessed 30 July 2014).
Vance-Sherman, Anneliese. 2013. King County profile. Employment Security Department Washington State. https://fortress.wa.gov/esd/employmentdata/reports-publications/regional-reports/county-profiles/king-county-profile (accessed 30 July 2014).
Vanderbeck, Robert & Cheryl Morse Dunkley 2003. Young people's narratives of rural–urban difference. *Children's Geographies 13(5)*, 241–259.
Viereck, Wolfgang. 1986a. Dialect speech areas in England: Orton's lexical evidence. In Dieter Kastovsky & Aleksander Szwedek (eds.), *Linguistics across historical and geographical boundaries*, 725–740. Berlin: Mouton de Gruyter.
Viereck, Wolfgang. 1986b. Dialect speech areas in England: Orton's phonetic and grammatical evidence. *Journal of English Linguistics* 19(2). 240–257.
Viereck, Wolfgang. 1992. Prince Louis-Lucien Bonaparte and English dialectology. In G. Gotzon Aurrekoetxea & X. Videgain (eds.), *Nazioarteko dialektologia biltzarra: Agiriak*, 17–30. Bilbo.
Viereck, Wolfgang & Heinrich Ramisch. 1991. *The computer developed linguistic atlas of England, Vol. 1*. Tubingen: Niemeger.
Viereck, Wolfgang & Heinrich Ramisch. 1997. *The computer developed linguistic atlas of England, Vol. 2*. Tubingen: Niemeger.
Wales, Katie. 2000. North and South: An English linguistic divide? *English Today* 16(1). 4–15.
Wales, Katie. 2002. 'North of the Watford Gap'. A cultural history of Northern English (from 1700). In Richard Watts & Peter Trudgill (eds.), *Alternative histories of English*, 45–66. London: Routledge.
Wales, Katie. 2006. Dialects in mental Contact: A Critique of Perceptual dialectology. In Herbert Grabes & Wolfgang Viereck (eds.) *The wider scope of English*, 57–66. Frankfurt: Peter Lang.
Watt, Dominic. 2002. "I don't speak with a Geordie accent, I speak, like, the Northern accent': Contact-induced levelling in the Tyneside vowel system. *Journal of Sociolinguistics* 6(1). 44–63.
Weijnen, Antonius A. 1946. De grenzen tussen de Oost-Noordbrabantse dialecten onderling]. In A. A. Weijnen, J. M. Renders, & J. van Ginneken (eds.), Oost-Noordbrabantse dialectproblemen. Bijdragen en Mededelingen der Dialectencommissie van de Koninklijke Nederlandse Akademie van Wetenschappen te Amsterdam 8. 1–15.
Weijnen, Antonius A. [1968] 1999. On the Value of Subjective Dialect Boundaries. Reprinted in Dennis R. Preston (ed.), *Handbook of perceptual dialectology*, Vol. 1, 131–133. Amsterdam: John Benjamins.

Weinreich, Uriel, William Labov, and Marvin I. Herzog. 1968. Empirical foundations for a theory of linguistic change. In Winfred F. Lehmann and Yakov Malkiel (eds), *Directions for historical linguistics*, 95–188. Austin and London: University of Texas Press.
Weck, Françoise. 2008. *Putain d'accent ! Comment les Méridionaux vivent leur langue*. Paris: L'Harmattan.
Wells, J. C. 1982. *Accents of English 1. An introduction*. Cambridge: Cambridge University Press.
Wells, J. C. 1986. *Accents of English 2: The British Isles*. Cambridge: Cambridge University Press.
Whelan, Christopher. 1995. Class transformation and social mobility in the Republic of Ireland. In Patrick Clancy, Sheelagh Drudy, Kathleen Lynch, & Liam O'Dowd (eds.), *Irish society: Sociological perspectives*, 324–357. Dublin: Institute of Public Administration.
Whelan, Christopher. 2003. Social class. In Brian Lalor (ed.), *The Encyclopedia of Ireland*. Dublin: Gill and Macmillan.
Wiesinger, Peter. 2009. Die Standardaussprache in Österreich. In Eva-Maria Krech, Eberhard Stock, Ursula Hischfeld, & Lutz Christian Anders (eds.), *Deutsches Aussprachewörterbuch*, 229–258 Berlin/New York: Mouton de Gruyter.
Williams, Frederick. 1973. Some research notes on dialect attitudes and stereotypes. In Roger Shuy and Ralph Fasold (eds.), *Language attitudes: Current trends and prospects*, 113–128. Washington: Georgetown University Press.
Williams, Angie, Peter Garrett, & Nikolas Coupland. 1996. Perceptual dialectology, folklinguistics, and regional stereotypes: Teachers' perceptions of variation in Welsh English. *Multilingua* 15(2). 171–199.
Williams, Angie, Peter Garrett, & Nikolas Coupland. 1999. Dialect Recognition. In Dennis R. Preston (ed.), *Handbook of perceptual dialectology*, Vol. 1, 345–358. Philadelphia: John Benjamins.
Williams, Raymond. 1973. *The Country and the City*. New York: Oxford University Press.
Wodak-Leodolter, Ruth. 1975. *Das Sprachverhalten von Angeklagten vor Gericht: Ansätze zu einer soziolinguistischen Theorie der Verbalisierung*. Kronberg/Ts: Scriptor.
Wodak-Leodolter, Ruth & Wolfgang U. Dressler. 1978. Phonological Variation in colloquial Viennese. *Michigan Germanic Studies* 4(1). 30–66.
Woods, Michael. 2006. *Rural geography*. London: Sage.
Woods, Michael. 2011. *Rural*. New York: Routledge.
Wood, Sidney. 2005. Praat for Beginners. http://swphonetics.com/praat/ (accessed 22 April 2013).
Wright, Joseph. 1905. *The English dialect grammar*. Oxford: The Clarendon Press.
Wright, Peter. 1986. *Notts natter: How it is spoke*. Skipton: Dalesman Publishing.
WWAMI Rural Health Research Center. 2005. [Rural-Urban Commuting Area Codes (RUCA) Data]. http://depts.washington.edu/uwruca/ruca-data.php (Retrieved June 30, 2011).
Yaeger-Dror, Malcah. 1998. Factors influencing the contrast between men's and women's speech. *Women and Language* 21(1). 40–46.
Yi, Kigap. 1998. Sonam pangon (Southwest dialects). *Sae Kugo Saenghwal* (Special Edn. on Dialects) 8(4). 95–110.
Yi, Sanggyu. 1995. *Pangonhak* (Dialectology). Seoul: Hagyonsa.
Yi, Sanggyu. 1998. Tongnam pangon (Southeast dialects). *Sae Kugo Saenghwal* (Special Edn. on Dialects) 8(4). 111–32.
Yim, Young-Cheol. 1993. Zainichi, zaibei Kankokujin, oyobi Kankokujin no Gengo Seikatsu no Jittai (Language life of Koreans, Korean-Japanese and Korean-Americans). Tokyo: Kuroshio.

Yonezawa Morris, Midori. 2010. Regional stereotypes and the perception of Japanese vowel devoicing. In Dennis R. Preston & Nancy Niedzielski (eds.), *A Reader in Sociophonetics*, 191–202. Berlin & New York: Mouton de Gruyter.

Index

FSC
www.fsc.org
MIX
Papier | Fördert
gute Waldnutzung
FSC® C083411

Kolibri 360 GmbH
Ferdinand-Jühlke-Straße 7
99095 Erfurt, Deutschland
produktsicherheit@kolibri360.de